PATHOGENS FOR WAR

Biological Weapons, Canadian Life Scientists, and North American Biodefence

Pathogens for War explores how Canada and its allies have attempted to deal with the threat of germ warfare, one of the most fearful weapons of mass destruction, since the Second World War. In addressing this subject, distinguished historian Donald Avery investigates the relationship between bioweapons, poison gas, and nuclear devices, as well as the connection between bioattacks and natural disease pandemics. Avery emphasizes the crucially important activities of Canadian biodefence scientists – beginning with Nobel Laureate Frederick Banting – at both the national level and through cooperative projects within the framework of an elaborate alliance system.

Delving into history through a rich collection of declassified documents, *Pathogens for War* also discusses the contemporary challenges of bioterrorism and disease pandemics from both national and international perspectives. As such, readers will not only learn about Canada's secret involvement with biological warfare, but will also gain new insights into current debates about the peril of bioweapons – one of today's greatest threats to world peace.

DONALD AVERY is an emeritus professor and adjunct research professor in the Department of History at Western University.

Pathogens for War

Biological Weapons, Canadian Life Scientists, and North American Biodefence

DONALD AVERY

UNIVERSITY OF TORONTO PRESS
Toronto Buffalo London

Toronto Buffalo London
www.utppublishing.com
Printed in Canada

ISBN 978-0-8020-8971-7 (cloth)
ISBN 978-1-4426-1424-6 (paper)

Printed on acid-free, 100% post-consumer recycled paper with vegetable-based inks

---

Library and Archives Canada Cataloguing in Publication

Avery, Donald, 1938–
Pathogens for war: biological weapons, Canadian life scientists, and North American biodefence / Donald Avery.

Includes bibliographical references and index.
ISBN 978-0-8020-8971-7 (bound). ISBN 978-1-4426-1424-6 (pbk.)

1. Biological weapons – Canada – History – 20th century. 2. Bioterrorism – Canada – Prevention – History – 20th century. 3. Communicable diseases – Canada – Prevention – History – 20th century. 4. Life scientists – Canada – History – 20th century. 5. Canada – Military policy – History – 20th century. I. Title.

UG447.8.A94 2013 358'.3882097109045 C2012-908541-3

University of Toronto Press acknowledges the financial assistance to its publishing program of the Canada Council for the Arts and the Ontario Arts Council.

University of Toronto Press acknowledges the financial support of the Government of Canada through the Canada Book Fund for its publishing activities.

This book has been published with the help of a grant from the Canadian Federation for the Humanities and Social Sciences, through the Awards to Scholarly Publications Program, using funds provided by the Social Sciences and Humanities Research Council of Canada.

For my four grandchildren – Brendon, Alexander, Olivia, Kaitlyn – and their furry friend Zep.

Contents

Acknowledgments

This book is the product of approximately twenty years of scholarly work in the fields of biodefence, public health, defence planning, and arms control / disarmament studies. While the focus of this study is Canada's response to the threat of biological weapons, terrorism, and disease epidemics since the Second World War, it also explores the broader dimensions of these themes within the prevailing international context. Of particular importance are the many dimensions of Canadian-American cooperation in dealing with disease as an instrument of war throughout the Second World War, the Cold War, and the more recent war on terror.

In preparing this study for publication I received assistance and support from many sources. I owe a special debt to my wife, Dr Irmgard Steinisch, who provided insightful but compassionate criticism as the manuscript gradually evolved. Scholarly assistance was also forthcoming from my three UWO colleagues Robert Murray, Grant McFadden, and Bhagirath Singh, who patiently explained the complexities of biological agents and disease outbreaks. Equally important were the insights of biodefence scholars John Ellis van Courtland Moon, Mark Wheelis, Brian Balmer, Julian Perry Robinson, and Malcolm Dando, who carefully assessed the author's contribution to the interdisciplinary project *Deadly Cultures: Biological Weapons since 1945* (Harvard University Press, 2006). Because of the nature of this subject, valuable information was often available only through personal reflections, and the author was fortunate in being able to conduct many quality interviews. High on the list of scientific confidants were Clement Laforce, Cam Boulet, Joan Armour, and Kent Harding of DRDC Suffield; Ronald St John, Frank Welsh, and

Marc-Andre Beaulieu of the Public Health Agency of Canada; and Frank Plummer, James Strong, and Yan Li of the National Microbiology Laboratory.

Readers will note that chapter 1 is partly based on sections of *The Science of War* (University of Toronto Press, 1998), while chapters 2 and 3 draw on some of the arguments developed in the author's article "The Canadian Biological Weapons Program and the Tripartite Alliance," in *Deadly Cultures* (2006). Another consideration is the quality and quantity of source material available in different sections of the book. For the first six chapters, most of the information has been drawn from diverse archival collections in Canada, the United States, and the United Kingdom. In contrast, chapter seven, with its focus on biodefence developments since 9/11, draws extensively on contemporary government reports, newspaper accounts, personal interviews, and secondary sources.

Over the years, my research has been facilitated by the staff of many archives and libraries. Within Canada, the following institutions provided special assistance: Library and Archives Canada (Ottawa); Connaught Medical Research Laboratories (Toronto); McGill University Archives; and Queen's University Archives. While the author consulted a number of US archival collections, the most important were located at the National Archives and Records Administration (College Park), the Archives of the American Microbiology Society (Baltimore), and the Archives of the National Academy of Sciences (Washington). Equally valuable were the government records and private papers housed at the National Archives of the United Kingdom (Kew) and the Archives of the North Atlantic Treaty Organization (Brussels).

The broad range of my research would not have been possible without the generous support I received from the Social Science and Humanities Research Council, the University of Western Ontario, the Department of National Defence, and a Hanna history of medicine grant. In addition, assistance in the publication of this book has been provided by the Social Science Federation's Aid to Scholarly Publication Program. I would also like to thank a number of people at the University of Toronto Press: Bill Harnum (retired) for providing the original contract; Len Husband, for shepherding the manuscript through its various stages; and Frances Mundy and James Leahy for their valuable editorial assistance. My grandson Brendon Avery / Quick also enhanced the book by preparing a fine index.

PATHOGENS FOR WAR

Biological Weapons, Canadian Life Scientists, and North American Biodefence

Introduction

Since the tragic events of the 1972 Munich Olympics, there have been fears that terrorist groups might exploit these high-profile international sporting events "to make a grandiose and symbolic statement with a potential for mass casualties."[1] This was certainly the mindset of Canadian security officials in 1976 when they had to deal with the possibility that radical organizations would attack the Montreal Summer Olympics with-chemical, biological, radiological, or nuclear weapons (CBRN).[2] Fortunately, this did not occur, but fears about such assaults resurfaced during the 1990s with the emergence of ever more ruthless terrorist organizations. This sense of vulnerability about the Olympic milieu was certainly evident in Atlanta (1996), Salt Lake City (2000), and Athens (2008), with more than 50,000 security personnel being involved in the latter event.[3] Given this legacy, it is not surprising that organizers of the 2010 Vancouver Winter Games decided to implement elaborative counterterrorist measures while trying "to avoid the appearance of a 'Fortress BC.'"[4]

Overall, they achieved their goals, and there were no CBRN incidents. But safeguarding the Vancouver Olympics challenged the capabilities of the Integrated Security Unit, composed of specialists from the Royal Canadian Mounted Police, the Canadian Security and Intelligence Agency, the Canadian Armed Forces, and local police forces. Of special importance were the secret BW monitoring activities of the special Microbiological Emergency Response team from the National Microbiology Laboratory, who maintained close contact with the BC Centre for Disease Control and military specialists at DRDC Suffield.[5] The effectiveness of this integrated threat assessment was enhanced by the use of new CBW technology such as the Vital Point Biological Sentry System, which provided detection, alarm, and identification capabilities for upwards of one kilometre, allowing its operators an opportunity to differentiate "between local background biological activity and threat activity in real time."[6] In addition, timely assistance was provided by the US Department of Homeland Security, the Pentagon, and the Centers for Disease Control and Prevention.[7] Yet despite these elaborate security measures, on 13 January the US State Department advised American sport fans travelling to Vancouver to be aware of the possibility of terrorist activity, given "Al-Qaeda's demonstrated capability to carry out sophisticated attacks."[8]

In reality, this warning was symptomatic of pervasive American concerns about the possibilities of terrorist WMD attacks, and particularly bioterrorism, since 9/11. In January 2010, for example, the United States Commission on the Prevention of Weapons of Mass Destruction Proliferation and Terrorism informed President Barack Obama that his administration deserved failing grades for its inability to provide a "rapid and effective response to bioterrorism."[9] These deficiencies were deemed particularly serious since the Commission claimed "that a weapon of mass destruction (WMD) will be used in a terrorist attack somewhere in the world by the end of 2013 … [and] that weapon is more likely to be biological than nuclear."[10]

Significantly, the US and Canadian governments were just recovering from another disease related crisis – the global outbreak of H1N1 influenza. Indeed, in June 2009, Dr Margaret Chan, Director General of the World Health Organization, had warned the international community that the Swine Flu pandemic could threaten the lives of millions of people around the world: "For five long years, outbreaks of highly pathogenic H5N1 avian influenza in poultry, and sporadic frequently fatal cases in humans, has conditioned the world to expect an influenza pandemic, and a highly lethal one … [but] the new influenza A (H1N1) strain, has emerged from another source on another side of the world … [which] spreads easily from person to person."[11]

These warnings about mutating strains of influenza assumed a sinister new perspective in December 2011 when the US National Science Advisory Board on Biosecurity, a high-status oversight agency for high-risk scientific research, recommended that two controversial articles on Avian Influenza (H5N2) be substantially revised before publication.[12] Of special concern were several innovative experiments that suggested ways in which this deadly virus, with its 60 per cent fatality rate, could be genetically altered so that it "could potentially spread quickly among humans … [establishing] a blueprint for the creation of a biological weapon."[13] Significantly, even the usually restrained *New York Times* was so concerned about these developments that it published a lead editorial entitled "An Engineered Doomsday."[14]

While these four incidents exemplify the types of challenges facing contemporary Canadian and American health security officials, these issues have a long historical legacy. Indeed, one of the major goals of *Pathogens for War: Biological Weapons, Canadian Life Scientists, and North American Biodefence* is to examine the impact of bioweapons, bioterrorism, and pandemics on the global community within the historical and contemporary context. More specifically, it seeks to explain how Canada has responded to the threat of disease as an instrument of war and terror, with special emphasis on the role life scientists have assumed in the country's biodefence strategies since the outbreak of the Second World War.

Adopting the Canadian perspective has a number of advantages. First, throughout these years Canada had a well-developed medical science research community, along with an evolving national health system, which by the 1970s had established medical coverage as one of the rights of citizenship. As a result, the country was able to withstand a series of major infectious disease outbreaks, such as the influenza pandemics of 1957 and 1968, while mobilizing resources for the national eradication of the twin scourges of smallpox and polio by the 1970s. Second, Canada maintained an active involvement in the military aspects of biological weapons during the Second World War and the Cold War, both in terms of its national research and testing programs, and through its ongoing cooperation with the United States and the United Kingdom in developing the offensive aspects of this weapon system.[15] Third, since Canada was forced to deal with the threat of an enemy attack with weapons of mass destruction, civil defence planners developed extensive protective measures that included major biodefence programs, often in cooperation with their counterparts in the United States. Fourth, fear that enemy saboteurs or international terrorists might use biological weapons against Canada's urban centres is another theme that has a long legacy, even if some pundits believed the threat only emerged after the 9/11 terrorist attacks. And finally, despite Canada's important historical and contemporary role in responding to the threat of biological warfare and bioterrorism, this subject remains virtually unexplored in the scholarly literature, in part because of the tendency of Canadian military historians to ignore the subject and, in part, because of the veil of secrecy that has shrouded this aspect of the country's national defence and international relations.[16]

Despite its extensive use of government records, *Pathogens for War* is not a narrow institutional study. On the contrary, the main players in this longitudinal analysis of Canada's BW experiences are the defence scientists themselves, many of whom were prominent medical professionals, both in developing national biodefence strategies and in working with their counterparts in the United States and United Kingdom. This diverse group, numbering less than 150, consistently provided high-quality research and testing standards for Canada's own needs and in cooperation with its two major allies. By way of contrast, since 1941 the United States has employed thousands of BW scientists, with 1,800 being involved in BW research at Camp Detrick alone by the end of the Second World War.

Definitions and Conceptual Issues

Biological warfare can be defined as a form of warfare that uses living organisms and natural poisons (toxins) to produce death and debilitation in humans, animals, or plants.[17] While biological weapons are often equated with nuclear and chemical weapons, in reality they are quite different since they are composed of or derived from organisms, which can replicate themselves inside the host, thereby allowing an attacker "to use a small amount of a biological weapon to inflict mass casualties."[18] Another major difference is the diversity of pathogenic microorganisms and toxins that can be used as biological weapons, particularly given the enormous advances in the biosciences during the past forty years. As a result, there has been a major change in the concept of bioweapons, which traditionally meant "a warhead with massive quantities of refined agents that were specifically designed for instant and catastrophic release ... [while] now a biological weapon might be merely a test tube of pathogens that are capable of wide replication or a tiny device that can carry a pathogen through the body."[19]

While biological warfare was only integrated into the military strategy of nation states during the Second World War, it has a long historical legacy. Indeed, hostilities between states have always been accompanied by an increase in infectious diseases, because of intensified problems of sanitation, poor nourishment, overcrowding, and the carnage of the battlefield. Indeed, until the twentieth century, deaths from disease usually far exceeded the numbers of killed on the battlefield, a situation that encouraged rival armies to exploit disease outbreaks as part of their military operations. This was evident, for example, during the 1346 siege of

Caffa when the Mongol army catapulted thousands of plague-infected cadavers into the Genoese city, causing a serious outbreak of bubonic plague. Another widely cited example was the 1763 decision of the British commander Sir Jeffrey Amherst to distribute smallpox-infected blankets among a number of Native American tribes who had been involved with the so-called Pontiac Rebellion against European settlers in the Ohio Valley, apparently with devastating results, given their lack of immunity against this dreadful disease.[20]

Yet despite these and other examples, biological warfare was not regarded as a serious threat by the Hague Conventions of 1899 and 1907, which attempted to establish civilized norms for the conduct of war. Nor did germ warfare occur on European battlefields during the First World War, although German agents engaged in a crude form of veterinary BW when they attempted to use anthrax and glanders against American war horses destined for the western front. Yet according to microbiologist Mark Wheelis, this German activity did establish an important precedent since it was "a) the first national programme of offensive biological warfare; (b) the first biological warfare programme of any kind with a scientific foundation ... [and] was directed against neutrals not belligerents, and targeted animals not humans."[21]

Several factors about the strategic and tactical use of biological weapons should be considered.[22] Above all is the advantage of achieving surprise through covert first use, since BW agents are very difficult to detect because of their physical properties and the similarity of symptoms with natural disease infections. Another argument in favour of using biological weapons is as a "force multiplier" within the context of a conventional war, as well as a follow-up device after a nuclear exchange, when the enemy's health care services would be in disarray. Recently, there has been considerable discussion about the possibility that terrorist groups might use bioweapons to attack urban targets, given the relative ease of acquiring BW devices and the capability of causing high casualties. While all of these scenarios are a source of concern, the historical reality is that unlike chemical and nuclear weapons, biological weapons have not been used in a major war. This, of course, could change as ever more deadly biological weapons are created through genetic engineering and genomic synthesis.[23]

Research in the life sciences has greatly influenced the development of biological and toxin weapons. This trend has been particularly pronounced since the late 1960s when scientists in Western countries and the Soviet Union were able to utilize new research techniques associated with recombinant DNA techniques and gene replication.[24] As one commentator observed in 1968, "We are now, though we only dimly realize the fact, in the opening stages of the Biological Revolution – a twentieth century revolution, which will affect human life far more profound than the great Mechanical Revolution of the nineteenth century or the Technological Revolution through which we are now passing."[25] Moreover, since the 1970s accelerated advances in the life sciences have influenced the development of new bioweapons and the biodefence technologies to deal with these new pathogens and toxins.[26] In many ways, this virtual explosion in biotechnology research has been one of the defining characteristics of modern society, as scientists moved "to the frontier where the outer edges of genetics, biochemistry, and microbiology were merging alongside a flood of new technologies, such as electron microscopy, crystallography, cell culture, and virology ... and steeply rising capabilities for information storage and analysis."[27]

Popular and Scholarly Depictions of Biological Warfare and Disease Pandemics

The image of biological warfare has often been equated with the terrible disease outbreaks of the past, notably the Black Death (bubonic plague), the scourge of smallpox (variola major), and the 1918 Spanish influenza pandemic, which killed over 50 million people globally. In addition, all these pathogens have been considered as biowarfare agents, along with other deadly diseases such as cholera, typhus, tularaemia, and, above all, anthrax. While *Bacillus anthracis* has not been responsible for global pandemics, the bacterium has certain qualities that make it an ideal biological weapon: its high lethality, its ease of production, its contained application, and its ability to develop a protective spore covering which "could last and penetrate where no bacillus could survive, producing new germs where the conditions were right."[28] Similar arguments have been made for using Botulium toxin, an extreme poison, which often became the second most common BW agent in national arsenals

Protecting the military and civilian population against bioagents and infectious disease outbreaks was closely associated with the development of vaccines and antibiotics / antivirals. Indeed, it is often argued that the advent of the antibiotic era began during the 1940s with the discovery of penicillin and related drugs. These developments appeared to usher in a new era of public health whereby infectious disease outbreaks could be prevented and many pathogens eradicated. This did not happen. While international public health officials enjoyed considerable success in virtually eliminating the twin scourges of smallpox and polio, its record against other pathogens, notably malaria and tuberculosis, has been disappointing. Even worse, emerging infectious diseases such as HIV/AIDS, which began its global rampage during the 1980s, have killed tens of millions of people during the past thirty years. These ominous trends have continued during the first decade of the twenty-first century when pathogens such as severe acute respiratory syndrome (SARS), and avian / swine influenza threatened the global community.

There is a vast body of scholarly literature on the above subjects. Of particular value are Charles Rosenberg's analysis of epidemics, both past and present trends, and the sweeping study by Paul de Kruif of the great microbe hunters of the nineteenth century.[29] Equally important are studies on the social impact of disease outbreaks such as the nineteenth-century cholera epidemics, the influenza pandemics of the twentieth century, and the AIDS/HIV saga.[30] Closely related are the descriptive accounts written by medical practitioners of their personal involvement in international and national campaigns to eradicate smallpox, polio, malaria, and tuberculosis.[31] And finally some authors have attempted to explain how disease has influenced global power relationships, as exemplified by Jared Diamond's Pulitzer prize-winning book *Guns, Germs and Steel: The Fates of Human Societies.* As part of his analysis Diamond points out that one of the major reasons why European powers could conquer the Americas was their ability to export "all of history's most lethal killers: smallpox, measles, influenza, plague, tuberculosis, typhus, cholera, malaria and others."[32]

In contrast with the large number of publications on natural outbreaks of infectious diseases, there are relatively few scholarly works dealing with biological warfare. Moreover, while much of the earlier work focused almost exclusively on the military dimensions of the subject, during the past twenty years the literature has become increasingly sophisticated and multi-disciplinary.[33] Another major change has been the growing appreciation that germ warfare, while it shares certain characteristics with chemical warfare, has many unique characteristics in terms of tactical use and casualty-causing potential.[34] Significantly, during the 1990s there was growing awareness about the formidable killing power of bioweapons, in part because of the biotechnology revolution and, in part, because of the proven ability of the Soviet Union to develop more sophisticated BW munitions and delivery systems.[35] These concerns were reinforced by the appearance of a number of memoirs by former Soviet BW weaponeers. The first Gulf War (1990–1) also demonstrated that rogue states such as Iraq could acquire a bioweapons capability, sufficient to cause large number of battlefield casualties and threaten neighbouring urban centres.[36] But it was the events of 9/11, and the subsequent anthrax letter attacks, that had the most powerful impact, serving as a catalyst for massive biodefence preparations in the United States, Canada, and the United Kingdom. Indeed, during the past decade not only has there been a wave of publications on biological weapons, scholarly and otherwise, but the field of bioterrorism / biodefence has become institutionalized with specialized research centres and journals.[37] Hollywood has also helped reinforce these fears about dangerous new pathogens, through such powerful films as *Outbreak* (1995), which explores the connection between a sudden outbreak of the haemorrhagic disease Ebola in the United States and a sinister biological warfare project under the direction of a fanatical army general (Donald Sutherland).[38] Fortunately, disaster is averted at the last moment through the heroic actions of two dedicated and resourceful US virologists (Dustin Hoffman and Renee Russo), who manage to develop an effective vaccine.[39] Many of these themes were also featured in the thriller *Contagion* (2011). In this case, a deadly new pathogen spreads rapidly throughout the world, killing millions and causing widespread panic and civil disorder. Despite its formula of dramatic action and a happy ending, the film has received considerable praise for its scientific accuracy in reconstructing the dynamics of a disease pandemic and the challenges of finding effective medical counter-measures.[40]

Microbiology as a Science: Major Trends

The emergence of bacteriology as a separate and important discipline has been an integral part of the triumph of scientific medicine. There were a number of reasons for this phenomenon. First, there were the major discoveries during the nineteenth century by that outstanding scientific trinity: Louis Pasteur in France, Robert Koch in Germany, and Joseph Lister in the United Kingdom. Through their individual and collective achievements, the isolation, diagnosis, and treatment of many bacterial diseases were now possible. Second, there was the development of a more demanding scientific methodology that stressed the value of focusing on specific bacterial pathogens while at the same time recognizing the complex relationship between the "germ" and the immune system of the human or animal host. Third, under the so-called Koch postulates, a rigorous procedure was initiated to determine the identity of pathogens: detection and isolation from an infected subject; growth in a controlled laboratory environment; and replication of the disease symptoms by using inoculated animal trials. This methodology was increasingly standardized. Fourth was the emergence of an expanded international medical research system of cooperation that dated back to the work of the early microbe hunters, with Pasteur, Koch, and Lister all insisting that the implications of their research should surmount national boundaries and be shared with scientists throughout the world.[41]

Throughout the Western world, acceptance of the "germ theories" proceeded at different stages, reflecting the dynamics of specific national medical cultures. By 1900, however, there was a general consensus that most major diseases were caused by bacteria, or by pathogens not yet identified, a form of diagnosis that reinforced the "popular military analogy of invading germs in conflict with the body's defences."[42] In addition, there was general agreement that "laboratory-based ideas and practices were important resources in the reshaping of prevention, diagnosis, treatment and patient management ... [and] ongoing movement of medicine to an expert, science based profession."[43] Of special importance was the recognition that laboratory investigations provided the basis for standardized analysis, avoiding "the messy clinical reality of an infectious disease, with its variety of symptoms and individual manifestations, its regional, seasonal and environmental variations, and its often unpredictable outcomes."[44] But it was a slow process.[45] Indeed, by the First World War, the level of laboratory experimental technology remained relatively primitive, with all viruses and many groups of bacteria still "beyond the capture of microscopy and the new laboratory techniques."[46]

Given these technical and conceptual limitations, it is not surprising that a series of controversies arose within the evolving fields of bacteriology and immunology. One of these involved the cause and prevention of epidemics during the late nineteenth century, notably those associated with the outbreaks of cholera (1881–2), smallpox (1884–5), and bubonic plague (1896).[47] Each of these epidemics presented common problems in isolating, identifying, and categorizing specific pathogens and in determining "their physical, chemical and biological effects on the body."[48] A related set of questions involved the relative importance of pathogens, heredity, and environment in causing disease, a debate that extended into the 1920s when W.W.C. Topley of the London School of Hygiene and Tropical Medicine stressed the importance of studying the effect of epidemics on collectivities rather than individuals, seeing these outbreaks as a herd phenomenon of susceptibility.[49] Immunological theories about the body's ability to produce antibodies were another important area of medical research, although at the time little was known about the specific process of molecular interaction.[50] Nor was much progress made in developing chemotherapy drugs for diseases such as pneumonia, diphtheria, scarlet fever, tuberculosis, or the bacterial pathogens that caused streptococcal and staphylococcal infections.[51] Indeed, it was not until the late 1930s that the sulpha drugs (sulphanilamide) became available, which, in turn, were followed by the development of penicillin and related antibiotics during the Second World War.[52]

Controlling Infectious Diseases during War and Peace

There is considerable historical debate about whether wartime conditions advance or delay fundamental scientific research.[53] In the case of military medicine, however, there are powerful arguments for a positive correlation. As many scholars have pointed out, during both world wars important medical advances sharply reduced the number of battlefield deaths from infection and disease. Indeed, this trend was already obvious during the Anglo-Boer War (1899–1902), which witnessed a sharp decline in the mortality rate from battlefield wounds because of greater use of antiseptic methods and more rapid treatment of casualties. As well, British military officials enjoyed some success in dealing with enteric fever, typhoid, and dysentery.[54] The challenge of reducing battlefield deaths was much greater during the First World War, when the possibility of a short, low-casualty conflict was dramatically replaced by a grim war of attrition in which machine guns, artillery, and poison gas turned the western front into a massive killing ground.[55] Although none of the belligerents was prepared for this scale of battlefield carnage, remedial measures were soon adopted that helped reduce the ratio of deaths from battlefield injuries and war-related diseases. On the western front, for example, typhoid cases were fewer than 10 per 1,000; tetanus, 1.47 per 1,000; and dysentery, 0.79 per 1,000.[56]

For Canada's medical scientists their experience with the horrors of trench warfare strongly influenced their professional and personal lives. This was certainly the case for the group of life scientists who would subsequently direct the country's biological warfare program during the next world war: Frederick Banting, Everitt Murray, Guilford Reed, James

Craigie, and Charles Mitchell. Banting, for example, in his September 1918 assignment to the No. 13 Canadian Field Ambulance casualty clearing station "proved the worth of his training as a doctor, and found it exhilarating to know that he was literally saving men's lives as he tended to their wounds."[57] This sense of confidence continued into the postwar years, when Banting became one of Canada's foremost public figures because of his 1922 Nobel prize for the discovery of insulin and his status as an outstanding medical researcher at the University of Toronto. In contrast, Murray's wartime experience was with the British Army Medical Corps, which gave him many opportunities to utilize his skills as an academic bacteriologist in dealing with an outbreak of meningococcal meningitis on the home front and then carrying out an extensive investigation of Shiga dysentery among British troops in the Middle East.[58] After demobilization in October 1918, as a physician at London's St Bartolomew hospital, Murray experienced first hand the ravages of the deadly H1N1 influenza pandemic that killed upwards of 50 million people globally.[59] What made the situation even more frustrating for front-line doctors was that the viral causative agent, influenza H1N1/A, had not been identified.[60]

After 1930, the Canadian careers of these two outstanding medical researchers would intersect, when Murray was recruited from the United Kingdom to become chair of the Department of Microbiology and Immunology at McGill University.[61] Given their many shared interests, Banting and Murray cooperated on a number of important projects, including the establishment of the National Research Council's Associate Committee on Medical Research (ACMR).[62] Indeed, under Banting's direction the ACMR not only carried out an extensive survey of Canada's medical research facilities, but also developed a range of contacts with the major British and American medical organizations, which were becoming increasingly involved with war-related research.[63]

Japan's 1931 assault on Manchuria, Italy's 1935 attack on Ethiopia, and fascist involvement in the 1936 Spanish civil war provided stark evidence that the world was speeding towards another world war. For many scientists this was a frightening prospect, given their grim experiences during the Great War and their awareness of how technology had intensified the destructiveness of modern warfare. These sentiments were aptly summarized in the June 1936 edition of *The Scientific Worker* (UK): "Bravery, physical fitness, unselfishness and nobility of character are of no avail against the methods of large scale mechanical slaughter developed by engineers, of burning and maiming developed by physicists, and of asphyxiation and blinding developed by the chemists. Soon the bacteriologists may surpass the other scientific workers and produce methods of 'destroying the morale of the civilian populations.'"[64]

This fear of germ warfare was largely based on developments during the First World War. Although neither side deployed biological weapons on the European battlefields, there was an active German BW sabotage operation in the United States. Yet despite this military innovation, neither the Versailles Treaty of 1919 nor the various postwar disarmament conferences made any reference to bioweapons. By the mid-1920s, however, there was a major change in how the international community viewed biological and chemical warfare. During the June 1925 League of Nations meetings in Geneva, for example, the delegates decided that, in addition to outlawing "asphyxiating, poisonous and deleterious gases," they should "extend this prohibition to the use of bacteriological methods of warfare."[65] Unfortunately, the Protocol did not prevent the development of either weapon system, and throughout the 1930s Germany, Japan, and Russia began exploring the military possibilities of germ warfare.

Overview of Conceptual Approaches

This study addresses a number of key questions about Canada's response to the threat of biological weapons, bioterrorism, and pandemics since 1939.[66] The book is divided into seven chapters, arranged chronologically, each with a specific thematic focus. The first section deals with the Second World War and the early Cold War, with special emphasis on the following questions: What measures were adopted by the government of William Lyon Mackenzie King throughout the war years to deal with the possibility that the Axis powers (Germany, Japan, Italy) would use biological and toxin weapons against Canada's armed forces and civilian population? How effective was Canada's wartime BW cooperation with the United States and the United Kingdom, and why did it continue after 1945? Why did successive federal governments try to conceal Canada's involvement with offensive BW research, despite the country's active involvement with the Tripartite Chemical and Biological Warfare Military Agreement of 1947? And finally, how did Canadian defence planners differentiate among threats emanating from biological, chemical, and nuclear weapons?

The second section of the book examines the major changes that took place in Canada's BW operation after US President Nixon's 1969 decision to unilaterally abandon the extensive American biological warfare program and support the establishment of the Biological Weapons and Toxin Convention (BWC) (1972–5). It also explores how Canadian defence scientists and diplomats supported the disarmament goals of the BWC while at the same time working to improve the country's BW defensive measures because of evidence that the Soviet Union was violating the Convention. During the 1970s attention was drawn to the challenge of bioterrorism and the effects of disease pandemics on Canada's public health system. Here again, a series of challenging questions is addressed. To what extent has Canada's professed commitment to prevent the spread of bioweapons been challenged by its alliance commitments, both before and after the operation of the Biological and Toxins Weapons Convention (1975)? In what ways did the civil defence measures of the Cold War period include protection from biowarfare, and to what extend did the lessons of this experience influence Canada's response to the threat of bioterrorism in the twenty-first century? How did the challenge of dealing with natural disease pandemics, such as the influenza outbreaks of 1957 and 1968, along with Canada's active involvement in the WHO's smallpox eradication program, facilitate the development of more comprehensive biodefence strategies, both at that time and in subsequent years? And to what extend did advanced research in the life sciences and biotechnology encourage the BW arms race during the Cold War and after 1991?

Lastly, the book examines questions that relate primarily to contemporary concerns over bioweapons, bioterrorism and pandemics: Are existing Canadian biosafety and biosecurity safeguards sufficient to prevent either natural disease outbreaks or acts of bioterrorism? What lessons were learned during the 2009 swine flu pandemic by the World Health Organization and Canadian health officials? Is the present controversy over the threat of bioterrorism – from either traditional BW weapons or "designer" germs – exaggerated or should the warning of organizations such as the United States Commission on the Prevention of Weapons of Mass Destruction Proliferation and Terrorism be taken more seriously?

1 Canada's Role in Allied Biological Warfare Planning in the Second World War

The highly secret BW unit at Grosse Isle is well underway. It is a co-operative enterprise between the United States and Canada and for administration is under Dr Maass as director of Chemical Warfare. This is another of the many projects which Sir Frederick Banting was keen about in the early days of the war. When I think of all the projects which Sir Frederick started and observe now how many of them have proven eminently successful, the stark tragedy of his early death bears heavily upon me.

(C.J. Mackenzie to General McNaughton, 2 January 1943;
cited, Thistle, *Wartime Letters*, 123.)

Canada's biological warfare program during the Second World War was part of a broader system of Anglo-American military cooperation that included a number of weapon systems – radar / sonar, radar / proximity fuze, explosives / propellants; poison gas / military medicine, and the atomic bomb. This defence science coordination, within the North Atlantic triangle, began well before the United States formally became a belligerent and was driven by mutual fear of Nazi Germany. In June 1940, for instance, it was the possibility of imminent invasion that forced the British government to seek assistance from the United States as well as to call upon its major military ally, the Dominion of Canada, to make every possible sacrifice to save the mother country. The advent of formalized British and American military cooperation was highlighted by the 1940 bases for destroyers deal and the creation of the ABC-1 defence arrangement of March 1941. Meanwhile, Canada and the United States had, through the Ogdensburg Agreement of August 1940, established the Permanent Joint Board of Defence with a mission to cooperate "in the broad sense the defence of the north half of the Western Hemisphere." Eight months later, the Hyde Park Declaration created a common North American market for defence production.[1]

These developments had an enormous impact on the organization and performance of Canadian defence science. During the first year of the war, the National Research Council (NRC) and the Department of National Defence (DND) were granted only very limited access to top-secret information about British weapons and virtually no entry into the world of American military technology. All this changed after the British scientific mission of September 1940, led by Sir Henry Tizard, one of the country's leading defence scientists. Now the British connection made it possible for Canadian scientists to become involved in the most sophisticated of American military projects. The fact that the United States was a non-belligerent until December 1941 also worked to Ottawa's advantage since information about various British and American weapons was often channelled through Canadian agencies.[2] As Dean C.J. Mackenzie, acting head of the NRC, would later admit, "if America had declared war in September, 1939, instead of twenty eight months later, the status of Canadian science would have been quite different from what it is today."[3]

There was a direct relationship between Allied chemical warfare and biological / toxin (BTW) warfare activities between 1939 and 1945. In all three countries the administrative structures and physical plants, initially developed for chemical warfare research purposes, were subsequently used by the biowarfare program. In addition, most of the BW delivery systems such as the 30-lb bomb, the 4-lb bomblet, and the 500 cluster bomb were adaptations of chemical munitions. Moreover, while the use of either biological or chemical weapons was outlawed by the 1925 Geneva Protocol, this was really not a deterrent for either Allied or Axis defence scientists, if military success could be achieved by using these devices. Yet despite the similarities of the two weapon systems, there were also significant differences. Unlike poison gas development, biological warfare research was secret and hidden, its practitioners shadowy and self-conscious. Most were medical researchers, not chemists, committed professionally to saving, not taking lives. But their weapons included some of the world's most terrifying diseases: bubonic plague, typhus, typhoid, dysentery, yellow fever, tularaemia, brucellosis, and anthrax, as well as the deadly botulinum toxin. And, unlike chemical weapons, these were primarily living pathogens whose battlefield use raised many alarming questions. Could they be controlled, or would both sides be crippled in a biological warfare exchange?

Canada's participation in biological warfare research, development, and planning in 1939 was related to four major factors. First was its involvement with the British military establishment through the Committee of Imperial Defence and, scientifically, through the British Medical Research Council. Second, Canada was fortunate in having many talented medical scientists and bacteriologists such as Frederick Banting and J.G. Craigie of the University of Toronto; Everitt Murray and J.P. Collip of McGill University; and Guilford Reed at Queen's University. A third factor was the institutional links which were established between the various university laboratories and the two patrons of this research: the National Research Council (NRC) and the Department of National Defence (DND).[4] After 1940, Otto Maass, one of Canada's most eminent chemists, assumed a major role in coordinating exchanges between these different wartime stakeholders, given his role as chairman of the Directorate of Chemical Warfare and Smoke (DCWS) and special assistant to NRC president C.J. Mackenzie.

Exploring the Bioweapons Threat, 1937–1942

Canada was one of the original thirty-eight states parties that signed the 1925 Geneva Protocol outlawing the use of chemical and biological weapons.[5] Despite its long-term importance in establishing a moral commitment against the use of germ and gas warfare, the creation of the Protocol attracted little public attention in 1925, largely because most Canadians felt that they should emulate American isolationism and avoid being involved in future European wars.[6] These anti-militarism viewpoints were reflected in drastic cuts in the budgets of the Department of National Defence throughout the interwar years, but particularly during the Great Depression. It was not surprising, therefore, that neither the Canadian Chiefs of Staff nor the Canadian public showed much interest in theoretical debates about different weapon systems.[7]

On the other hand, several disturbing accounts about the future threat of biological weapons appeared during the early thirties. One of these was the March 1933 article "Bacterial Warfare: The Use of Biological Agents in Warfare" by Major Leon Fox of the United States Army Medical Corps. Overall, Fox believed that this type of warfare faced overwhelming technical problems since most BW agents would either be destroyed by environmental factors (sunlight / wind) or by the impact of explosives, if placed in shells and bombs. And even if infectious disease agents were successfully sprayed, he argued, this offensive strategy could cause an uncontrollable epidemic that would decimate the aggressor's own armed forces and civilian population. Fox did admit, however, that anthrax spores had the potential of being an effective military pathogen, since the agent met all of the logistical requirements: "It must be quick acting, highly virulent, and capable in causing disease in small quantities. It must be highly resistant, capable of surviving outside the body under the most adverse conditions ... The causative organisms must be able to force its entrance through all the avenues of infection."[8] In 1934, an even more sensationalist report about biological warfare appeared when Wickham Steed, a British journalist, claimed that he had obtained secret German documents outlining plans to use anthrax and other biological agents against London's subway system. While the veracity of Steed's accusations was challenged by government experts, they became a catalyst for a series of frightening predictions about imminent biological warfare. In Canada, for example, the *Toronto Daily Star* ran a series of revelations under the lurid title, "Disease Germs Going to Flood Cities When Next War Comes."[9] But it was not until 1937 that Canadian defence scientists began to consider seriously the threat of an enemy BW attack, largely through the efforts of Nobel laureate Sir Frederick Banting.[10]

Banting's concerns first found public expression on 11 September 1937, when he warned NRC chairman General Andrew McNaughton that scientific and military evidence demonstrated that Germany was prepared to initiate germ warfare. With McNaughton's encouragement, Banting then prepared a detailed analysis of the various airborne, water-borne, and insect-borne agents which he believed could be used militarily.[11] For Banting the threat was serious and imminent: "An epidemic of disease would have a paralysing effect on both the military and civilian population ... By the invention of an automatically air-cooled shell, bacteria such as gas gangrene, tetanus and rabies could be added to the danger of the wounds inflicted, so that even a scratch would be deadly ... In the same way botulinus toxin could be introduced with deadly effect." Banting also expressed great concern that since the filtered water supply of Canadian cities was stored in open reservoirs, "if an enemy aeroplane were to drop a ton of live bacteria or water-borne diseases such as typhoid, cholera or dysentery, an epidemic would be a certainty."[12] Nor did he feel that the 1925 Geneva Convention against bacteriological and chemical warfare would deter either Germany or Italy since "scientists of these countries are carrying out research on practical utilization of these deadly weapons."[13] Banting's prestige, combined with McNaughton's endorsement, meant that the memorandum was passed on to the Canadian Army Chiefs of Staff and the British War Office.[14]

But biological warfare also had its British and Canadian detractors. One of the most prominent was the Cambridge University geneticist J.B. Haldane, who in a October 1937 public lecture, "Science and Future Warfare," declared that he was "very sceptical" about the possibility of using either bacteria or viruses as weapons.[15] Similar reservations were forthcoming from Dr N.E. Gibbons of the NRC, who had been given the task of analysing Banting's memorandum.[16] Banting, however, was not deterred by these naysayers and resumed his BW campaign during the 1938 Munich crisis.[17] In a letter to General McNaughton on 16 September 1938, he emphasized the need for Canada to obtain "adequate supplies of anti-serum and vaccines, particularly anti-tetanic serum, typhoid vaccine and supplies for active and passive immunity, both of soldiers and civilians."[18] McNaughton was quick to respond since many of Banting's warnings had also been raised by the Committee of Imperial Defence (CID) and its Subcommittee on Biological Warfare.[19]

Yet significantly, the final report of CID Sub-Committee emphasized the value of adopting a public heath approach through expanded emergency bacteriological laboratories, rather than creating a military-focused biodefence system.[20] This was also the opinion of the Canadian Army Chiefs of Staff, who decided to transfer the Banting file to the Department of Pensions and Health on the grounds that it was best suited to deal with all disease related issues.[21] After the outbreak of war in September 1939, however, Banting was able to convince the leadership of the National Research Council to support his personal mission to the United Kingdom with the goal of convincing Britain's defence scientists of the urgent need for BW preparedness.[22] In December 1939, at General McNaughton's request, he prepared a detailed report, "Memorandum on the Present Situation Regarding Bacteria Warfare." One of its key recommendations was the importance of developing an immediate retaliatory capability so that "British Forces should be in a position to give back in a ten fold measure any attack that the Germans may attempt." Banting also emphasized that effective BW defence could only be achieved "by means of a study of offensive warfare." He was not impressed with the arguments of W.W.C. Topley and other British consultants who considered germ warfare only in terms of naturally occurring diseases "since under the new conditions of bacteriological warfare, the disease producing bacteria may be scattered anywhere – and ... as long as the bacteria gain entrance to the body they produce disease."[23]

Banting's greatest challenge, however, was convincing Lord Hankey, chairman of the newly established Cabinet Committee on Bacteriological Warfare, about the seriousness of the enemy threat.[24] From Banting's perspective their January 1940 meeting went rather well, and he was pleased with Hankey's compliment that he was the first "scientist of standing" who had presented a "definite proposal."[25] But Hankey's feigned interest was more an exercise in Commonwealth diplomacy than an acceptance of Banting's "somewhat alarmist view" of the danger of a German biological weapons attack.[26]

Nor was the Canadian government prepared to support a major biodefence program without British support.[27] Yet despite these setbacks, Banting continued his efforts to create a core group of Canadian BW researchers, starting with bacteriologists P.H. Greey and Donald Fraser of the University of Toronto, along with virologist James Craigie, of the world-famous Connaught Medical Research Laboratories.[28] One of the earliest goals of this small BW team was to investigate the capabilities of different agents and methods of dissemination, an exercise that included dropping a payload of coarse sawdust from a bush plane over Ontario's Balsam Lake, designed to simulate a bioweapons aerial attack, by either bombs or insect vectors.[29] Banting also recruited other Canadian scientists into the project such as Dr Charles Mitchell, animal pathologist with the Department of Agriculture, an authority on rinderpest; and Guilford Reed of Queen's University, one of the country's leading experts on anthrax and botulinum toxin.[30] Another participant was the noted McGill microbiologist Everitt Murray, who had already been involved in biological research as a member of the subcommittee on gas gangrene. He and Banting soon became confidants, given their shared belief that Nazi Germany was poised to adopt germ warfare strategies and that "the only safe defensive position against any such weapon is afforded by a thorough understanding which can only be gained by a complete preparation for the offensive use."[31]

These suspicions appeared vindicated in May 1940 with the dazzling German military successes in Holland and France, a situation that Banting partly attributed to the use of biological weapons. Even worse, he was convinced that the Germans would undoubtedly use germ warfare "if England … is invaded."[32] As a result, Banting renewed his efforts to convince the National Research Council and the Department of Defence to begin large-scale experimental work on both the defensive and offensive dimensions of biological warfare.[33] It was a hard sell.[34] But in the end, C.J. Mackenzie managed to provide Banting with about $50,000 of seed money out of the so-called "Santa Claus Fund" until British support was forthcoming.[35] Indeed, Banting's 21 February 1941 tragic airplane accident, off the east coast of Newfoundland, occurred while he was en route to London to lobby to advance his biodefence project.[36]

Significantly, by this stage in the war the British Cabinet Committee on Bacteriological Warfare had already moved the country towards an institutionalized response to germ warfare. Of crucial importance was Lord Hankey's ability to convince Prime Minister Churchill, through somewhat devious means, of the need for a bioweapons offensive capability.[37] As he explained: "Broadly speaking my own view is against initiating frightfulness of this kind … [but] the right course is to investigate the practicability of the proposals … in case they are used against us … [whereby] we put ourselves in a position to protect our own interests, if they are attacked and to retaliate if the Cabinet should decide."[38] For his part, Churchill's support for an aggressive bioweapons policy was not surprising, since he had pushed hard for the offensive use of gas warfare during the summer of 1940 if a German invasion should occur, despite arguments from some of his military advisers that Britain would be "throwing away the incalculable moral advantages of keeping our pledges … [Geneva Protocol] for a minor tactical surprise."[39]

Chemical and Biological Warfare Preparedness, 1941–1943

Throughout the early stages of the Second World War, Canada provided substantial support for the British chemical weapons program in the fields of agent research and the development of munitions and delivery systems.[40] In 1941, however, Porton Down officials had another request. Could Ottawa supply a large-scale experimental testing facility where British and Canadian scientists could improve the operational efficiency of existing chemical munitions and develop even more deadly war gases?[41] Once the Suffield, Alberta site had been selected, there were sustained negotiations between the British and Canadian governments about shared costs, administrative responsibilities, and the long-term consequences of this shared project. Under the final agreement of 1 March 1941 it was stipulated that Experimental Station Suffield (ESS) "would consist of two branches, one for research and development, and the other for field trials comparable with the actual use of gas in warfare."[42] It was also agreed that the costs of land, buildings, equipment, materials and maintenance would be shared equally,[43] while the Alberta government agreed to lease the 700,000-acre site to Ottawa for ninety-nine years at a cost of one dollar per year.[44] Responsibility for administering the station would rest, however, with the Department of National Defence, a commitment the Army Chiefs of Staff reluctantly agreed to assume.[45]

Canada's chemical warfare operation was greatly enhanced by a number of administrative changes during the spring and summer of 1941. Of particular importance was the establishment of the Directorate of Chemical Warfare under the jurisdiction of the Master General of Ordnance (Army), with McGill chemist Otto Maass as its full-time director.[46] And under Maass's watchful eye Experimental Station Suffield gradually evolved, with the construction of laboratories, toxic storage facilities, and residences for its 200 employees. Arrangements were also made with the Canadian Pacific Railway to build a spur line to the Station, while the Alberta government agreed to construct an all-season road and establish telephone connections between ESS and the outside world.[47] Protecting the Station from enemy agents was another priority, with the local RCMP detachments and military police assuming responsibility for safeguarding the perimeter. But the most important development was the arrival of a cadre of capable and respected scientists.[48] Among the original scientific "stars" were University of Toronto chemist Dr H.M. Barrett, who was appointed superintendent of research, and Major J.C. Paterson of the Royal Canadian Army Medical Corps, who was named head of the physiological and pathological section.[49] The Superintendent was E.L. Davies of the Porton chemical warfare laboratories, who brought a wealth of research and administrative experience to his challenging new job.[50]

On 29 December 1941 the Directorate of Chemical Warfare issued a warning that since "the possibility of the outbreak of C.W. in the Pacific sphere of operations would seem very likely ... there is every chance that it would spread to other zones."[51] As a result, plans were made to equip Canadian troops, both at home and abroad, with protective equipment and gas munitions for retaliatory purposes.[52] But the most controversial issue was whether Canadian Forces should actually initiate the offensive use of chemical weapons or, more specifically, use mustard gas against Japanese assault forces landing on the coast of British Columbia.[53] This policy was vigorously promoted by the CW Inter-Service Board at its 22 March meeting. Fortunately, no invasion occurred.[54]

Canada's chemical warfare policies were also influenced by the 10 May 1942 warning of Prime Minister Winston Churchill that German use of poison gas against the Soviet Union, Britain's wartime ally, would be regarded "as if it were used on ourselves ... and we will use our great and growing air superiority in the west to carry gas warfare on the largest possible scale far and wide upon the towns and cities of Germany." On 5 June President Roosevelt gave a similar warning to Japanese authorities: "I desire to make it unmistakably clear that if Japan persists in this inhuman form of warfare against China or against any other of the United Nations, such action will be regarded by this government as though taken against the United States, and retaliation in kind and in full measure will be meted out."[55] In response to these two important proclamations, the Canadian War Committee attempted to bring its policies into line with those of its British and American allies, particularly with growing evidence that chemical warfare might erupt in either the European or Pacific theatres.[56]

Significantly, in his June 1942 address President Roosevelt made no reference to the Japanese BW threat, despite evidence that these weapons had been used against Chinese troops and civilians.[57] But at this stage in the war the level of American biological warfare preparedness was seriously underdeveloped because of limited funding, the military's general distrust of CBW weapons, and serious jurisdictional disputes between the Chemical Warfare Service and the Surgeon General's Office.[58] Because of this leadership deficit in the US Army, most of the original ideas for biological agent development came from civilian bacteriologists and virologists, who operated out of the so-called WBC Committee, under the chairmanship of Professor Edwin B. Fred of the University of Wisconsin.[59] The importance of this organization was evident in February 1942 when the WBC submitted its first report to Secretary of War Henry Stimson that called for a comprehensive study of biological warfare, with a special warning that the Japanese might deploy these weapons operationally in the Pacific theatre and possibly against the American mainland.[60] In turn, Stimson informed President Roosevelt about the importance of decisive action: "Biological warfare is, of course, 'dirty business,' but in light of the Committee's report, we must be prepared."[61]

Similar discussions were taking place in the United Kingdom. In late December 1941 Lord Hankey, chairman of the Bacteriological Warfare Committee, notified the War Cabinet that the Porton research team under bacteriologist Dr Paul Fildes had been carrying out experiments with biological agents and toxins, which could be deployed in a retaliatory attack, although their only operational weapon was "the use of anthrax against cattle by means of infected cakes dropped from aircraft."[62]

Impressed by Hankey's arguments, the Chiefs of Staff agreed that the Biology Section at Porton should be the focal point of Britain's offensive BW research and development program. By May 1942, Fildes was able to justify the military's confidence by providing a preliminary report on his attempts "to dispense anthrax spores by explosion in such a way as to create an invisible cloud of spores travelling down wind, in the hope that anthrax will be produced by inhalation."[63] Plans were also made to invite "gifted" American bacteriologists to Porton's Biology laboratory, as a means of improving bilateral relations and impressing the leadership of the US Chemical Warfare Service "with the scientific quality of the work which was being done, the thoroughness and vision with which it was planned ... for prompt retaliation with this type of warfare, should it first be instituted by the enemy." And it was a good investment. In April 1942, James Defandorf, one of the visiting CWS scientists, informed his superiors about Porton's move towards an offensive BW capability, including the use of anthrax cattle cakes and "a promising but untested method of using anthrax against man ... [and] a highly potent botulinus toxin."[64]

Canadian scientists were also important participants in this evolving Anglo-American biowarfare alliance. Indeed, the groundwork for cooperation with US specialists had been established in December 1940 when Banting had carried out extensive correspondence with Dr M.V. Valdee, chief of the Division of Biological Control, National Institute of Health, one of the key members of a secret US group that was considering "the possibilities of the use of bacteria and other similar substances in warfare."[65] These contacts continued after Banting's death, largely through the efforts of McGill's Everitt Murray, who took the initiative to attend the December 1941 meetings of the WBC Committee in Washington. After establishing their many common interests, Murray invited American BW specialists to visit Ottawa for extended discussions about possible BW threats and how the two organizations might pool their scientific resources.[66] This liaison system was greatly enhanced by the fact that E.B. Fred and Murray soon became close confidants, based on professional respect, warm personal relations, and a shared awareness of the seriousness of the bioweapons threat.[67] Both men also appreciated that their respective national programs desperately needed British assistance, given Porton Down's achievements in developing suitable BW agents and delivery systems.[68]

On 19 December 1942 a crucial meeting in Ottawa of the newly formed biodefence organization (code named the M-1000 Committee) took place with Everitt Murray in the chair.[69] The discussion was lively and the recommendations numerous.[70] High on the list of concerns was the possibility of enemy agents using biological weapons against Canada's civilian population through acts of sabotage.[71] Another serious problem was the vulnerability of Canada's 8 million cattle to the rinderpest virus, which was known to spread "rapidly, extensively and disastrously." Since the disease was virtually unknown in North America, it was recommended that Canada develop its own vaccine supply, hopefully with US assistance.[72] But the most intense discussion focused on the question of whether biological warfare would occur, and, if so, which pathogenic or toxin agents would the Germans and Japanese deploy?[73] Botulinum toxin was Guilford Reed's choice, because of its extreme toxicity and because large-scale production was "very easy." There were, however, other fearsome candidates such as the viral diseases yellow fever and Rift Valley fever, with the latter deemed particularly dangerous since the virus was "hardy and the disease difficult to diagnose."[74] But in order to develop a more comprehensive approach, the Committee recommended the preparation of an inventory of likely BW agents, with each of the eleven participants being asked to prepare an analysis of specific pathogens and toxins for review at the next M-1000 meeting.

How best to organize biowarfare research was another subject of debate. As chairman, Murray took the lead in outlining the advantages of having long-term contractual arrangements with both Canadian universities and private companies. This approach was strongly endorsed by Sir Henry Dale, representing the British Medical Research Council, who claimed that the British model of biodefence research had profited from "the decentralization of laboratories," combined with an effective "register of pathologists, bacteriologists, building and equipment." Both Dale and Murray also stressed the importance of developing an effective exchange system for the efficient transmission of BW documents "from the English Committee to this and the American W.B.C."[75]

The most contentious issue was the proposal for a joint Canada-US rinderpest research station at Grosse Ile. For Murray, the advantages of the site were obvious and compelling: the island was isolated, yet only thirty-five miles from Quebec City; it was uninhabited, but had a useful physical infrastructure left over from its days as an immigrant quarantine station; and above all, it was available, and could be immediately brought under the control of the Department of National Defence. Many other delegates agreed, including Captain Stephenson of the US Navy Medical Service, who pointed out that the yellow fever virus had been "produced in the middle of New York City," and wondered "if the virus of rinderpest was really any more difficult to handle than that of yellow fever." This line of argument was, however, challenged by veterinarians E.A. Watson and Charles Mitchell, who claimed that "four miles of water did not afford sufficient barrier against the spread of Rinderpest."[76] In the end, there was a consensus that the Grosse Ile project should proceed, with the notable exception of Sir Henry Dale, who made it clear that British BW experts were not concerned about enemy use of the rinderpest virus.[77]

Murray's next task was to convince the hierarchy of National Defence that they should support a comprehensive Canadian biodefence system, in general, and the Grosse Ile rinderpest project, in particular.[78] Initially there was a hostile response. What was the purpose, the generals asked, for Canada to pursue this odious form of warfare when neither Britain nor the United States had indicated they would "engage offensively in Bacteriological Warfare."[79] Faced with this opposition, Murray turned to Washington for support.[80] Would it be possible, he asked WBC officials, for Secretary of War Henry Stimson to endorse the Grosse Ile undertaking since it "would carry great weight in deciding the question"?[81] This request was quickly granted.[82] Given this clear evidence of US support, the Canadian Cabinet War Committee soon endorsed the establishment of the War Disease Control Station (WDCS).[83]

Under the final agreement of August 1942, the rinderpest project was placed under the jurisdiction of the Joint US-Canadian Commission of eight members, chosen by Otto Maass and General William Porter, head of the US. Chemical Warfare Service (CWS).[84] Financial support was also divided: the Canadian government provided $300,000, along with the physical site, while the US Army paid 75 per cent of all costs associated with new buildings, equipment, and general operating expenses.[85] Administratively, a number of initiatives were undertaken during the fall of 1942 to ensure that the rinderpest production facility would soon become operational. One of the most important was the appointment of Dr Richard Shope, a highly regarded virologist from the Rockefeller Institute and Princeton University, as scientific director.[86] It was also decided that Charles Mitchell of the Department of Agriculture would be the vice-director of research, while Colonel J. Dickinson of the Canadian Army was named base commander. In October 1942 other important decisions were worked out, such as the location of the rinderpest laboratory and pilot plant, when the Joint Commission visited the island for a second time. On this occasion, it was also decided that Shope would personally select the scientific personnel for the project and that essential equipment would be obtained through the auspices of the CWS and the US Army Medical Corps.[87] Yet, despite its formal dual character, most of the policies affecting the War Disease Control Station came from Ottawa not Washington.

Canadian and Allied BW Research and Development, 1942–1944

By the spring of 1943 the administration of Canada's biological weapons program had become integrated into the Department of National Defence. As a result, ultimate decision-making authority over BW matters rested with the minister, Colonel J.L. Ralston, and his advisers, the Chiefs of the General Staff. Under this system, once a policy had been determined, instructions would be issued through the Master General of Ordnance to the Directorate of Chemical Warfare and Smoke, headed by Dr Otto Maass. In reality, however, this hierarchical system was a myth, largely because the leadership of the Canadian Army rarely showed any interest in monitoring Canada's biological warfare developments throughout the war years. Consequently, most of the essential decisions came from the triumvirate of Otto Maass, Everitt Murray, and Guilford Reed.

On the research side, the most important facility was Reed's Queen's University laboratory, occupying the top floor of the Medical Science Building "of the New Medical Science Building, along with the adjoining New Animal House, a 2664 square ft. specially designed brick building to contain "offensive odors ... and [maintain] a sanitary condition."[88] It was here that Reed's small research team, including medical bacteriologist John Orr, J.J.R. Campbell, J.W. Stevenson, and his nephew Roger Reed, assessed the BW weapons potential of dangerous pathogens such as *Yersina pestis* (plague), *Malleomyces mallei* (glanders), yellow fever virus, *Bacillus anthracis* (anthrax), and botulinum toxin (botulism).[89] Ultimately, Reed decided that anthrax and botulinum toxin had a decided edge over the other candidates, because of their high killing power, ability to maintain their virulence, resistance to environmental degradation, and good storage capability. In addition, the Kingston laboratory developed two innovative methods for the dissemination of pathogens, namely the use of finely ground peat dust and the use of insect vectors.[90] On the administrative side, Reed assumed an important role in coordinating the anthrax project at Grosse Ile and the field testing facility at Suffield, while at the same time maintaining close contact with Paul Fildes at Porton. He also accompanied Murray at numerous meetings with US defence scientists in Washington, DC and at Fort Detrick, the nerve centre of the American biowarfare program..

During the summer of 1942 there were major organizational changes in the American biological warfare operation.[91] One of these was the appointment of pharmaceutical mogul George Merck as chairman of the newly created War Research Services (WRS), a secret civilian organization which was given responsibility for coordinating BW developments in the United States and maintaining close contact with British and Canadian research teams.[92] Indeed, in October, after meeting with Murray in Washington, Merck decided to arrange a special visit to Quebec City and Ottawa. It was a great success.[93] According to NRC president C.J. Mackenzie, the new American BW chief was "an extraordinarily fine man ... [who] is many times a millionaire ... [and] is a personal friend of the president's."[94] Even more important, Canadian officials were impressed with Merck's questions about the progress of the GIR-I project, his positive response towards their role in the evolving tripartite biological warfare alliance, and his knowledge about Anglo-Canadian plans for the weaponization of anthrax.

Canada's involvement with the development of anthrax bombs was directly associated with the November 1942 mission of Dr Paul Fildes and Dr David Henderson to North America.[95] Their first stop was Washington, where they attempted to achieve three major goals: "To encourage the formation of an effective BW service; To seek help in specific items of large-scale production; To suggest certain lines of research which we could not undertake."[96] But there were a number of obstacles preventing these goals from being realized, particularly the jurisdictional disputes between the War Research Service, who were mandated to carry out the research and development work on BW agents, and the CWS, which was authorized to produce the munitions. In contrast, the Canadian triumvirate of Otto Maass, Everitt Murray, and Guilford Reed appeared competent and willing to assist Britain's BW program. And, as Fildes assured his superiors, the Dominion had much to offer:

> On the field trial side the C.W. experimental station at Suffield is available. Suitable laboratories are in course of erection, and these will probably be administered by Major J.C. Paterson, R.C.A.M.C. ... Before I left America I heard that negotiations for permitting the use of dangerous materials at Suffield had been concluded successfully ... Production in a (N) semi-scale plant has been underway for some time at the University, Kingston,

> and a very large plant already existing on Grosse Isle in the St. Lawrence river has been taken over.[97]

In his meeting with Canadian defence scientists, Fildes made a strong case for the battlefield potential of anthrax spores, based on recent laboratory experiments and cloud chamber tests at Porton.[98] His most important revelation, however, was the detailed account of the 1942–3 field trials that had been carried out at Gruinard Island, in northern Scotland, which, he claimed, showed that exploding anthrax bombs could create a spore cloud "effective against a man one mile distant."[99] This briefing was followed by an impassioned request from Fildes: would Canadian scientists help Porton develop sufficient quantities of anthrax munitions (code named N) for a retaliatory response against a possible German germ warfare attack? Predictably, the response was positive, and an agreement was quickly signed.[100] Nor did the C-1 Committee waste any time in advancing this novel undertaking. On 25 November, Murray and Reed made an emergency trip to Grosse Ile to arrange the "most satisfactory plan for reconstruction of the [anthrax] building." Of even greater importance, Murray was able to convince Otto Maass that $40,000 was required for the purchase of vital scientific equipment "which must be bought immediately in order that the 'N' project can be put into production quickly."[101] This brief also praised the potential of the anthrax (GIN) project:

> It is entirely Canadian; It is designed to provide "N" for the British authorities immediately and in suitable quantity. If practice comes up to our calculations we hope to produce 300 lbs of "N" spores per week; this would provide for 1500 of the 30 lb. bombs per week. Should the U.S. wish to join us in the "GIN" Project, CI would not raise any objections, but there are no indications that they are prepared to do so immediately.[102]

As it turned out, Murray's scepticism about American intentions proved unfounded. In reality, Merck and his WRS colleagues were greatly interested in the GIN project, and during the spring of 1943 the United States became a full partner in this operation. Indeed, under the subsequent agreement, the War Research Services assumed three-quarters of GIN's operating costs, while providing additional technical support and equipment.[103] American bioweapon officials, however, adopted quite a different position towards the British BW program, despite extensive lobbying by Edgar Lord Stamp, who became Porton's official envoy in

North America in January 1943.[104] Frustrated with this lack of progress, in April 1943 Fildes proposed an elaborate division of scientific labour.[105] As he told officials of the Chemical Warfare Service, "if you make the bombs and the charging, it will obviously be better for many reasons to fill on your side ... since ... these materials are more likely to be used in Japan than in Germany." Fildes was adamant, however, that Porton should continue to provide the improved anthrax seed stock (Vollum Strain), design the delivery system, and supervise the necessary field trials. At the same time, he acknowledged that Porton still had not resolved a number of serious technical problems:

> We have as yet no bomb which is satisfactory for dispersing N. We are at work ... trying different designs filled with bacteria ... [which] are harmless but otherwise comparable to N ... I suggest that an area is required for experiments with weapons of this sort, namely where bombs up to 30 lbs can be detonated and sampled ... We hope that similar work will be undertaken in U.S.A. on parallel lines in order that an adequate bomb may be available for filling when your N production comes into operation."[106]

Grosse Ile Production Plants for Rinderpest Vaccine and Anthrax Spores

Overall, the two biological warfare projects at the War Disease Control Station at Grosse Ile made considerable progress during 1942–3. In the case of GIR (rinderpest), Richard Shope's small research team of eight scientists (six Americans, two Canadians), along with technicians and support staff, pushed ahead with their challenging research project. Throughout this difficult period, continuous contact was maintained with Everitt Murray, who was appointed full-time coordinator by the Joint US-Canada Commission. Fortunately, these two eminent microbiologists worked well together, given their extensive laboratory experiences and their mutual interest in developing a viable rinderpest vaccine "which could be produced more economically than known vaccines."[107] This was a daunting undertaking on several counts. First, there was the difficulty of converting an old immigration medical screening centre into a modern microbiology laboratory while maintaining an acceptable level of safety. In part, this was achieved by building ten completely isolated research units, along with an anteroom for infected animals. The most formidable challenge, however, involved the development of an innovative scientific technique for the production of an inactivated egg-cultivated virus which would produce an effective and safe rinderpest vaccine. The system worked as follows: "The calves were inoculated subcutaneously either with egg culture virus or bovine splenic virus and killed ... on the second or third day of fever ... Final storage of the vaccine was in 500-cc serum bottles capped with vaccine-type rubber stoppers ... with full retention of its potency for one year."[108]

At the end of December 1942 Murray provided Otto Maass with a positive assessment of the GIR project, which he claimed was largely due to Shope's leadership and scientific creativity. He did, however, note that the scientific director was often "unduly impatient and magnifies difficulties out of proportion because of his anxiety to get started."[109] What Murray failed to mention was that Shope had good reasons for being frustrated, given the physical isolation of the island, the problems of obtaining vital laboratory supplies, serious flaws in the military support system, and deplorable living conditions. As Shope bitterly recorded in his report of 20 July 1943, "If our Station were in Labrador or up in Hudson Bay ... I could see some sense to being completely cut off for 5 months. But here we are within 35 miles of the city of Quebec where there are facilities ... [and] supplies. This extremely unpleasant experience did little to boost morale here ... because you get too damned hungry." On the other hand, despite these obstacles, Shope's research team achieved amazing results. As he explained, "the colloid mills seem to be working satisfactorily and the resulting vaccine makes a nice looking product ... Things at the laboratory are going marvelously well and we are having an extremely interesting time of it. Believe that after the war, it can be justly said the 'R' Project, in addition to serving as a War measure, paid its own way as a strictly scientific Project."[110] Given these encouraging results, the US-Canada Joint Commission attempted to rectify the situation by recruiting additional scientific workers, by improving the system of obtaining laboratory supplies and test animals, and, above all, by establishing an all-season ferry service with Quebec City. Another key initiative was the appointment of Colonel A.E. Cameron as superintendent of the Station, with a mandate to establish a more orderly and disciplined work environment. For Shope this was a welcome change, and he applauded Cameron's management style: "He talks my own language ... and I feel that I can go to him with some of my laboratory problems and get a sympathetic ear."[111] But it was a relatively brief working relationship. In October 1943 Shope was informed by his US Navy superiors that he would be transferred to another American biological warfare project, where, it was reasoned, his Grosse Ile experience would be a useful asset.[112]

During the winter of 1942–3, Grosse Ile became the site of another top-secret research undertaking: the production of weaponized anthrax spores. The project began in November 1942 with the Anglo-British agreement that Canada's C-1 Committee would supervise the development of facilities, at the War Disease Control Station, to ensure that sufficient quantities of anthrax spores were produced for British experimental purposes.[113] During its early days the GIN project was closely monitored by Paul Fildes, who expedited its work by forwarding large numbers of Porton technical reports to Guilford Reed's Queen's University laboratory, which was carrying out experiments to increase the virulence of the Porton Vollum 98 anthrax strain.[114] Fildes also arranged for the GIN project to receive a sophisticated anthrax spore charging machine, to streamline filling the empty 4-lb bombs, which would be shipped from Porton to Grosse Ile, with the "N" munitions being subsequently transported to Suffield for the October 1943 field trials.[115]

Murray strongly endorsed this arrangement, since in his opinion the Anglo-Canadian anthrax project had reached the stage "where weapons development and Field Trials are essential."[116] The problem was that the GIN operation could not meet these expectations due to a series of logistical difficulties. First, there were technical challenges in constructing the production facility, since it was necessary to pump air through the corn steep growth medium "in order to get the anthrax to sporulate."[117] Second, Murray's request for an anthrax research team of fifteen scientists, including three qualified bacteriologists, did not materialize, and the laboratory remained short-staffed for the duration of its existence.[118] Third was the failure of local military officials to provide the necessary supplies, notably the 5,000 Bakelite trays required for mass producing N within the autoclaves or sufficient numbers of laboratory animals. But the essential problem was that the GIN operation was based on outdated technology, with limited productive capacity, particularly when compared with the large-scale fermenters being constructed at Camp Detrick for US anthrax production. Even worse were the biosafety issues connected with producing liquid spores (called N mud), which were described as being "thick grayish white material not unlike oatmeal gruel [and] is essentially a mass of N spores with a few vegetative forms and debris."[119] In October 1943, for example, Murray reported a problem with leaking gaskets in the three aged autoclaves (built in 1893), which, he claimed, were only kept in operation because of the pressure "to produce sufficient 'N' for the experiments at Suffield this fall."[120] Nine months later the possibility of anthrax infection became a reality when some of the scientific staff had to be hospitalized in Quebec City "with an unexplained fever."[121] Fearing a serious outbreak on the island, Murray ordered the immediate requisition of special protective clothing and masks, as well as special equipment to supply fresh filtered air. For a time these measures were effective, but serious health hazards continued to disrupt the GIN operation.[122]

By August 1944, there seemed little point in undertaking major renovations at the production facility since Washington had given notice that the Canadian-American anthrax contract would be cancelled at the end of the month.[123] As a result, the C-1 Committee decided that all Canadian GIN personnel would be transferred either to Suffield or Kingston, along with most of the equipment.[124] The exception was the anthrax charging machine, which was personally delivered by Major Duthie to Camp Detrick, along with a 100-cc sample of "high quality" anthrax.[125] But despite its problems, during its eighteen months of operation the Grosse Ile anthrax project had produced 382 litres of the Vollum 98 strain of anthrax spores, "equivalent to some 2,000 Type F 4-lb bombs," along with 439 litres of the Bacillus globigii (U) simulant.[126]

Even more impressive was the large-scale production of rinderpest vaccine at the War Disease Control Station prior to June 1945, when the project was terminated. According to the final report of the Joint Commission, the integrated team of Canadian and American scientists successfully produced over 1 million doses of the vaccine for military and civilian purposes, through the innovative techniques implemented by Richard Shope and the dedicated work of the small US-Canadian research team. As the report pointed out,

> the mission ... was of a two-fold character. In the first place it called for the prompt development of ways and means for the safe production and stocking of a type of rinderpest vaccine ... [and] it was likewise highly important to institute research studies to determine whether an effective vaccine could be produced without having to utilize cattle as the source of virus ... After a number of attempts, success was attained in establishing the virus in developing chick embryos ... Aside from the insurance provided by the rinderpest investigations ... the findings and knowledge gained from this work will be of great post-war value.[127]

Biological Weapons Testing at Experimental Station Suffield

Canada's other major biological warfare facility was Experimental Station Suffield (ESS), with its approximately 1000-square-mile testing range.[128] By the end of the war, ESS would accommodate 584 employees, including 50 professionals, 100 technicians, and 60 field staff "trained in the exacting techniques employed in carrying out large scale experiments."[129] In 1943, however, Suffield's BW operation was much smaller, with only limited laboratory facilities and a slowly evolving field trial capability. But the C-1 Committee was determined to rectify this situation, in cooperation with Superintendent E.Ll Davies and with the assistance of visiting Porton experts.[130] In May 1943, for instance, Lord Stamp provided the British Biological Warfare Committee with details about Suffield's BW testing area:

> The proposed site of the enclosed square mile ... is some thirty miles to the N.E. of the Station. About four miles to the S.W. there is a lake near which is the proposed site of the camp. A mile or so from the proposed enclosed area there is a spring of fresh water, near which it is proposed to make an enclosure for the infected animals. Some thirty to forty miles to the N.E. is the boundary of the main Suffield enclosed area, so there is a safety zone of some thirty to forty miles with the prevailing S.W. wind ... It remains to be seen whether the material [anthrax] for this summer's trials will be forthcoming ... The ordering of all the equipment for the trials appears to be well in hand and while we were at Suffield a preliminary dropping trial of peat powder and fly bait was carried out.[131]

There were a number of reasons why the C-1 Committee was determined that the proposed "N" field trials should be carried out expeditiously. First, there was continuous pressure from British officials warning Murray about the desperate need for a retaliatory capability because of the anticipated German BW attack. Second, the project had the support of the American WRS Committee, which gave assurances that it would "collaborate with our Canadian friends on 'N' munitions."[132] And finally, there was the opportunity of preparing ESS scientists for large-scale bioweapons field trials, a learning curve that was enhanced by the use of a small bursting chamber which provided a flow area of 224 cubic metres for the study of "weapon produced aerosols under controlled conditions ... of cloud coagulation and stability and for assessment of the destructive effects of bursting munitions on labile biological agents."[133] Equally important was Major J.C. Paterson's innovative research on the survival rate of bacterial pathogens when exposed to intense heat, excessive drying, detonation waves, and ultraviolet light.[134]

Porton scientists also helped prepare their Suffield colleagues for the forthcoming "N" field trials.[135] Of particular value were the reports of the 1943 anthrax stationary and bombing trials at Gruinard Island, which demonstrated the effectiveness of the improved anthrax seed stock (Vollum strain) and the advantages of using the 4-lb bomb (F-4 Mark 1), filled with 320 mL of anthrax spore suspension.[136] Logistically, it was decided that about fifty of these small devices (bomblets) would be packed into the 500-lb cluster bomb, which would subsequently be "released from 10,000 ft. and ... fuzed to open at 3000 ft, " producing a substantial lethal aerosol cloud.[137] The second level of planning involved the use of two aircraft outfitted with high-precision Sperry bomb sights and photographic equipment to record the impact of the bomb burst on the group of sheep tethered along the projected bomb explosion route.[138] According to Porton experts, once they were exposed to this aerosol cloud the test animals would experience "death from anthrax septicaemia ... confirmed ... by microscopic examination of a smear of peripheral blood."[139] In the final stages of the experiment, all the sheep would be slaughtered and cremated in order to prevent anthrax contamination at the ESS target site and neighbouring farmlands.[140]

Yet despite this meticulous planning, Fildes remained apprehensive about the ESS trials since, as he told Otto Maass, "the scale of the operation is ... much larger than we have previous attempted, but we hope you will be able to bring it off."[141] This would not happen. At the very last minute, Fildes sent an urgent telegram to the C-1 Committee advising that the cluster bomb test be cancelled, in part because it represented a much more dangerous situation than the previous Gruinard trials since it was "the first time in which a bacteriological weapon had been tried out on the full scale." This apprehension was reinforced by reports of a disease outbreak in northern Scotland, where, as Fildes related, "a more or less inexplicable case of anthrax occurred 10 miles away from the explosion of the [Gruinard] 4-lb bomb. According to our meteorological advisers, we can hardly be responsible for it, but I am not myself so certain." Given this alarming situation, Porton's chief BW scientist reasoned that "if a single 4-lb bomb is responsible for killing a cow at 10 miles, the idea of letting off fifty 4 lb. bombs at Suffield, with only 20 miles of safe distance, would seem to be associated with considerable risk, and I must say that I would hesitate before staging such a trial." In his concluding arguments, Fildes recommended that the C-1 Committee make arrangements with US biological warfare officials "for using Camp Wise [Dugway], and I hope that this can be done if you decide that the risk is too great for Suffield."[142]

Not surprisingly, Canada's BW team were stunned by these developments, but their responses were quite different. For Guilford Reed, the cancellation had major policy implications since it demonstrated "serious deficiencies in [ESS] laboratory equipment for anything but very small and simple field trials." This situation was complicated by the Station's lack of qualified bacteriologists, thereby supporting Reed's concerns "that a man who does not understand the nature of the agent he is handling is not likely to give that agent the best treatment." This critique brought an immediate rebuttal from Superintendent Davies, who denounced Reed's scientific elitism and his narrow Ottawa / Kingston perspective.[143] While Otto Maass disapproved of this squabble between key members of his BW team, he quickly defused the situation by reminding his scientific colleagues about the importance of keeping Canada involved with this important weapon system. To ensure further progress, Maass proposed a new set of guidelines. First, since production of weaponized anthrax spores was taking place at both Kingston and Grosse Ile, he decreed that Guilford Reed should assume responsibility for "the filling of bombs for experimental purposes ... [or] small scale experiments ... at Suffield." Second, he strongly endorsed Fildes's recommendation that the anthrax trials be shifted to Horn Island, along with the ESS experimental team, on the grounds that this would provide not only a suitable testing area, but also a catalyst for having Canadian and US scientists "working together for a period in order to get mutual experience of methods of technique and assessment which are used in the two countries." As a result, in December E.G.D. Murray began negotiations with Colonel M.S. Chittick of the US Chemical Warfare Service about possible joint trials in the gulf coastal region:

> Mr. Davies is not worried about carrying out the entire procedure outlined (in Fildes Bio / 2461, Sept 27th, 1943) but the Committee were not satisfied that it could be done with reasonable safety at Suffield. The contamination of a very large area would be involved and must be regarded as permanent for soil and climate existing at Suffield and are liable to be spread by the large numbers of gophers and antelope there. Adequate decontamination

of the area is not considered possible. It was thought the area now available at Granite Peak might be considered not large enough for an experiment of this scale, though soil, climate and animal life are more suitable than Suffield. Your views on this question would interest us very much.[144]

The CWS hierarchy were intrigued with Murray's proposal, and plans were made for Suffield and US scientists to carry out a number of joint trials at Horn Island during the winter of 1943.[145] For the most part, this involved bacterial simulants that replicated the behaviour of BW agents such as anthrax, plague, and brucellosis. The exception were the small-scale "hot" trials with bolutinum toxin (X), since there was no danger of contamination or an infectious disease cycle.[146] Another project of common interest was the use of insects as BW vectors since Reed's Kingston Laboratory had already made considerable progress in breeding and studying the *Musca domestica* (common housefly) and *Aedes sollicitans* (salt-marsh mosquito), demonstrating that these vectors could be used to carry plague and other infectious diseases.[147] While a joint Canadian-US team of entomologists and microbiologists was established at Horn Island, their trials enjoyed only limited success. Nor did this vector program make much progress during the remainder of the war, a trend that was subsequently criticized as being "one of the great shortcomings of Allied biological warfare planning."[148]

Considering the Offensive BW Option

In December 1943 Paul Fildes prepared a comprehensive assessment of Porton's anthrax bomb project for the British Cabinet Biological Warfare Committee.[149] He began his report with the observations that during the previous two and a half years the Porton Biology group had been engaged "in the development of a project for dispersing N spores by A/C bombs to produce an airborne cloud of spores capable of killing man by inhalation."[150] Considerable success, Fildes claimed, had also been achieved in the production of anthrax N spores, including methods for increasing virulence, improving stability in storage, and enhancing the casualty-producing capability of "N" bombs (4 lb. H.E. /Chem. Type F, Mark I). On the negative side, Fildes expressed his disappointment that meaningful assistance from the United States had not been forthcoming. And unless Washington changed its policy, Fildes warned, there was little possibility that an Allied retaliatory BW weapon would be available "if required, in this war."[151]

This pessimistic outlook was not shared by his US colleague Lord Stamp. In February 1944, for example, he informed Fildes that the leadership of the American BW program were becoming quite interested in the British retaliatory doctrine "to try to be ready to retaliate effectively in as short a time as possible in the event of the enemy commencing B.W. if such a policy was decided on by the higher command."[152] Equally important were the enormous changes that were taking place at Camp Detrick, notably the construction of an anthrax pilot plant, capable "of producing about 25,000 lbs of 4% N slurry per month sufficient to fill some 25,000 bombs ... [with] new facilities expected to produce some 25 times that amount." Stamp's boosterism was not misplaced. By the spring of 1944 Detrick had become the world's leading facility for the mass production of pathogenic agents and toxins, with advanced large-scale fermenters and separation plants, along with sophisticated air filtration systems and negative pressured sealed laboratories. Equally important was an elaborate division of scientific labour, whereby each major bacterial or toxin agent was developed in its own building in order to avoid cross-contamination.[153]

But would anthrax remain the Allied BW weapon of choice? While Porton scientists remained firmly committed to the "N" cluster bomb, there was considerable discussion in Canada and the United States about using other munitions. This subject was, in fact, the focus of Murray's memorandum of 15 February 1944, when he informed the Canadian Chiefs of Staff that while a number of possible biological warfare agents had been examined by the C-1 Committee, there was a consensus that "X (bot tox) takes first place as a BW weapon," given its killing power, long term stability, and ease of dissemination. Murray also praised several innovative BW delivery systems such as the use of finely ground peat (LP) that could be used "for the distribution of pathogenic bacteria maintaining their numbers, viability and virulence ... [and] disseminated as a cloud or as ground contamination." Even more promising was the use of insect (LMP) vectors for transmitting "pathogenic bacteria; typhoid, paratyphoid, food-poisoning, dysentery." Murray concluded his presentation by pointing out that while the Grosse Ile rinderpest project was primarily defensive, there were also "possibilities of using this virus as an offensive weapon."[154]

British reaction to the C-1 Committee report was divided. On the negative side was the predictable critique by Paul Fildes, who dismissed Murray's claim that Canada had provided critical assistance to the UK program "in terms of producing agents / munitions, developing new delivery systems, or in establishing effective defensive measures (X-toxoid / rinderpest vaccine)." According to Fildes, the C-1 Committee's exaggerated sense of importance was based on their unfortunate tendency to ignore the strategic reality "that research and development is a waste of time without producing approved weapons."[155] The UK Cabinet Committee on Biological Weapons, however, adopted a more conciliatory approach, both because it wanted to ensure future Canadian scientific assistance and because of growing concerns that the UK program, with its emphasis on anthrax, had put "far too many 'eggs in one basket.'"[156] Consequently, the committee's report of 19 April provided a positive assessment of past and present Canadian BW projects:

1. X. [Bot tox] "Work should be continued in Canada to determine the best field method of using X and its value when so used; further work towards large scale production is also desired."
2. N. [Anthrax] "We naturally want work on all aspects of N to be continued. As soon as we hear that the trials of the experimental bombs which the Americans have in hand are successful, an order will be placed through the usual Lend / Lease channels for the early delivery of half million. For the filling of these bombs, P.F. wants a material five times as virulent as the present standard."
3. LP. [Peat as BW vehicle] "We feel that though research should be continued there is not at present a case for further production."
4. LPM [Infected bait dispersal system] "This project was regarded as interesting but not of vital importance."
5. R [Rinderpest] "The Panel was unable to reach a decision regarding the need for stocks of vaccine either in this country or in other parts of the Empire."[157]

While these discussions were taking place, Canadian bioweapons scientists were charging ahead with attempts to develop botulinum toxin (type A) into a battlefield weapon.[158] In August 1944, for example, Murray advised E.L. Davies that if Suffield could "bring off satisfactory trials and prove the ESS 4 lb (bomb) can be a satisfactory weapon, we shall have X nearly as far advanced as the U.K. have N," and that Detrick had "no (X) weapon near production."[159] Murray's scientific boosterism was partly vindicated during the fall of 1944 when ESS scientists carried out a series of trials with a newly designed weapon: a 30,000 X tipped cluster projectile, exploded from a 500-lb aerial bomb. According to Superintendent Davies, "De-clustering in all cases occurred at about

3000 ft ... In every case about 10,000 square yards were contaminated to a density of 1 to 5 darts per square yard ... [with] penetration through battledress ... of 6 inches or more."[160]

During the latter stages of the war, Canadian BW officials also worked closely with their counterparts in the United States in the development of other bacterial agents such as *Brucella suis* (brucellosis), *Francisella tularensis* (tularaemia), *Yestina pestis* (bubonic plague), *Burkholderia mallei* (glanders), *Burkholderia pseudomallei* (melioidosis), and *Chlamydid psittaci* (psittacosis), as well as viruses such as yellow fever and the deadly paralytic toxin produced by shellfish.[161] But in operational terms, most of the attention was focused on the potential of *Brucella suis* because of its infectivity, its stability, and its ability to incapacitate rather than kill, and because, as one British expert remarked, it was "100 times more potent than N or X."[162] In December 1944 a series of joint trials were carried out at Area E Suffield to improve the delivery system of brucellosis munitions. A Detrick technical report described the results: "Two types of bombs, the Mark I and the shot-gun shell, were fired statically during the Canadian trials and the resulting clouds sampled by means of impingers and animals stationed on area of 50 to 100 yards from the site of the bomb-burst ... From 10 to 35 percent of the organisms were recovered from the clouds, more than sufficient to cause animal infection."[163] But for *Brucella suis* to make the grade as battlefield weapon "friendly" troops would need to be protected by either an effective vaccine / toxoid, or other medical therapies. Unfortunately none existed. This deficiency was highlighted in January 1945 when a serious brucellosis outbreak occurred among laboratory workers at Camp Detrick, with seventeen requiring extensive medical treatment.[164] The following month US officials advised Otto Maass of these problems, including the offer that Detrick could send one of its medical safety officers to Suffield "for the purpose of treating the personnel at the station and giving instruction for such further treatment as may be necessary."[165]

Preparing for Offensive Biological Warfare

On 11 February 1944 a high level meeting of Canadian, British, and American biological warfare experts took place at Camp Detrick. At the top of the agenda was the British request for 500,000 "N" bombs in order "to retaliate effectively in as short a time as possible in the event of the enemy commencing B.W." Although Canada was not a voting member of this select Anglo-American planning group, Otto Maass worked hard at being the "helpful fixer," by proposing that "400 of the first delivery [of bombs] should be sent to Canada for experimental purposes, presumably to be tested at Experimental Station Suffield.[166]

British efforts to accelerate the development of the anthrax cluster bomb were affected by several factors.[167] First, there was mounting evidence that Germany was preparing to launch a BW attack, either through its V-1 rocket campaign or in opposing the June 1944 Allied Normandy invasion. This could only be prevented, Fildes argued, if the British armed forces could retaliate in kind, since all BW defensive measures were futile.[168] Second, there was overwhelming scientific proof that biological weapons were vastly superior to their CW counterparts, with the "N" bomb estimated as having "from 5 times to 20 times the effectiveness of phosgene ... [since] there is no warning with N and the fatal dose is inhaled in a few seconds, thus a respirator is likely to be useless."[169] A third factor was whether the United Kingdom should obtain its initial half-million anthrax bombs from the United States, despite Churchill's preference for having these weapons produced at home.[170] Finally, there were questions about the strategic use of anthrax cluster bombs against enemy targets. As Fildes explained,

> One cluster projectile containing 106 4-lb bombs produces a lethal ct. 50 over an area of 500,000 square yards reaching 1750 yards downwind of the point of burst ... The ground surrounding the burst bomb becomes heavily contaminated with N spores and by subsequent impingement of the spores during the passage of the cloud. The ground so contaminated cannot be occupied with safety ... The N bomb cannot therefore be used in actual combat; it takes 3 or more days to become effective, the cloud being odourless and invisible. The weapon is therefore only applicable to bombardment from the air of cities and islands which it is not proposed to occupy.[171]

Problems of BW Intelligence and the "Bot Tox" Scare of 1944

During the Second World War one of the most challenging aspects of biological warfare was obtaining accurate intelligence reports about enemy intentions and capabilities. Part of the problem was the lack of trained field agents and operational analysts, which meant that American and British intelligence officials often had to rely on reports by exile scientists or resistance groups in occupied Europe. In March 1943, for example, a report from the Polish underground stated that "Hitler and Himmler had conferred ... on the subject of bacteriological warfare and concluded to employ this ... when it appears the tide is definitely turning against them."[172] Because of these ominous warnings, the US War Research Committee decided that it needed more reliable information and established close links with US Army Intelligence (G-2). As a result, in April 1943 Colonel James Defandorf, an accomplished biodefence scientist, was assigned to the London headquarters of the US Army's European Theatre of Operations with instructions to obtain evidence "which might indicate an intention by the enemy to employ B.W. against Allied Forces."[173] Although Defandorf concluded that none of the rumours about Nazi BW activities had any substance, these assurances did not satisfy officials of the US Office of the Surgeon General (OSG), who demanded increased surveillance on the grounds that they were "responsible for the medical protection and treatment of the [US] military forces in this theatre."[174] In particular, they recommended making greater use of the US Army's system for the analysis of blood samples, taken from prisoners of war (POWs), since this could provide evidence that the enemy was preparing its troops for biological warfare.

Significantly, one of the earliest proponents of this scientific technique was Canada's Everitt Murray, who in July 1942 had recommended carrying out a blood analysis of the eight German secret agents who had been arrested in the northeastern United States trying to establish a sabotage operation. According to Murray, there was the chance "that these men were prepared to use infectious agents and in that case might have been immunized ... [also] the finding of unwarranted antibodies might well reveal the type of organism the Huns are prepared to use."[175] But in operational terms, it was not until the 1943 campaign of the US Army in North Africa that a more systematic method of doing blood tests on enemy POWs was adopted. This approach was later refined and expanded in the European theatre of war, with impressive results. By May 1945 over 326 blood specimens of German POWs had been examined (plus 40 Italians), with special attention being focused on the three strains of botulinum toxin (408), anthrax (220), and yellow fever (297).[176] Curiously, neither Porton's BW scientists nor the British medical services participated in this project, on the grounds that their "depleted medical services could hardly be asked to undertake a job so unlikely to be productive."[177]

However, the greatest challenge for Allied intelligence came in December 1943 when there was a report from the Office of Strategic Services (OSS) that Germany was "preparing to use long range rockets and biological toxins ... [which] if used suddenly on a large scale the effect would be devastating."[178] In building their case, the OSS experts referred to Canadian tests that had demonstrated how airborne botulinum powder was lethal even "in extremely great dilution," and how casualty rates could reach "70% mortality."[179] Reference was also made to the tactical advantages the enemy would obtain by using such a weapon:

a. This toxin is not contagious and its military use by the Germans, would, therefore, not backfire onto the continent.
b. The air-borne dust would have no odor or taste, and thus a population would have no reason to protect itself with gas masks. Symptoms do not develop until four or five hours after contact, when death invariably follows.
c. The cause of death is an embolism and would tend to bewilder medical opinion.
d. The toxin ... if mixed with high explosives might not be detected.[180]

Faced with this serious biowarfare threat, the United States Joint Chiefs of Staff decided to establish a special group of expert consultants, known as the 'Barcelona' Sub-Committee. Chaired by George Merck, this group included representatives from the Chemical Warfare Service, the Office of Surgeon General, and the Office of Scientific Research and Development (OSRD).[181] Despite an initial report that claimed the armed forces of the United States could "defend themselves against enemy use of biological warfare, and to retaliate in kind if necessary," the committee soon acknowledged serious weaknesses in the American biodefence program. Of particular concern were delays in the establishment of manufacturing facilities for the production, loading, and storage of biological warfare agents and munitions; failure to develop plans for the strategic and tactical use of BW agents; and an inability to appoint qualified BW scientists who would "be available for the instruction of combat personnel in all offensive aspects of biological warfare."[182] But the most serious question was whether the United States could "retaliate in kind if necessary." And, on this issue the committee could only speculate that a retaliatory capability might be possible *within a year*, with "the manufacture of 'X' and 'N' on a large scale."[183] But the Joint Chiefs of Staff remained unconvinced. And to ensure at least a minimalist US biodefence response, they adopted two contingency plans: to ship additional poison gas munitions to the European theatre of war and to immunize American troops against "bot tox" prior to the D-Day invasion.[184]

The advantages of the immunization option was strongly supported by Canadian BW scientists and their military allies.[185] This was evident on 17 December 1943 when Master General of Ordnance, General J.V. Young, notified Minister of Defence Ralston about the OSS report, the response of the US. Joint Chiefs of Staff, and the fact that Murray and Reed had visited Washington to support the plan "to provide immunization for American troops."[186] And the prospects of this defensive response appeared promising. In January 1944, for instance, the C-1 Committee claimed that Camp Detrick was preparing to ship 1,138,000 type A toxoid to the United Kingdom, sufficient for treating 300,000 military personnel, with another 46,000 units of the alternative type B toxoid being held in reserve.[187] Armed with information, Murray and Maass easily convinced the Canadian Chief of Staff that 60,000 units of the "X" toxoid (code named Esoid) should be produced at Reed's Kingston laboratory in order to meet the requirements of Canadian troops overseas.[188] This crash program had impressive results. By April 1944, Reed was able to announce that he could ship 25,000 man doses overseas, with another 150,000 available by mid-May.[189] Plans were also made to send Reed and Murray overseas so they could ensure that Canadian invasion troops were receiving the appropriate X-toxoid inoculations.[190]

For a variety of reasons British BW planners did not share Canadian and American concerns.[191] Predictably, Paul Fildes was the most outspoken critic of the December OSS intelligence report and of subsequent claims by Professor Hellmuth Symon, a refugee scientist in Switzerland, that the German military "may employ a mixture containing the causative agents of plague, typhus, psittacosis and glanders."[192] Nor was he impressed, unlike his Canadian counterparts, with the battlefield potential of "bot tox" munitions. But most important of all was Fildes's insistence that specific defence against unknown biological agents was impossible and "that the only defence was the power of retaliation."[193] The major test came on 18 April 1944 when Fildes manage to persuade the British Committee on BW Intelligence that large-scale immunization against "bot tox" was both foolish and dangerous.[194] Despite the opposition of Lord Rothschild, a prominent medical civil defence expert, who ridiculed Fildes for his "almost ghoulish scientific interest in the offensive potentialities of bacteria, the Committee's decision was a foregone conclusion."[195] In reality, what mattered most for the British Chiefs of Staff and the War Cabinet was that the country's leading BW expert had justified their anti-immunization policy.[196]

Quite a different approach was adopted by the Canadian scientific delegation, which arrived in England in early May. For Everitt Murray it was a wonderful homecoming, having left his academic position at the University of Cambridge in 1930 for a career at McGill. And he took the opportunity to contact many of his former scientific contacts, including Paul Fildes.[197] But his attempt to visit Porton was rudely rebuffed, an indication of how relations between these two biological warfare administrators had deteriorated during the previous six months. Instead, Murray and his three colleagues (Maass, Reed, and Brigadier General Brock Chisholm) went to London, where they attempted to convince British and US military authorities about the necessity of preventative immunization. But it was an uphill campaign, particularly when the British Committee on BW Intelligence did not recognize their credentials as official spokesmen for the Canadian government.[198] Nor did the situation improve when Lieutenant General Kenneth Stuart, Chief of Staff, Canadian Military Headquarters (London), was given a special briefing by the British War Office outlining the various reasons why the German military could not possibly launch a "bot tox" attack against Operation Overlord.[199] Stuart was impressed not only by the evidence, but also by the determination of British and American military planners to force Canadian compliance. This message was reinforced in subsequent SHAEF meetings, when Stuart was bluntly told that if Canada tried to immunize its troops, this would create "considerable adverse psychological effects on unprotected British and Allied troops" and, even worse, seriously jeopardize security for the Normandy invasion.[200]

Once the Normandy invasion was successful, the immunization controversy quickly subsided. The next question was whether the German armed forces would use biological weapons in a desperate defence of their homeland. While the British Chiefs of Staff remained sceptical of such a possibility, it did agree that the situation merited careful investigation. As a result, a special unit was set up in the War Office, including two Canadian liaison officers, to examine UK and US intelligence reports of German biological warfare activity.[201] Of particular value were the reports of the special British and American ALSOS teams, which fanned out across recently liberated regions of France, Belgium, and Germany in search of leading German biological and chemical warfare experts. Significantly, these debriefing sessions provided no evidence that Germany had developed a viable biological warfare program, despite its large numbers of gifted microbiologists and well-endowed research centres. But Everitt Murray was not convinced. Indeed, as late

January 1945 he warned the Canadian Chemical Warfare Inter-Service Board that "the enemy may have devised more effective bacterial warfare than we are aware of and these might contribute to a combination of weapons with terrible effect."[202] What he didn't realize was that in Nazi Germany "any offensive program was barred by [Adolf] Hitler's interdict against BW development."[203]

Administrative Changes and New Biological Warfare Guidelines

During the summer of 1944 military officials in Canada, Britain, and the United States began to explore the possibilities of using biological weapons against either Germany or Japan.[204] This major transition in policy, away from the defensive / reactive response, was largely due to the impact of the D-Day controversy over botulinum toxin weapons, which had forced war leaders of all three countries to address the challenges posed by biological warfare rather than leave this sensitive subject to their military subordinates. In August 1944, for example, the British Chiefs of Staff decided to transform the Porton BW operation into a military agency, under the administrative control of the Inter-Services Sub-Committee for Biological Warfare (ISSCBW), chaired by Air Marshall Sir Norman Bottomley.[205]

A similar process of centralization was under way in Washington.[206] In September 1944 the United States Biological Warfare Committee (USBWC) was established with sweeping authority to streamline the American BW program and improve communication with its two allies. Both these goals were quickly realized. Indeed, USBWC chairman George Merck made a point of insisting that both British and Canadian representatives should participate in the committee's deliberations, a gesture that was greatly appreciated in Ottawa.[207] In contrast, British officials were outraged that one of the Dominions was afforded equal status on such an important wartime body.[208] And a nasty confrontation occurred just before the USBWC's first meeting when Lord Stamp and Colonel Paget asked that Otto Maass be excluded as an official participant on the grounds that it would establish an unfortunate precedent for Commonwealth unity and complicate the work of the Combined Chiefs of Staff.[209] Neither Merck nor General William Porter, head of the CWC, was impressed with these arguments. And to ensure that Maass had a place at the table, on 2 December Merck advised US Secretary of War Henry Stimson that direct liaison between Canada and USBWC "was important and desirable," even if it meant offending British sensibilities.

He gave three reasons for his decision: "(a) We are engaged in important joint projects with the Canadians; (b) Our countries are contiguous and our b.w. policies should therefore be closely coordinated, especially in defensive matters; (c) The Canadians have made important progress in this field, considerably more than the British in some respects, and direct exchange of ideas with them should be facilitated in every practicable way."[210]

From the perspective of Canada's military hierarchy, this high-level US endorsement greatly enhanced the status of Canada's biological warfare program. It also facilitated increased cooperation between the Directorate of Chemical Warfare and Smoke and the Army's Chiefs of Staff, particularly after Colonel Wally Goforth's October assessment of Canada's BW programs.The report stressed three major themes:[211]

1. "For reasons that are self-evident, it is essential that defensive aspects of BW should take general precedence over offensive activities and preparations.
2. "On the offensive side ... it is research and assessment which should hold priority over procurement or even over improved weapon design.
3. "It is suggested that the possibility of an allied decision to retaliate with BW is so very remote as to render accumulation of operation reserves – as distinct from those needed for experiment and assessment – a matter of little or no strategic importance."[212]

Despite its controversial nature, Goforth's brief was well received by General Murchie, Chief of the General Staff, in part because it coincided with a request he had received from the Minister of Defence R.L. Ralston about how the reorganization of British and American BW programs would affect Canada.[213] In particular, Ralston wanted to know "why the planning and Service integration of BW matters has not made rapid progress, in view of the extensive research and consideration given to the matter in all three Countries." Unfortunately Murchie was ill-prepared to answer this important question, since his previous involvement with BW matters had been virtually non-existent. This lack of knowledge about the fundamental scientific and military issues associated with biological warfare was evident in his response to Ralston's inquiry. On the other hand, in trying to satisfy the curiosity of his minister, Murchie did raise two important questions: why all aspects of Canada's biological warfare activities were cloaked in such secrecy and the extent to which this approach had hampered Canadian research and development work. As he explained:

> For security and policy reasons BW research and development has been confined, up to very recently, to a small group of experts and appropriate senior officers of the Service. It is only in the past few weeks that it has been considered advisable that definite material, from these experts, should be made available to the Service Staffs on which the latter could attempt a proper strategic appreciation … It is obviously not a matter which can any longer be confined to the technical officers concerned, regardless of the very valuable work which they have done, and are still doing.[214]

Although Murray was one of those "insider" scientific experts, he was delighted that National Defence Headquarters was finally demonstrating some interest in BW issues, and even appreciating the achievements of the C-Committee.[215] And it came at an opportune time since he was becoming increasingly concerned that ISSCBW representatives in Washington were trying to undermine Canadian biowarfare priorities. In October 1944, for example, Lord Stamp unilaterally made arrangements for Camp Detrick scientists to conduct extensive trials with brucellosis munitions at the Suffield Experimental Station.[216] The problem was that while the C-1 Committee resented Stamp's high-handed actions, it could not cancel the trials without running the risk of alienating American BW administrators.

Another dimension of this Canadian-British power struggle was Murray's ongoing debates with Paul Fildes over the relative merits of botulinum toxin (X) and anthrax (N) weapons.[217] In the former case, Fildes had long maintained that "bot tox" had limited tactical or strategic value since "the dispersal of X for subsequent inhalation effects is valueless." By the fall of 1944, however, Murray was well prepared to refute this argument by pointing out that recent SES field trials, using dried forms of X, "had demonstrated considerable improvement in the creation of a lethal aerosol cloud." Another aspect of "bot tox" weaponization were the trials with "X"-tipped particles, which even Fildes conceded could have considerable battlefield utility "by increasing the number of hits per weapon and the number of deaths per hit." But the most contentious issue between the two BW administrators was the status of the anthrax cluster bomb. And while Murray did not directly attack the project, he raised serious questions about the lethality of anthrax munitions, based on the laboratory experiences at Grosse Ile and Camp Detrick: "Up to now more N has been produced by the Grosse Ile plant than in the U.S. and U.K ... The agent produced is satisfactory when tested on sheep and on guinea-pigs but it is not effective with cattle. Canadian opinion is not convinced that N would be as effective as some suppose. Man is one of the more resistant animals to that disease, possibly more resistant than cattle [and] considerable exposure at both the W.D.C.S. ... has resulted in a very few insignificant cases of human disease."[218]

Allied Planning for the Offensive Use of BW against Germany and Japan

On several occasions the British and American governments considered the strategic advantages of using biological or chemical weapons. One of the most controversial incidents occurred in July 1944 when the British Chiefs of Staff submitted a report to Prime Minister Winston Churchill outlining the relative merits of attacking German cities with chemical or biological munitions in retaliation for the V-1 rocket attacks on London. While the report predicted that phosgene and mustard gas would produce a casualty rate of between 5 and 10 percent, military experts estimated that a large-scale anthrax bombing attack would be even more devastating. Indeed, in their opinion anthrax was "the only Allied biological agent which could probably make a material change in the war situation before the end of 1945 ... [and] used on a large scale from an aircraft [it] might have a major effect ... [causing] heavy casualties, panic and confusion in the areas affected ... [leading] to a breakdown in administration with a consequent decisive influence on the outcome of the war."[219] Significantly, this feasibility study was never incorporated into British military strategy, in part because the Chiefs of Staff believed that an outbreak of biological or chemical warfare would impede rather than accelerate Allied military success. Not surprisingly, Paul Fildes disagreed. And in a lengthy memorandum of 19 August he outlined some of the dimensions of a possible bioweapons exchange between the United Kingdom and Nazi Germany:[220]

> He [Germany] is known to have experimented before the war with bombs filled with the same agent (N) and if he has developed it satisfactorily and uses it against us we shall suffer from ... [and] the weight of the attack can only be reduced by air defence, since specific defensive measures are at present of little value. If it were decided to retaliate (after the end of the year) we should have to do so with the same weapon, hoping that united

> resources and opportunities were greater than those of the enemy. On the other hand, if the enemy initiated BW with some other form of agent and we found it desirable to retaliate with the present project, the resources of Germany would be quite inadequate to develop a similar project in time to be effective in this war.[221]

Similar discussions were taking place in the United States between the Joint Chiefs of Staff and the leadership of Chemical Warfare Service, who were now responsible for the operational use of biological / toxin weapons. In this situation, the CWS bioweapons strategy was influenced by their previous chemical warfare experiences, and this linkage between the two weapons system had become more pervasive at this stage in the war, since the US did not have operational BW weapons "to retaliate in kind." Instead, the Joint Chiefs of Staff had decided that if Germany used biological weapons against US forces in Europe, the American military would immediately attack enemy cities and troop concentrations with poison gas. And while this represented a major expansion of their wartime role, what the hierarchy of the Chemical Warfare Service really wanted was authority to launch first-use CW attacks against either Germany or Japan. On 13 January 1944, Brigadier General Alden Waitt outlined the merits of this approach in an Ottawa meeting with the Canadian CW Inter-Service Board: "Properly used, gas can shorten the war. By the proper use, I mean 'overwhelming attack with gas,' with non-persistent as well as persistent use on a very large scale ... The U.S. figures 40 to 80 tons of mustard per square mile and 150 to 200 tons of phosgene per square mile. I think phosgene will kill more people per ton than high explosives ... I do not believe in harassing the enemy, I want their resistance eliminated completely."[222]

Despite its lack of operational capability, the American BW program made considerable progress during the fall of 1944, largely because of the determination of the United States Biological Warfare Committee (USBWC) "to ensure that the research and operational dimension ... was more effectively integrated."[223] But despite this progress, American BW scientists were still unhappy over a number of policy and logistical issues. Many of these concerns were outlined in an October 1944 report by Lieutenant Colonel Oram Woolpert, Commander of Camp Detrick. Woolpert began his critique with the observation that while the US bioweapons strategy emphasized a retaliatory response, "no policies have been laid down in respect to the types of agents which will be accepted for this purpose." More specifically, he asked for clarification of the following key questions:

(a) Will agents that are highly persistent or highly transmissible be acceptable to the War Department should research and development be carried to the point where their use appears to be practicable?
(b) Should attention be directed primarily to agents which can be used tactically against combat formations, or rather for strategic effects against civilian centres?
(c) Is there a requirement for agents which will quickly produce large numbers of fatalities or for agents which will cause only temporary or prolonged disability?
(d) Will the using Services accept agents which, because of their instability, will require special handling in storage and transportation, particularly in respect to refrigeration?
(e) Will an agent be accepted even though adequate protection of our Forces is not yet assured, as in the case of "N"?
(f) If an agent is to be used for retaliation, on what scale should preparations be cast?
(g) Should research installations confine their investigations strictly to authorized classes of agents, or explore in addition other possible agents, on the assumption that circumstances might later alter policies?[224]

Although Woopert's memorandum did not mention the anthrax cluster bomb, Detrick scientists were extensively involved in its development. Indeed, during the summer of 1944 the Anglo-American project gained considerable momentum after a consensus was reached on several important issues. First, the original British order of half a million 4-lb anthrax bombs, scheduled for production in the United States, would be insufficient for the sustained attack on Berlin, Stuttgart, Frankfurt, Hamburg, Aachen, and Wilhelmshafen.[225] Evidence for this assessment came from a special Porton report which demonstrated that since these cities covered 538 square miles, the Anglo-American bombing offensive would require a total of 4,277,100 four-pound bombs, or 40,350 clustered projectiles, "all of which can be carried by 2017 Lincoln aircraft."[226] On the logistical front, Fildes calculated that 4.5 million gallons of anthrax spore slurry would be required to fill one million 4-lb bombs, a process that would take approximately seven weeks of production time at the Vigo plant.[227] Another issue was the technical problems associated with the cluster bomb container, including the debate over whether the British or American models were best suited for this complex task. And finally, there was common agreement that before the "N" aerial bombs could be considered operational, they would have to be tested at either the proving grounds at Granite Peaks or at Experimental Station Suffield.

Despite these bold predictions, serious problems continued to plague the anthrax bomb project. One of these involved the 500-lb projectile (M-19), with its erratic declustering patterns and premature explosions.[228] Much more serious, however, were reports that the experimental anthrax spores produced at Camp Detrick had only a three-month storage life, far below the anticipated nine months of virulence associated with Porton's Vollum strain. In turn, this raised the crucial question of whether "N" munitions for the Anglo-American project would require frequent refilling in order to maintain effective virulence. And if this was the case, would the viability of the undertaking be undermined?[229]

Another set of issues related to the actual offensive use of the anthrax bomb projectiles once sufficient numbers were available for military purposes. And there were several options. If immediate deterrence was the major goal, it was recommended that the UK Air Ministry accept immediate delivery of the crated clusters for storage in the United Kingdom. In contrast, if viability and stability of the anthrax charging were the priority, then the munitions should remain in the United States. This latter option appealed to Paul Fildes because it would cover a number of scenarios: "The American plant could turn out 1 million bombs per week ... [which] could be delivered here within 3 months ... stable for immediate use ... It may also be taken into account that proof of the use of BW must take time, say 14 days after receipt of suspect material ... [and] if 14 days is allowed for diagnosis in the field, transmission delays and inter-Allied decisions, it may well be a month before the decision is taken ... [while] an order to America to proceed, based on suspicion, could be given in a few days after the incident took place."[230]

Finally, important questions were raised about proposed field trials at Experimental Station Suffield with operational "N" cluster bomb projectiles. When Otto Maass had originally proposed this option in February 1944 Allied BW planners had feared that a German "bot tox" attack was imminent. But by the fall of 1944 quite a different situation prevailed, and serious questions were being asked whether such dangerous field trials should be carried out.[231] For the Canadian C-1 Committee the answer was no. In September an official DCW&S report announced that "large scale trials with 50 bombs clusters charged N are not warranted at Suffield, because of the heavy contamination and possibility of spread by Gophers and Antelope." This decision was reinforced by a second directive: "Official clearance of safety distances for the Suffield area have not been forthcoming. This Most Secret project should, therefore be cancelled; it may be pursued in the United States.[232] It was also instructive that on 4 November 1944 Fildes declared that Porton was now committed "for doing [N] trials at sea, because they are obviously impossible on land, and if we cannot find a way to test things out it is not worth while continuing after the war."[233]

There were other reasons why the 'N" trials at Suffield never materialized. One of these was the lack of progress in obtaining either sufficient qualities of the weaponized anthrax spores or workable cluster projectiles because of serious procurement problems in the United States.[234]

Even more important was growing evidence that CWS officials had now decided that the choice of the anthrax spore agent, the filling of the munition, and the testing of the clustered projectiles would take place in accordance with American, not British, priorities.[235] One demonstration of US dominance over this "winning weapon" came in October 1944 when Secretary of War Henry Stimson created the so-called DEF Committee, composed of a group of prominent life scientists who were given the challenging task of evaluating the scientific and technical aspects of the anthrax cluster bomb project.[236] High on the list of concerns were safety standards at the Vigo plant, which was being transformed for the "mass production of agent 'N' ... and the filling of a munition designed for its dissemination" [Mk. I, Type F, 4 Lb bomb]. This program had two major goals: to protect scientific workers in the production plant and to implement emergency measures "in case of a large 'spill' to the surrounding community." Because of these health concerns, the DEF Committee decided that Vigo would have only a limited output – despite its twelve 20,000-gallon fermenters, capable of producing 240,000 gallons of anthrax slurry, or the simulant *Bacillus globigii* (U), which became the facility's first biological product – until contamination problems had been rectified. In addition, guidelines were established to ensure that the scientific and technical staff were sufficiently trained "to insert the newly produced anthrax into the 4-lb bombs ... being manufactured ... in a Detroit plant."[237] This was not an easy undertaking since the production system, using a noisy and dangerous assembly line, was expected to fill one shell with "N" every fifteen seconds, which was followed by the insertion of detonators in the bomb assembly facility.[238]

Another part of the DEF Committee's mandate was to determine the feasibility of using the Granite Peak (Dugway) proving grounds, "where munitions charged with 'N' are to be tested in the field." In developing their experimental protocols, DEF scientists made extensive use of the "lessons learned" at Camp Detrick, Grosse Ile, and Suffield in avoiding outbreaks of anthrax infection. This careful and incremental approach, however, soon encountered vigorous opposition from the hierarchy of the Chemical Warfare Service, who were determined that their anthrax strategic weapon become operational before the end of the war. And in their opinion, DEF scientists did not realize that while laboratory and cloud chamber tests were useful, the most crucial stage in the weaponization process was to determine what bioagents would do in the field, and what length of time anthrax suspensions would remain virulent, since "field tests so far ... on bombs charged with 'N' have been made on single bombs ... [so] it will be necessary to cluster the 4-lb bombs and to determine the performance of these clusters in actual serial-dropping trials."[239]

Eventually, the Chemical Warfare Service and its allies convinced Secretary of War Stimson that the anthrax weapon was crucial for American national security. And by March 1945 the Vigo plant was ready to produce limited quantities of *Bacillus anthracis* spores, with the expectation that the filling of the 4-lb bombs would reach industrial levels by November 1945. And once dropping trials at Granite Peak had been completed, these deadly anthrax cluster bombs would be ready for use in the Pacific theatre.[240] Of course this did not happen. In August 1945 the Second World War ended, with the US atomic bomb attacks on Hiroshima and Nagasaki.

Ironically, as the United States and its allies were making preparations for the invasion of Japan, there was evidence that the Japanese military was preparing to use biological weapons against the west coast of North America.[241] Between November 1944 and August 1945 over a thousand balloons were identified, many of which were armed with incendiary bombs.[242] What made the situation most threatening was the fact that these paper balloons, floating at altitudes of between 20,000 and 30,000 feet, could carry upwards of 400 lbs – more than enough to disseminate biological warfare agents such as plague, anthrax, rinderpest, and Japanese B encephalitis.[243] In the end, while no BW agents were detected, this incident provided a major catalyst for extensive biodefence cooperation between Ottawa and Washington, providing the basis for civil defence cooperation in the Cold War.[244]

Conclusion

By September 1945, the team of Maass, Murray, and Reed had advanced Canada's biological warfare operation to a level that Frederick Banting would have found incredible.[245] Most of this research, development, and testing was initially carried out at the request of the British Bacterial Warfare Committee and its successor, the Inter-Service Sub-Committee for Biological Warfare. By the fall of 1944, however, Maass and Murray were much more involved with Camp Detrick than with Porton Down, since the United States had become the world's only BW superpower. They were also gratified that Edwin Fred, George Merck, General R.A. Kelser, and General William Porter treated them as scientific equals, not as colonial technical assistants.[246] This sense of being appreciated was reinforced in September 1947 when the United States government honoured Murray, Maass, and Reed with the Medal of Freedom for their work on the Joint United States Canadian Commission. In contrast, no formal recognition of the wartime exploits of the C-1 Committee was forthcoming from either the British or Canadian governments.[247]

Neither the Canadian military nor its scientists were involved in planning a biological weapons attack against Germany or Japan. As a junior partner in the wartime Anglo-American alliance, Canada had neither the capability nor the responsibility of deploying these awesome weapons. This does not, however, diminish the important contribution that Canadian scientists made to the Allied military effort, both in the defensive and offensive dimensions of biological warfare. In the former case, the C-1 Committee directed the Grosse Ile rinderprest vaccine project; the development of the toxoid against weaponized "bot tox"; and civil defence preparations against Japanese warfare balloons. More controversial was the involvement of Canadian scientists in weaponizing botulinum toxin and brucellosis, both at Guilford Reed's Kingston laboratory and at Experiment Station Suffield, largely because the C-1 Committee believed these BW agents were better suited for battlefield use than anthrax. On the other hand, Canada did support the development of the Anglo-American "N" cluster bomb as the only available retaliatory weapon.

What evidence do we have that the United States was willing to use biological weapons against Japan? This question can only be answered if a distinction is made between anti- personnel and anti-crop agents. Certainly, the Chemical Warfare Service was investigating the use of anti-plant fungal agents such as potato blight, wheat rust, rice fungus, and chemical plant growth inhibitors and defoliants against Japanese crops.[248] This project gained greater legitimacy in March 1945 when the US Joint Chiefs of Staff decided that such weapons did not violate the Geneva Convention since "the compounds used are comparatively harmless to human beings."[249] Closely related was the May 1945 JCS decision for the strategic use of poison gas against Japanese island strongholds since "the character of the weapon was no less humane than phosphorous [*sic*] and flame throwers and need not be used against dense populations or civilians – merely against these last pockets of resistance which had to be wiped out."[250] In contrast, there were no strategic plans for the offensive use of lethal anti-personnel biological weapons against Japan. But what would have happened if the war had continued for another two years, when anthrax cluster bombs would be available? Historian John Moon has provided the best response to this hypothetical question: "Once a weapon is developed, of course, the options are broadened … [and] the United States was ruthless enough to use the A-bomb … [But] the two weapons are profoundly different in origin and purpose … [and] it is difficult to argue convincingly that dropping anthrax bombs all over Japan would have ended the war quickly."[251]

2 Bioweapons in the Cold War: Scientific Research, Civil Defence, and International Controversy

At the height of its development, the Special Projects Division of the Chemical Warfare Services of the [US] Army, which carried the main responsibility for the [BW] Program after June of 1944, had a total personnel of nearly 3900, of which some 2800 were Army personnel, nearly 1000 Navy, and nearly 100 civilian ... It should be emphasized that while the main objective ... was to develop methods of defending ourselves against possible enemy use of biological warfare agents, it was necessary to investigate offensive possibilities in order to learn what could be used for defense ... All evidence to date indicates that the Axis powers were behind the United States, the United Kingdom and Canada in their work on biological warfare.

(Report to the Secretary of War by Mr George Merck, Special Consultant for Biological Warfare, 3 January 1946)

With the advent of the Cold War, many nations were keenly interested in the potential of biological weapons. This was certainly the case for Canada, the United Kingdom, and the United States, who resumed their wartime biowarfare alliance through the 1947 Tripartite Biological and Chemical Weapons Agreement. Yet despite many shared interests, notably the importance of deterring Soviet use of biological weapons, the three countries had quite different views on how these military devices could be used. For British military planners, there was the awareness that since their country lived in a dangerous neighbourhood there were powerful strategic reasons for developing a biological and chemical weapons deterrent, at least until the British military acquired a nuclear capability. Predictably, most of this BW research and development work was carried out at the Microbiological Research Department (Porton Down), with close connections to the Ministry of Defence and the Foreign Office.[1] The program also received strong financial and administrative support from both Labour and Conservative governments within the framework of pervasive secrecy.[2]

For Canada, there were less obvious advantages in remaining involved with biological warfare research and testing, since the country did not face the same external threats as the United Kingdom. Nor did Ottawa intend to develop atomic weapons, despite the impressive wartime research achievements of the Montreal-based Anglo-Canadian Montreal Laboratory and its close connections with the US Manhattan Project. Yet despite its unwillingness to become directly involved with weapons of mass destruction, the Canadian government was prepared to assist the offensive BW programs of its British and American allies within the context of the early Cold War by joining the 1947 Tripartite CBW Agreement. This involvement was enhanced by several factors. First, Canadian scientists were highly regarded for their expertise, in both the offensive and defensive side of biological warfare. Second, the unique testing and research facilities at Experimental Station Suffield (changed to SES in 1950) and Grosse Ile were greatly appreciated by Canada's two allies, since neither possessed such assets after 1945. And finally, the advantages of assisting the Pentagon were an integral dimension of Canada's evolving postwar military relationship with the United States, as demonstrated by the work of the Permanent Joint Board of Defence and the Canadian-United States Military Co-operation Committee.[3]

After the Second World War the United States, with its vast scientific and industrial resources, was the dominant global power in the field of biological warfare. Yet ironically, this fact did not necessarily impress the American military hierarchy, who continued to insist that containment of the Soviet Union was impossible without the US nuclear deterrent, and that other WMD systems had only marginal utility. This situation, however, changed dramatically in 1948 when Secretary of Defence James Forrestal launched a full review of the strengths and weaknesses of the US biological warfare system. The results were bleak. According to the report of the Biological Warfare Sub-Committee, the United States was vulnerable to possible enemy attack since "the current research and development program for biological warfare is grossly inadequate in view of the potentialities of the present international situation."[4] This blunt warning had an immediate impact on the White House and Congress, which, in turn, resulted in additional funding for BW research, development, and testing.

Most of this work was carried out at Fort Detrick, where scientists made considerable progress in improving the major biological agents of the Second World War: *Bacillus anthracis* (anthrax), *Yersinia pestis* (plague), *Brucella suis* (brucellosis), *Francisella tularensis* (tularaemia), and botulinum toxin. In addition, there were innovative efforts to weaponize a number of viruses, with Venezuelan equine encephalitis (VEE), variola major (smallpox), and yellow fever being considered the most promising candidates. But research without large-scale agent production still left the US without battlefield bioweapons. This problem was largely rectified in 1950 when Congress authorized the construction of a huge BW manufacturing plant at Pine Bluff, Arkansas, which was completed three years later at a cost of $90 million.[5] With its 858 employees, and its capability of producing 1 million 4-lb cluster bombs per month, Pine Bluff soon became the world's most formidable bioweapons production facility. Closely related were plans for large-scale field trials, using a range of munitions and delivery systems, at both the Dugway Proving Grounds and the Suffield Experimental Station.

How to contain disease outbreaks associated with natural pandemics was another formidable challenge for Western defence scientists during the Cold War. In Canada, for example, while the federal government's civil defence program was initially focused only on the threat of atomic bombs, by 1949 it was expanded to include germ warfare. In developing their CBRN preparedness programs, Canadian civil defence organizers worked closely with their counterparts in the United States on developing a number of joint programs during the next decade. In addition, the World Health Organization provided valuable medical information and public health assistance in helping Canada deal with the challenge of infectious diseases, either synthetic or of natural origin. This was not surprising since the WHO's first director general was Dr Brock Chisholm, former director general of medical services of the Canadian Army and a steadfast opponent of biological warfare.[6]

The involvement of high-profile scientists with weapons of mass destruction during the Cold War has attracted considerable scholarly attention. Until recently, however, most of the emphasis has been on elite nuclear scientists such as Robert Oppenheimer, Edward Teller, and Ernest Lawrence, who moved effortlessly between the worlds of academe and military strategic planning. One of the earliest studies of this phenomena was Herbert York's book *The Advisers* (1976), which examined the critical debate during the Truman administration over whether the United States should continue its focus on tactical atomic weapons or develop the world-destroying hydrogen bomb.[7] York's work also acted as a catalyst for other scholars, and during the past twenty-five years the literature on this important subject has mushroomed.[8] In contrast, the world of biological warfare scientists remains relatively unexplored by either military or scientific historians. There are, however, two notable exceptions. One of these was Brian Balmer's imaginative analysis, based on rich archival sources, of how Britain's leading microbiologists helped guide the country's bioweapons program during the Second World War and the early Cold War.[9] Another useful contribution came from Russian historian Nikolai Krementsov, who developed the intriguing thesis that the post-1945 arms race between the USSR and the US would have been different if biological, rather than nuclear, warfare had been the dominant strategic factor. As he argued, "the U.S. detonation of the atomic bomb boosted the importance of physics and the authority of physicists … [but] had the Cold War military competition between East and West concentrated upon the development of biological weapons (say, a superpathogenic virus) … biology rather than physics would have vastly expanded its institutions."[10]

The Legacy of the Second World War

Atomic Bombs and Biological Weapons: The Debate

During the Second World War the development of atomic bombs and biological weapons took place within separate scientific and military environments, each surrounded by extensive security measures. But after 1945 a symbiotic relationship evolved between the two weapon systems. Initially, most of the discussion focused on the awesome potential of nuclear weapons, given the devastation at Hiroshima and Nagasaki, where more than 160,000 people perished.[11] At the time, one of the more useful analyses of these attacks was carried out by the British Bombing survey team, including Canada's Omond Solandt, which in September 1945 examined the physical damage and casualty levels at the two target cities.[12] On the technical side, the Smyth Report, issued by the newly formed US Atomic Energy Commission in the fall of 1945, provided a comprehensive overview of the major features of the Manhattan Project. Not surprisingly, since they had not been used, there was little information about Allied bioweapons programs until October 1945, when George Merck submitted his final report to Secretary of War Robert Paterson. In this brief document, Merck outlined the main aspects of the US biological warfare program during the Second World War while also acknowledging the valuable assistance the US received from its Canadian and British allies. He also pointed out that "unlike the development of the atomic bomb ... the development of agents for biological warfare is possible in many countries, large and small, without vast expenditures of money or the construction of huge production facilities ... [since] the development of biological warfare could very well proceed in many countries, perhaps under the guise of medical or bacteriological research."[13]

The availability of the Smyth and Merck report helped stimulate discussions within the American defence science community about the possibilities of international control of atomic and biological weapons. According to a January 1946 report by physicist Robert Oppenheimer, former scientific director of the Manhattan Project, while both weapon systems posed a threat to US national security, he was confident that the United Nations Atomic Energy Commission would establish a workable system of controls to prevent a nuclear arms race. On the other hand, he believed that "any attempt ... to establish effective safeguards against the use of biological warfare will be illusory ... with little possibilities of reliable detection." Part of the problem was technical: BW agents could be produced in small-scale plants, with the munitions being filled in secret locations. But the most serious problem, Oppenheimer noted, was that aggressor countries would develop various strategies to conceal their programs since "the effective surprise use of biological weapons depends upon keeping a potential enemy ignorant of one's plans."[14] Significantly, Secretary of War Robert Paterson forwarded Oppenheimer's report to Bernard Baruch, chairman of the US delegation to the Geneva-based atomic energy talks, whose mandate involved international control over all weapons of mass destruction.[15]

Meanwhile defence officials in the United Kingdom were tackling many of the same issues. This was evident at the December 1945 meeting of the Chiefs of Staff Committee, which attempted to evaluate the "efficiency" of biological warfare in comparison with the demonstrated capabilities of the atomic bomb.[16] In particular, the discussion focused on what factors were required to elevate the status of BW as a weapon of mass destruction.[17] These included: "(a) The probability of tremendous increase in lethality in BW agents which may be produced after further research; (b) The comparative ease of production; (c) The possibility of BW being complementary to operations with Atomic Bombs."[18] The

Chiefs of Staff brief concluded with the observation that the use of temporary disabling agents, such as brucellosis and Q-fever, might convince the British public to become less hostile towards germ warfare by demonstrating that this form of bioweapon was "very humane indeed by comparison with atomic weapons."[19]

There were no such debates in Canada because the government of Mackenzie King had no intention of developing either atomic or biological weapons. In some ways this was surprising since Canadian defence scientists had been actively involved with both forms of warfare during the previous four years. In the nuclear field, for example, important research and development work was carried out at the Anglo-Canadian Montreal Laboratory, with considerable success being achieved in producing plutonium through the heavy water process. Indeed, because of its scientific contributions, Canada was represented on the powerful Combined Policy Committee, which determined Allied atomic strategy during the latter stages of the war. Canadian military officials also carefully monitored the actions of the US Manhattan Project, including the dramatic August atomic bomb attacks on Japan.[20] In September 1945 another dimension of Canada's involvement with nuclear research occurred when the Chalk River Laboratories began operations with the goal of making Canada one of the world's leaders in the peaceful use of atomic energy.[21] Diplomatically, Canada also participated in the tortuous international discussions about the postwar use of atomic energy, with General Andrew McNaughton assuming an active role as Canada's representative on the United Nations Atomic Energy Commission (1946–7).[22]

But while discussions about the future of atomic devices were carried out in public view, this was not the case for biological weapons. Instead, the King government made a determined effort to conceal the offensive BW activities of the C-1 Committee, with all of the press releases being focused on either the Grosse Ile rinderpest project or the 1944 production of the botulinum toxoid for protecting Canadian soldiers overseas. Nor was there any Canadian debate, either in Parliament or in the media about the ethical dimensions of Canadian medical scientists becoming involved with biological weapons research in the first place. Strangely, even the Canadian Medical Association remained silent on this subject. In contrast, on 25 July 1945 the British Medical Research Council issued a statement that it would not support the country's postwar bioweapons policy since its responsibility was "to promote by research, knowledge conducive to the saving of life, whereas the primary object of BW is destruction."[23]

Safeguarding Suffield and Grosse Ile for Canada's BW Program

Canada's decision to maintain its biological and chemical warfare facilities at Experimental Station Suffield was based on several factors. First, there was clear evidence that the British government wanted to utilize these facilities after 1945, given its commitment to develop an offensive BW retaliatory capability during the immediate postwar years. Second was the sustained campaign by Otto Maass that continuing experimental work on biological and chemical weapons at ESS was essential for national security and was the only such facility "in the whole of the British Empire."[24] He also stressed that with the discovery of the German nerve gases (tabun / sarin / soman), poison gas warfare had been transformed, paralleling developments in the field of germ warfare.[25] On 2 August 1945, despite the formidable financial burden, Ottawa made arrangements for the postwar operation of Suffield, including negotiations with the Alberta government of Ernest Manning for retention of the site for another ninety-nine years.[26] This decision was strongly supported by the Canadian CW Inter-Service Board, which pointed out that the Station would continue to attract British and American bioweapons specialists given its "excellent laboratory equipment, an adequate animal colony and most important, a well trained staff for the carrying out of trials with pathological organisms ... so powerful that they could not be overlooked."[27] The prospects of continued cooperation between ESS and Camp Detrick were greatly enhanced in September 1945 when US Secretary of War Stimson recommended that "a reasonable" program of BW research and development be continued by the US Army since "this form of attack still constitutes a threat which must be guarded against."[28]

While most of the postwar attention focused on the future of Experimental Station Suffield, several outstanding problems associated with the War Disease Control Station also had to be resolved. One of these were the ramifications of the serious riot of 4 August 1945 between members of the Station's military garrison, most of whom were French Canadians, and Canadian and American rinderpest scientists. While order was quickly restored by a contingent of military police, who came to Grosse Ile during the middle of the night, Defence officials were initially determined that all of the perpetrators should be severely punished. Fortunately cooler heads prevailed. In part, this was due to Everitt Murray's sage advice that the incident should be viewed as an unfortunate aberration since for three years American scientists and support staff had been "living side by side with ours at that Station ... [and] have made a happy family with the Canadian(s)."[29] Significantly, news of this incident was not reported in Canadian newspapers, largely became of Otto Maass's ability to censor any stories about Canada's biological warfare program in general and the Grosse Ile operations in particular.[30] And there was much to hide. In August 1945, for instance, the Directorate of Chemical Warfare issued instructions for the dismantlement of the GIN production plant and the destruction of the remaining stockpile of anthrax spores. A vivid account of this process was provided by Thomas Stovell, one of the scientists seconded for this dangerous task: "I was asked to clean up the anthrax ... because I done a lot of work in bacteria. They needed somebody to go in there and sterilize the whole lab ... And mop the floor and everything with chemicals ... With anthrax they [put] ... formaldehyde in it for a while ... and then put it in the St. Lawrence and dumped it."[31]

The Defence Research Board and Canada's BW Operation

In 1946 administrative control of Canada's CBW programs were transferred to the newly created Defence Research Board (DRB), with Dr Omond Solandt as its first Director General. A respected medical scientist and skilful administrator, Solandt recognized the importance of re-establishing the wartime exchange system, which would give Canadian scientists continued access to top-secret American and British BW research.[32] In developing the DRB research operation, Solandt greatly profited from the operational and administrative skills he acquired while serving as Superintendent of Operational Research for the British Army, reinforced by his involvement with the 1945–6 British Atomic Bombing Survey Team in Japan.[33] This impressive CV was certainly an asset during Solandt's initial years as Director General, when he tried to advance several key policies: to maintain DRB autonomy within the Department of Defence system; to secure reliable and generous financial support; and to recruit the best scientists available. In pursuing these goals, he received valuable assistance from Lt General Charles Foulkes, Chief of the Army General Staff, who believed that the Defence Research Board should be a broadly based organization composed of key government administrators, senior university scientists, and prominent industrialists. Solandt and Foulkes also recognized the advantages of creating quasi-independent advisory committees that would not only provide information about specialized weapon systems but also assist in the complex task of administering DRB research establishments. Another important innovation was the formation of a permanent liaison with US and UK defence organizations "for the exchange of scientific results of benefit to military operations and to defence industry."[34]

During this important transition period, one of Solandt's closest confidants was Professor B.J. Schonland of South Africa, his wartime mentor in operational research, and a key member of the postwar Commonwealth Defence Science Committee.[35] Of particular value was Schonland's advice that the DRB should concentrate its efforts in attracting well-qualified scientists, who often preferred positions in Canadian universities rather than with government agencies.[36] Another useful contact was the British defence scientist Dr O.H. Wansborough-Jones, who was an influential member of the Biological Research Advisory Board (BRAD), which helped guide Britain's postwar biological warfare policy.[37] Solandt also benefited from the counsel of Otto Maass, who became chairman of the joint DRB Committee on biological and chemical weapons, a position he held until the mid-1950s.[38]

Solandt also appreciated the political skills of C.J. Mackenzie, whose ability to manoeuvre in Ottawa's competitive government structure was legendary. Significantly, their first working relationship occurred during the 1946 Espionage Enquiry, when twenty-two Canadians, including seven scientists, were accused of spying for the Russians and eleven of whom would later be sent to jail.[39] But what most concerned Mackenzie and Solandt were the frantic inquiries they had received from Washington and London asking about what military secrets had been compromised. For US officials the most pressing question was whether the Soviet spy ring had obtained "any samples of uranium or plutonium ... [or] any plans or drawings showing [atomic] bomb construction."[40] In his official response, Mackenzie reassured Washington that important US nuclear secrets could not have been compromised since Canada was not directly involved with "the construction, method of assembly, and operating features of an atomic bomb."[41] Nor was he concerned about critical the biological and chemical weapons systems, given Otto Maass's assurances that the spy ring could not have obtained any useful information in these fields, since the security measures adopted by the Directorate of Chemical Warfare were so extensive "that it excited the admiration of the United States Chemical Warfare Service."[42]

Postwar CBW Allied Cooperative Systems

As Director General of DRB Solandt assumed a major role in the formation of several CBW arrangements. The most important of these was the Canadian, British, and American Tripartite Military Agreement on Chemical and Biological Warfare, which was created in August 1946.[43] The founding conference took place at Fort Detrick, Maryland, with an impressive array of Allied defence scientists in attendance. The meeting began with David Henderson, director of Porton Down's Microbiological Research Department (MRD), providing an overview of Britain's biological warfare problems: limited numbers of qualified scientists, minimal research facilities, and the problems of reconciling military and scientific goals.[44] In contrast, Solandt's description of the Canadian CBW situation was quite positive, particularly since the Chiefs of Staff had promised generous financial assistance for "fundamental and basic research at Kingston and field experiments at Suffield ... with BW munitions." In his concluding remarks, Solandt made a special plea for a continuation of the wartime pattern of trilateral bioweapon research, with the observation that "close scientific cooperation ... [was] more important than cooperation at the policy level." This viewpoint was strongly endorsed by General Waitt, head of the US Chemical Warfare Service (CWS), who claimed that the postwar development of biological weapons was essential for American national security "on a basis comparable to the Atomic Bomb."[45] He also expressed considerable interest in having American bioweapons tested at Suffield, since the wartime facilities at Granite Peak and Horn Island facilities were being decommissioned.

While the DRB welcomed US interest in its programs, it also wanted to expand the level of Anglo-Canadian BW cooperation. This was evident in June 1946 when Solandt attended the first meeting of the Commonwealth Advisory Committee on Defence Science. Under its terms of reference, the committee was expected " to consider and review major items of the defence research programme of the Commonwealth countries and ... [encourage] liaison in defence science, including methods of exchange of scientific information and interchange of scientific staff."[46] High on the list of priorities was the postwar use of biological weapons since, in the opinion of the committee, "they provided a possible solution to some important strategic problems confronting the United Kingdom ... [and] offered the possibility of achieving victory without the total destruction of the economic life of enemy countries."[47] This commitment to the offensive BW option was evident in the establishment of the so-called "Operation Red Admiral," which promised "to bring into service by 1957 biological weapons comparable in strategic effect with the atomic bomb, and defensive measures against them."[48] This was not wishful thinking. In November 1947, for example, the Air Ministry submitted a request for a 1,000-lb cluster bomb, "intended for strategic use against industrial targets ... [which] should contain the most effective biological agent for the incapacitation of workers."[49] Significantly, this interest in incapacitating agents such as brucellosis and Q-fever, rather than lethals, represented a major departure from the previous agenda of Paul Fildes, for whom the anthrax cluster bomb was the British weapon of choice. It was also clear, as was the case in Second World War, that scientists at Porton's Microbiological Research Department (MRD) could not develop an effective BW retaliatory capability without American and Canadian assistance.[50]

The DRB and Biological Warfare Developments, 1947–1950

Under his organizational model, Solandt insisted that biological and chemical warfare planning be carried out by specialized committees and panels of the Defence Research Board. This was evident in June 1947 when there was an important meeting in Suffield of the newly established Special Weapons Advisory Committee, chaired by Dr Otto Maass. It was a high-profile event featuring representatives from the three armed forces, the DRB, the National Research Council, along with two prominent guests: Dr David Henderson, Director General of MRD at Porton Down, and Dr O.C. Woolpert, Scientific Director of Camp Detrick.[51] During this two-day event discussion on CBW issues was wide-ranging and contentious. While there was brief reference to the military potential of ricin (W) and compound Z, which Canadian scientists had developed during the Second World War, the most promising work was associated with the German nerve gases (tabun, sarin, soman), given their ability to cause high casualties, either on the battlefield or against civilian targets. But for the most part, the delegates concentrated on the military capabilities of biological weapons. There were, for example, detailed reports by Suffield scientists on improved use of the BW agents anthrax, brucellosis, and "bot tox," particularly given new techniques for increasing aerosol levels.[52]

Despite these imaginative studies, research at the Suffield Experimental Station (SES) was hampered by the fact that its overall manpower resources had declined by 30 per cent from the 584 level of June 1945, with 80 per cent of the scientific contingent (50), along with 50 per cent of the technicians (100) and field staff (60) having found other jobs.[53] On the other hand, demands from Canada's two allies for large-scale trials provided justification for the appointment of additional scientists at SES in order to expedite research on the dispersion of pathogen particulates and the development "of more efficient munitions for dispersing this class of substance." Much of this work was made possible by upgrades in the bursting chamber, allowing it to be safely used "with pathogenic bacteria and the highly toxic proteins."[54]

During the summer of 1947 the DRB leadership asked Guilford Reed to prepare a report on the major trends in offensive bioweapons development. For Reed, this turned out to be a challenging task because of "the meager information that is available in many phases, particularly methods of dissemination, of controlling the spread of induced epidemics and of detecting B.W. agents rapidly."[55] Despite these obstacles Reed prepared a thorough overview of the contemporary BW situation, at least from the perspective of the bioweapons specialist. This did not, however, satisfy Omond Solandt, who complained that Reed's approach had been too theoretical, and did not address "the practical problems of defence nor outline any plan for attacking these problems. "[56] As a result, in September 1947 Otto Maass was asked to prepare a more user friendly study on 'special weapons' for the Chiefs of the General Staff.[57]

The Maass report began with the assumption that Canada would adhere to its Geneva Protocol obligations on the use of war gases and biological agents, which meant that while Canada would not initiate biological warfare, "the Government might decide to employ these agents as a retaliatory measure." Another major theme were the differences between defensive and offensive BW requirements. In the former category, emphasis was placed on the need for effective respirators, vaccines, prophylactic drugs, protective clothing, and devices for the detection of biological agents. But Maass was most concerned about the offensive dimensions of germ warfare, notably the importance of developing more lethal agents, and improvement in BW delivery systems. As Maass noted, "It is the opinion of the General Staff that research into methods of dispersion is of prime importance, and that in this field, profit would be obtained by research work being carried out in Canada in addition to that proposed by the United States and the United Kingdom."[58]

Two years later, as the Cold War became more intense, the Defence Research Board carried out an even more comprehensive analysis of how the Canadian Forces should deal with atomic, biological, and chemical (ABC) warfare. In the February 1949 report three major themes were highlighted. First was the reiteration of the familiar policy position that while Canada had no intention of developing offensive ABC equipment, its research and development policies would continue "to be closely integrated with the programmes of the UK and USA." Second, a distinction was made between atomic weapons, which the armed forces unequivocally renounced, and the offensive potential of germ and poison gas warfare.[59] As the report noted,

> The new chemical [nerve gases] and biological agents possess astounding lethal and destructive potentialities which cannot be fully exploited with conventional weapons and munitions. To achieve full strategic and tactical value, methods of dissemination of these new agents must be evolved and perfected. Basic research must continue in search of more powerful and effective agents ... The Canadian Army does not contemplate the use of biological agents in conventional field army munitions. There is, however, a requirement for BW agents for use in operations by a supporting air force or as means of sabotage. There is, therefore, a requirement for the development of munitions designed to take advantage of the highly infectious and toxic characteristics of these agents.[60]

Reaction to this DRB report varied. Representatives of the Naval Services, for example, made it clear that they were only interested in CBW defensive equipment and had "no requirements for offensive weapons."[61] Surprisingly, the Royal Canadian Air Force adopted the same position, despite the fact that their American and British counterparts were pushing hard for aerial BW cluster bombs. This left the Canadian Army as the only sponsor of CBW weapons, a situation that would prevail throughout the Cold War. At this stage, however, the Army's major priority was that its ground troops be equipped with "non-persistent [gas] agents – preferably of the G type [nerve gas] when we get them in quantity production."[62] In contrast, operational use of bioweapons remained uncertain, at least until an April 1949 study by Colonel Wallace Goforth. A former high-ranking CBW specialist, Goforth began his report by drawing on his Second World War experiences:

> It is recalled that Professor E.G.D. Murray told the writer (when he wrote the first General staff appreciation of BW in 1943–44), BW tends to spread outwards, from the central infected area, in a diminishing wave, both in time and distance; also that most BW agents can be expected to have a micro-fractional ratio of actual infection (i.e. of human beings or animals, except in very special cases) to the volume used for 'infecting' an area; "N" [anthrax] is a notable exception of this rule.[63]

Convinced that Canadian cities, not troop concentrations, would be the primary target of a Soviet BW attack, Goforth strongly recommended that Ottawa adopt a range of defensive measures, including assistance for the newly established civil defence system. In his opinion, Major General F.F. Worthington, Director General of the Civil Defence Coordinating Committee, was doing an outstanding job in mobilizing "large stocks of immunizing agents in advance of an emergency." On the other hand, Goforth had serious concerns whether these civil defence initiatives would work, since there was always the possibility "of possible mutations in the micro-organisms of BW … [which] might nullify immunization procedures, as now established."[64]

During these years, the Defence Research Board developed a number of programs to ensure that its CBW programs met the needs of its major client – the Canadian Army. On the educational side, there was the annual DRB research symposium, held at its Ottawa headquarters and featuring prominent American and British defence specialists, along with technical papers from biowafare scientists at Suffield, Kingston, and Ottawa (Shirley's Bay).[65] Closely related was the operation of the Atomic, Biological and Chemical Defence School at Camp Borden that provided junior officers of the three armed forces with general information about ABC forms of warfare, including the warning that biowarfare agents were "theoretically many million times more dangerous, weight for weight, than the most effective poison gas known."[66] While free discussion was encouraged at these briefing sessions, this situation would soon change as fears of Communist subversion and espionage mounted. This chill was evident in January 1949 when Director General Solandt issued a series of new security guidelines. One of these proposed more rigorous selection procedures in hiring new scientists and technicians "to ensure that no one with a record of disloyalty has been employed." Another directive focused on the protection of top-secret documents since, as Solandt pointed out, during the Second World War "the majority of leaks … were not through the deliberate passing of classified documents to enemy agents; but rather to the bits and pieces that were let drop in the unguarded moments."[67]

While DRB headquarters established the broad structures of Canada's biowarfare system, the Bacteriological Warfare Research Panel was responsible for the specific weapons projects. During its early years, the panel was dominated by the familiar foursome of Everitt Murray, Guilford Reed, Charles Mitchell, and Otto Maass because of their collective expertise and their range of scientific contacts at home and abroad.[68] But despite these assets, the panel was hard-pressed to deal with so many pressing BW issues. At its meeting of 31 January 1948, for example, they discussed whether the Defence Research Board should participate in the forthcoming British BW maritime trials (Operation Harness), which were ambitious attempts to assess the performance of anthrax, plague, brucellosis, and tularaemia munitions on the open seas, thereby minimizing environmental and health problems. After some deliberations, they reached a consensus that microbiologist A.J. Wood of the University of British Columbia should join Harness's core group of scientists, thereby providing a liaison between Porton and the DRB.[69] The next set of meetings (20 January 1949) had an even more demanding set of challenges. One of these involved the ramifications of the October 1948 re-opening of the Grosse Ile research facility because of concerns that the Soviets might deploy the rinderpest virus agent against North American livestock, thereby creating a need for a new and more effective "R" vaccine. But the most pressing issue facing the panel was to determine which of the various offensive BW agents should be prioritized for possible weaponization and use. It was pointed out, for example, that botulinum toxin (X) was Canada's most potent bioweapon because of its lethality as an aerosol and its use in the poison dart cluster bomb. Close behind was melioidosis, a non-sporulating bacterial agent which was praised for its long storage life and the fact that its "mortality in man is also very high."[70] On the other hand, the Panel acknowledged the value of using highly infectious non-lethal agents such as brucellosis, both because of its psychological impact on enemy troops and because it caused many non-lethal casualties, thereby overwhelming the enemy's battlefield medical facilities.[71] The meeting concluded with a discussion of Guilford Reed's proposal that a BW pilot plant be established at Barriefeld, approximately ten kilometers from Queen's University, not only to provide badly needed laboratory facilities, but also to ensure that Reed could supervise the production of select BW agents. But before the project could proceed, three questions had to be resolved. Did the DRB have sufficient funds to launch the pilot plant? Could qualified scientific and production staff be recruited? And would Queen's University be comfortable having one of its faculty members involved with such a "visible" military operation?[72]

The outbreak of the Korean War in June 1950 produced intensified interest in Canada's biological warfare program. As a result, the Defence Research Board once again asked Guilford Reed to provide an update of the major scientific and technology developments that had occurred in the biowarfare field, based on open and classified sources. In his report, Reed concentrated on several key issues. One of these was the progress that the Tripartite Group had achieved in carrying out laboratory and field trials of five major anti-personnel agents: *Bacillus anthracis, Brucella suis, Francisella tularensis, Yersinia pestis,* and botulinum toxin. On the negative side, Reed observed that Canada had not followed the lead of Camp Detrick and Porton in producing large quantities of these BW agents, despite "the opinion of the bacteriologists in this country who have studied the matter that pilot plant production is essential to progress in the field."[73] Another major problem, he noted, was that the DRB had not yet developed an effective BW dispersal system, although several options had been explored:

> Bombs. In a long series of field trials the concentration of agent per cubic meter of air per minute over target areas has been measured and the infecting or killing dose for animals determined. Several types of bombs have been tested in the field ...
>
> Insects. Reasonably accurate estimates have been made of the rate at which house and fruit flies distribute bacteria of the enteric and dysentery group of bacteria from contaminated baits to human or animal foods.
>
> Other means. Several other means of dispersal are possible, particularly sprays, but no adequate field trials have been carried out.

Reed did not feel that Canadian biodefence measures were sufficient to meet the challenge of a Soviet BW attack. High on his list of deficiencies was the lack of vaccines and other immunizing agents "against infections of toxemia likely to be used in BW agents operations, notably ... Brucella, Tularemia, Anthrax." Even more serious, however, were problems in surveillance and detection. In fact, Reed concluded his report with the recommendation that the DRB should encourage the training of more specialists in the field of aerobiology, who could be appointed "in two or three strategic points in provincial, military and veterinary laboratories, for the improved detection of biowarfare agents.[74]

Defence Research Establishment Kingston (DRKL)

In April 1946 a deal was negotiated between the Department of National Defence and Queen's University for the extended use of Guilford Reed's biodefence laboratory, located on the fourth floor of the New Medical Science Building, along with the top section of the adjoining Animal House. This arrangement, Reed assured Principal Wallace, would facilitate the fusion of fundamental and military-related microbiological research, with "no sharp line of demarcation between strictly University work or equipment and that of the new organization."[75] Funds were also allocated for the salary of the director (Reed) and employment of eight microbiologists and technicians. In his semi-annual report of 1947, Omond Solant explained why his organization welcomed the opportunity of securing the services of Canada's most prolific defence scientist.[76] He began by listing the thirty biological warfare reports that Reed's team had submitted since 1942. On the offensive side, this included the production of botulinum toxin; house flies as vectors of pathogenic bacteria; the viability and virulence of non-sporulating bacteria in peat; and the use of fine ground peat as a carrier of spores. Defensively, the major projects were "production of XA toxoid ... [and] mixing XA and XB toxoids"; along with his analysis "of Japanese balloons for Bacterial Warfare agents."[77] After the war, Reed continued to pursue many of these projects, both in his Kingston laboratory and as a key member of the Bacteriological Warfare Research Panel, and to develop a number of important BW research projects.[78] His own research work was enhanced in April 1953 when, after considerable delay, the Defence Research Kingston Laboratory (DRKL) moved to its new permanent buildings at Barriefield Military Camp.[79] While Reed was named DRKL's full-time superintendent, he did maintain an affiliation with Queen's University, with special provision for "the granting of Ph.D.'s to DRB bacteriology fellows working ... at DRKL ... provided that the candidates reach the required standard."[80] Unfortunately, Reed's dual career as an academic and defence scientist ended abruptly on 21 February 1955 when he had a fatal heart attack after returning home from biodefence meetings in Washington.[81] Without its dynamic leader, Kingston's biological research operation was gradually phased out, its equipment and personel moved to either Shirley's Bay (Ottawa) or the Suffield Experimental Station.[82]

During its heyday, the Kingston laboratory was highly regarded by both American and British BW experts. Of particular interest was Reed's work on the bioweapons potential of poisonous mussels and clams, located along the shores of Nova Scotia and New Brunswick, that often reached toxicity levels of over 400 units of poison per 100 grams.[83] In order to weaponize this natural toxin, the Defence Research Board worked out an arrangement with the federal Department of Fisheries for the large-scale harvest of 'red tide' shellfish, which were then processed at the Kingston laboratory.[84] By July 1948, an even more efficient system was established by having the shellfish poisons concentrated at a special plant in Digby, Nova Scotia, with the toxic product then being flown to Camp Detrick's secret airfield at Frederick, Mayland.[85] From the perspective of the US Chemical Corps this arrangement had enormous advantages both in maintaining its own bioweapons arsenal and in assisting the Central Intelligence Agency to develop special assassination weapons for its covert global operations.[86]

The use of insect vectors for transmitting infectious biological material was another promising Kingston project. Although Reed had briefly worked on vector research during the Second World War, it was not until 1949 that he launched his special study "Possible Use of Insects as BW Vectors – Scientific Intelligence Aspects." This undertaking was strongly endorsed by the BW Research Panel on the grounds that it advanced three major Canadian health security goals: "(a) public health precautions in peace-time when dissemination of disease is mainly due to 'natural' causes; (b) Defence against B.W. attack in wartime when the existing non-lethal insect infestation might be used to spread more or less lethal diseases (human, animal, or agricultural); (c) For purpose of B.W. attack in areas outside Canada having an insect infestation similar to certain Canadian areas."[87] This latter goal had an important offensive bioweapon dimension which Reed believed would provide an incentive to establish a special team of Canadian, British, and American bacteriologists and entomologists to carry out a comprehensive survey of possible BW vectors.[88] For a variety of reasons this did not occur. In February 1950, for example, the DRB liaison officer in London reported that Porton Down was "not doing any work on insect vectors of B.W. agents and is not proposing to do so."[89] Detrick officials had a somewhat different response: while they informed the DRB that the US insect vector program had been terminated in August 1949, they also expressed strong support for Reed's project, promising "to direct all possible information and intelligence on the subject to Canada."[90]

Grosse Ile Research Establishment

In 1948 the Grosse Ile research station was reactivated because of Canadian and American concerns that the Soviet biological warfare program could target the North American cattle industry with rinderpest or other agro-biological weapons.[91] The task of creating an effective "R" vaccine was, however, complicated by the fact that the wartime activated virus "seed stock" was no longer suitable for large-scale immunization. Nor was it easy to develop a suitable alternative. Indeed, it was not until

December 1950 that DRB scientists were able to announce "that they could fully protect cattle against highly virulent (R) material ... [and] prevent an exceedingly grave attack."[92]

Significantly, by this juncture the Grosse Ile research agenda now featured an offensive dimension, largely because the US Chemical Warfare Corps wanted to carry out "basic research in the field of anti-animal agents." More specifically, American microbiologists and veterinarians requested permission to carry out laboratory tests at the station with a number of exotic and dangerous animal pathogens, since these facilities were not available in the United States. This wish was quickly granted, with one reservation: Dr Charles Mitchell, representing the DRB, would have "full authority over the trials."[93] While the defensive aspect of rinderprest research had initially been the major focus, during the early 1950s attempts were made to determine how the virus could be used against enemy's cattle. This project involved a series of experiments with twenty-two different strains of the virus to determine whether they met the necessary criteria: ease of production, storage capacity, stabilization as an aerosol, "and methods for dissemination."[94] Canadian and American scientists were also successful in developing several other anti-animal pathogens such as weaponized forms of African swine fever, Venezuelan equine encephalomyelitis, Newcastle disease, fowl plague, hog cholera, rabies, and Rift Valley fever.[95]

But were these projects successful? Yes, was the assessment of the October 1954 report of the DRB Biological Warfare Research Panel. It noted that four viruses had been developed by the Joint US-Canada anti-animal program as efficient agents "for the destruction of ... food-bearing animals.[96] These included, "1. Cattle (Rinderpest); 2. Swine (African Wart Hog and African Swine Fever); 3. Chickens (Fowl Plague)."[97] Yet despite these scientific successes, the station's operation would soon be terminated, largely because the US decided to phase out its anti-animal program. DRB officials were not disappointed over these development, because of the high costs of the GI facility and because of growing criticism that its biological warfare facilities were responsible for mysterious outbreaks of disease. Of particular importance was the legacy of the devastating foot and mouth (FMD) epidemic that ravaged western Canadian cattle herds during the early 1950s.

This FMD outbreak was detected in southern Saskatchewan during November 1951; it lasted until August 1952, when the Canadian Department of Agriculture finally announced the epidemic was over.

But the consequences of the disease had been devastating, with over 1,300 cattle and 300 swine being destroyed as part of a containment strategy that ultimately cost western ranchers over $800 million.[98] An even more challenging aspect of this agro-biodefence campaign was to convince the US Department of Agriculture that Canada was FMD free and that Washington should lift its total ban against Canadian cattle entering the United States.[99] While most of the credit for the containment campaign was given to provincial and federal agricultural officials, the DRB's Bacteriological Warfare Research Panel had provided vital assistance behind the scenes in analysing the FMD viral strain and in monitoring the animal quarantine system. But it was a gruelling and stressful experience. Guilford Reed, whose laboratory was in the forefront of this research, reputedly quipped that "if a solution was not quickly found for the epidemic, his Kingston operation might be shut down."[100]

Suffield Experimental Station

Throughout the Cold War the Suffield Experimental Station assumed a major role in carrying out biological weapons trials for American and British military planners.[101] Most of these tests involved botulinum toxin (X), *Francisella tularensis* (UL), and *Brucella suis* (US), with the latter being favoured by British and American BW planners because of its high infectivity and ability to incapacitate enemy troops.[102] The actual trials went through several stages. Initially, small-scale, controlled tests were carried out in the SES bursting chamber.[103] The operation then shifted to Area E of the proving ground, where a variety of animals (goats, guinea pigs, monkeys) were exposed to aerosolized pathogens within the large-scale wind shed because it provided protection against the severe cold and because its facilitated the recovery of the bioagent. And finally, the aerial munitions trials, using the US Army's M-17, B-E1, and E-48 cluster projectiles, were separately assessed by the respective Canadian and American scientific teams.[104]

Given their shared interests in biological and chemical weapons trials, scientists from Suffield, Detrick, and the Dugway Proving Ground (DPG) worked on a number of joint projects.[105] These cooperative arrangements featured the exchange of information, equipment, and BW agents, along with shared technical meetings where scientists discussed which research facility was best suited for specific weapons trials. The criteria for this site selection process were described in the following

1954 US report: "Fort Detrick had a large number of chambers to do highly specialized operations. Suffield had a rather exclusive wind shed with a facility for doing other specialized operations. Dugway had no chambers or wind shed but they had a large amount of available space."[106] Not surprisingly, given previous concerns over anthrax contamination, Suffield officials insisted that these trials should take place at Dugway while, in return, they would carry out tests with non-persistent *Brucella suis* and *Pasteurella tularensis* agents.[107]

With the outbreak of the Korean War in June 1950, biological warfare cooperation between Canada and the United States increased dramatically.[108] This was evident in October 1950 when the US Joint Chiefs of Staff drafted a plan for the strategic use of weapons of mass destruction, which included the observation that in its plans to develop a biological warfare capability, the US Army had "received valuable assistance from our Canadian ally."[109] And this pattern of collaboration continued to evolve. In December 1951, for example, Omond Solandt was informed that the United States Air Force (USAF) wanted to carry out BW bombing trials in Canada and were preparing an inventory of possible sites "where user tests of live agents can be undertaken." There was particular interest in the Cold Lake (Alberta) base of the Royal Canadian Air Force "since the climate there should be similar to many parts of European Russia." Other possible candidates were the joint US-Canadian base at Churchill, Manitoba, and the Suffield Experimental Station (SES).[110] While Canadian officials tried to be accommodating, they were adamant that US trials should not take place anywhere other than SES "without very good reason." This message was subsequently conveyed to the USAF, along with an extensive description of Suffield's advantages as a bioweapons testing facility: "It is a completely self-contained unit with living quarters, animal houses, decontaminated centre, and a mile-square area fenced with rodent proof fence for the conduct of BW trials ... a bursting chamber of about 200 cu.m. capacity which will handle shells up to 25 pd. size. It has a vacuum system and lines to allow sampling of dispersed agents from six positions in the chamber. There is also an animal exposure chamber attached."[111]

During this period Suffield scientists also worked closely with David Henderson and his colleagues at Microbiological Research Department (Porton). Of particular importance was SES support for a series of British BW maritime trials carried out in the Caribbean and North Sea between 1948 and 1956. The first of these trials, Operation Harness, was a large-scale undertaking located in the Antigua-Leeward island region that involved over 450 scientific and support personnel. Scientifically, the major goal was to determine whether, as the experts claimed, "the most effective way to distribute pathogenic bacteria was to produce airborne particles containing the virulent organism ... [and] by the direct inhalation of such clouds a percentage of casualties in man or animals would follow."[112] Harness was, however, a risky venture since there were numerous tests involving "hot" pathogens such as *Bacillus anthracis, Yersinia pestis,* and *Francisella tularensis.* Overall, the operation achieved its basic goals: indeed, of the twenty-two toxic tests attempted, fifteen were regarded as completely successful.[113] In addition, Canadian scientists were involved with subsequent maritime trials: Operation Hesperus in 1952–3 (Hebrides), Operation Ozone in 1954 (Nassau), and Operation Negation in 1955 (Bahamas). They were particularly interested in this latter trial since it involved the testing of viral agents such as Venezulan equine encephalitis (VEE), and vaccinia, used as a stimulant for the much-feared variola virus (smallpox).

On the policy side, the Defence Research Board monitored ongoing debates in the United Kingdom about the relative importance of biological and nuclear weapons. There were several stages in this competition. Until October 1952, when the UK exploded its first atomic bomb, supporters of the bioweapons option could successfully argue that BW was crucial for the country's national security, particularly since the Air Ministry was actively promoting the development of a 1,000-lb BW cluster bomb. As a result, MRD Porton received considerable financial support to expand its laboratory facilities, hire new scientific staff, and make plans for the mass production of bulk pathogens in a large-scale pilot plant. On the other hand, there was abundant evidence that Britain's military hierarchy regarded bioweapons as a rather dubious stopgap measure, until the country acquired a nuclear deterrent. As one report to the British Chief of Staff pointed out, "lest over-optimistic results should be expected from [BW] retaliation on a massive scale, it should be borne in mind that the large-scale use of biological warfare from the air as a weapon of war has not been tried."[114] Although Porton's BW operation continued to function after 1952, it was completely eclipsed by the rapidly expanding American program, which, as David Henderson noted, "had become very offensive minded ... [and] the Services had gained complete control of BW matters."[115]

Medical Internationalism and Civil Defence

Canada and the World Health Organization

In their attempts to prevent biological warfare, Canadian scientists also worked closely with the newly created World Health Organization. This unique organization was inspired by the internationalism associated with the emergence of the United Nations, reinforced by the dedicated work of a talented group of medical scientists, government officials, and members of humanitarian organizations.[116] One of the most important of these international medical leaders was Dr Brock Chisholm of Canada, who in 1947 became the first Director General of the WHO.

Chisholm brought impressive credentials to his new position, particularly his experiences as Director of Medical Services of the Canadian Army during the Second World War. In this role he took special interest in the problems of infectious diseases, both in terms of the biological weapons threat and the possibility of an influenza pandemic. In the former case, Chisholm had been part of the Canadian biodefence team that visited London in April 1944 in an attempt to convince Allied military planners to immunize all D-Day landing forces against botulinum toxin. Once he returned to Canada, the Army's Director of Medical Services had encouraged efforts by the Connaught Medical Research Laboratories of Toronto to develop an effective influenza vaccine capable of "meeting the needs of the armed services relative to control of this disease."[117] Convinced that an epidemic was imminent, during the summer of 1944 Connaught mobilized a team of thirty-six researchers under the leadership of virologist Ronald Hare to develop a vaccine for use against the anticipated A influenza strain.[118] Once the project was completed, over 200,000 doses were distributed to the Canadian Armed Forces, with the surplus being sent to the United States.[119]

During his six years as WHO Director General, Chisholm dedicated himself to the fundamental goals outlined in the WHO charter, namely that "the health of all peoples is fundamental to the attainment of peace and is dependent upon the fullest cooperation of individuals and States."[120] Because of the Cold War, this was not an easy undertaking, as was evident in February 1949 when the Soviet Union and its Communist satellites withdrew from the organization on the spurious claims that it was a tool of American foreign policy.[121] Nor were Chisholm's attempts to bridge this ideological divide successful, and the Communist bloc remained outside the organization for the next seven years.[122]

Despite this polarized international situation, the World Health Organization achieved many of its goals during the Chisholm years. These included a global campaign to control major infectious diseases such as malaria, smallpox, tuberculosis, plague, and cholera. Indeed, one of the early WHO successes came during the 1947 cholera epidemic in Egypt, when prompt action by Geneva health officials in mobilizing medical resources and vaccine supplies helped contain the outbreak. This incident also set the stage for a major debate about whether the WHO should concentrate on the eradication of the world's greatest disease threats or whether such campaigns were beyond its financial and manpower resources.[123] Closely related was the question of whether, if eradication was attempted, the WHO should start with smallpox, which was Chisholm's choice, or malaria, which was the US preference. In 1956 the World Assembly endorsed the latter position on the grounds that malaria killed more than 2 million people annually and because public health scientists erroneously elieved that extensive use of DDT and antibiotics would eliminate this ancient scourge.[124]

Another subject that preoccupied the Director General was the threat of biological weapons. And he spoke frequently on this subject. His most strident statement came in 1949, shortly after US Secretary of War James Forrestal announced that the Pentagon intended "to develop means for defence against possible enemy [BW] attacks."[125] On 9 September Chisholm added his own spin to this controversy when he provided an alarmist and exaggerated interpretation of the BW threat in an address to the World Union of Peace Organizations: "Biological warfare is not a new kind of war, it is just the latest step … Some seven ounces of a certain biological, if it could be effectively distributed, would be sufficient to kill all the people of the world." [126] This bold intervention into the CBW disarmament debate did not win him many friends in Washington, where his warnings were dismissed as irrational fear-mongering. Ironically, three years later some of Chisholm's fiercest US critics would be forced to ask the WHO for assistance in confronting Communist allegations that the Pentagon had used biological weapons in the Korean War. Unfortunately, his attempts to mediate the situation were unsuccessful since neither the Chinese nor North Korean governments sanctioned site visits by WHO experts on the grounds that the Geneva-based organization was a toady of the United States.[127]

In December 1952 Chisholm suddenly announced his resignation as Director General. According to his brief, it was a decision based on personal fatigue, along with his belief that an international organization such as the World Health Organization "should not have the same head for too long, particularly at the beginning of its history."[128] Behind the scenes, however, it was clear that the US government wanted a more malleable Director General who would be less inclined to criticize its strategic policies. In May 1953 they got their way when Dr M.G. Candau, a Brazilian malariologist, was nominated. Candu held this position for the next twenty years.[129]

Canada's Civil Defence System and Cooperation with the United States, 1948–1953

During the Second World War there were rudimentary attempts to develop a Canadian civil defence system. [130] This was certainly the case in 1940–1 when the threat of Nazi sabotage appeared imminent, and again in 1944, when Japanese BW balloons began drifting across the western sections of the country. But it was the advent of the Cold War, with its accompany threat of weapons of mass destruction, that convinced North American political leaders that civil defence was an essential defensive program. [131] In 1948, for example, there was extensive discussion on how Canada should prepare for a possible atomic bomb attack, with special emphasis on the need to safeguard the country's civil institutions and infrastructure. According to George Lawrence, a prominent nuclear scientist, Canada was particularly vulnerable for this type of attack "owing to the centralization of senior defence officers, government and senior civil servants – in short, those who must organize and coordinate every aspect of national defence – in an area of a few hundred acres in the city of Ottawa." [132]

Because of this controversy, Major General F.F. Worthington was appointed Civil Defence Coordinator and Special Advisor to the Minister of National Defence. [133] In August 1949 Worthington's influence over the country's civil defence system was further enhanced when he was given responsibility for a new umbrella organization, the Civil Defence Coordinating Planning Committee (CDCPC).[134] However, at this stage it was uncertain whether the federal government would develop the country's civil defence programs unilaterally or whether provincial and local authorities would also be involved with the planning process.[135] Many of these jurisdictional problems were resolved at the August 1950 dominion-provincial conference on civil defence when it was decided that while Ottawa would coordinate the overall system and provide the liaison with US and UK agencies, each province could pass its own civil defence legislation, appropriate for its local needs. By the end of 1950, it was estimated that over 224,000 persons nationwide were participating in civil defence activities, a third of whom were "full-time provincial and municipal personnel (police, fire, engineers, etc)."[136]

While the Civil Defence Coordinating Planning Committee was primarily concerned about atomic bomb attacks, there was also growing apprehension about other weapons of mass destruction.[137] In part, this was related to the growing interaction between the committee and CBW experts of the Defence Research Board.[138] In a 27 September 1950 memorandum, for instance, the veteran BW consultant Charles Mitchell outlined how a large-scale BW attack would overwhelm Canada's civil defence system "particularly hospitals, public health laboratories and agencies ... [which] would be first to feel the shock of biological aggression." But in Mitchell's opinion, an even more alarming situation would be a bioweapons assault on the country's livestock population:

> No human disease is comparable to two of our great animal plagues of cattle [rinderpest & foot and mouth] ... made to order for saboteurs. Moreover, each of these agents ... requires an infinitesimal amount to produce a primary outbreak ... Fifty active saboteurs each armed with as much material as could be held in a fountain pen could involve Canada and the United States in an outbreak which if uncontrolled would ... ruin the animal industry of the two countries ... Unfortunately, the public ... have little conception of the danger ... Another point which should be considered in regard to BW is the rapid identification of the agents.[139]

In a second memorandum of 16 February 1951 Mitchell concentrated on the broader aspects of Canada's biodefence system, which in his opinion was reasonably well organized. On the other hand, he lamented the fact that irresponsible demagogues such as the Reverend James Endicott, President of the Canadian Peace Council, were causing unnecessary public hysteria because of their scare-mongering diatribes about American BW intentions.[140]

In 1950, the Civil Defence Coordinating Committee was shifted from National Defence to Health and Welfare because of the increasing importance of medical countermeasures.[141] This transition of responsibility was accompanied by the publication of a series of civil defence manuals focused on the tactics of survival.[142] Of special importance was the September 1952 report of the Civil Defence Health Planning Group (CDHPG), which explored several biodefence scenarios.[143] First was the question of whether there would be a biowarfare incident "by sabotage before or after open war or as part of an overt attack ... [or] in conjunction with explosives or atomic munitions."[144] Second, it was recognized that the most deadly form of germ warfare came from "organisms which are capable of transmission by airborne clouds," particularly if the targeted population had not been immunized against this pathogen.[145] Somewhat reassuring, however, was the report's rejection of exaggerated fears about enemy use "of a new, super-agent capable of producing an uncontrolled epidemic"; instead, the authors insisted that Canadian health security responses would "be limited to known disease agents and their potential to produce disease."[146]

Another important trend was Ottawa's concerted effort to establish a North American civil defence system.[147] The first stage in this cooperative venture occurred in April 1949 when the Canadian and American governments endorsed the formation of a Joint Industrial Mobilization Committee, under the auspices of the Permanent Joint Board of Defence.[148] Further progress was achieved at the joint meetings, held in Ottawa on 21 February 1950, when it was agreed that first responders, trying to protect civilians from enemy attack, should proceed "as if there were no border."[149] The following month, the Joint United States / Canadian Civil Defence Committee was established with a mandate to develop cross-border exercises "for immediate warning and action in the event of an attack."[150] These initiatives were confirmed at the committee's April 1951 meetings in Washington, DC, where it was proposed that regional civil defence protocols be negotiated between US states and Canadian provinces, including arrangements for "entry and egress of civil defense forces ... of one country into the other."[151]

Canada and the Germ Warfare Controversy

The 1952–3 germ warfare campaign by the Communist bloc should have been anticipated.[152] Throughout the immediate postwar years, Soviet diplomats at the United Nations had lobbied hard for an international ban of both atomic and biological weapons, claiming that unlike Washington, Moscow was not developing any weapons of mass destruction. However, after the successful September 1949 test of its first atomic bomb, the Kremlin changed its position. Now the Soviet propaganda machine stressed that the greatest threat to world peace was the germ warfare program of the United States.[153] An integral part of this campaign was to exploit the fact that the United States had not yet ratified the 1925 Geneva Protocol banning bacteriological and chemical warfare and to claim that the US military, in advancing its postwar BW capabilities, had used the services of Dr Shiro Ishii and other Japanese scientists who had been guilty of war crimes.[154]

On 28 January 1952 Russian authorities formally charged that US military forces in Korea were guilty of "poisoning wells and spreading small-pox and typhus bacteria in various places in North Korea." On 22 February, Bak Hun Yung, North Korean Foreign Minister, registered a formal protest against American germ warfare aggression against his country. Two days later, Chou En-lai (Zhou Enlai), Foreign Secretary of the People's Republic of China, added his voice to the Communist choir with claims that his country's public health officials had conclusive proof of "the use of bacteria against the front lines and the rear by the Americans." Accusations of a US "germ warfare" campaign were not restricted to Communist regimes. On 8 March, the prominent French nuclear scientist Frédéric Joliot-Curie, President of the World Peace Council, issued a statement denouncing the use of biological weapons by the United States.[155] This was followed by similar allegations by the Commission of International Democratic Lawyers (31 March), the Canadian Peace Congress (30 April), the World Peace Council Commission (14 September), and the International Scientific Commission Report (2 October). And to reinforce the litany of allegations, Chinese officials produced a series of "confessions" by American POWs, who testified that they had been personally involved in US germ warfare atrocities.[156]

Canada's response to this international crisis assumed several forms. First, attempts were made to determine the reasons for the Russian, Chinese, and North Korean charges, and whether this campaign against the United States represented a dangerous escalation in the Cold War, as many US experts claimed.[157] Second, Canadian diplomats at the United Nations attempted to use their influence with other Commonwealth countries in countering Soviet attacks on the United States, particularly at the important UN Disarmament Commission.[158] Third, Ottawa strongly endorsed the US proposal that independent experts from the International Committee of the Red Cross or the World Health Organizations should be allowed to visit Chinese and North Korean sites where these germ warfare incidents ostensibly occurred.[159] At the same time, Canadian diplomats were often troubled by certain aspects of the US response to the crisis, particularly Washington's insistence that it would not "reconsider ratification of the Geneva protocol nor did it intend to be committed in any way not to use bacteriological weapons short of a comprehensive disarmament agreement."[160]

On 12 May 1952, Lester B. Pearson, Secretary of State for the Department of External Affairs (DEA), outlined Canada's official response to the germ warfare controversy. It consisted of a vigorous counter-attack in the Canadian House of Commons against the allegations of Reverend James Endicott, President of the Canadian Peace Congress (CPC) and a major critic of US military policies in Korea and China.[161] In his speech Pearson also defended Omond Solandt, Director General of the DRB, whom Endicott had denounced as a war criminal during his visit to China in April 1952.[162]

> As in the United States, there are agencies in Canada engaged in United States germ warfare preparations. So far as I know there is a big plant in the Alberta Province of Canada which turns out, on a large scale, infected insects harmful to men, animals and crops. The head of this organization, Dr. Solandt, in an article in the Montreal Standard openly declared "the future of death on a mass scale is very bright." This shows that these people are entirely devoid of humanitarianism. I am inclined to believe that perhaps many of the germ-carrying insects spread by the United States over north-east China are produced in Canada because in that area the weather is very cold for six months of the year. Such a climate is suitable for breeding bacteria carrying cold-resistant insects.[163]

In refuting Endicott's accusation, Pearson pointed out that Solandt was an outstanding Canadian public servant "whose reputation for service and sincerity and trust has never been challenged before." In contrast Endicott was described as an unscrupulous fellow-traveller who was "an undiscriminating supporter of everything that emerges from the Communist higher command ... [an] example of what this worship of this false god Communism, can do in the disintegration of an individual."[164]

The dilemma for the Canadian government was how to curb Endicott's anti-American activities without turning him into a martyr.[165] Although RCMP and Justice officials were eager to prosecute on the basis of Endicott's charges that Canada was an accomplice in the US germ warfare campaign, cooler heads prevailed. For the realists in the Cabinet there was an understanding that even if the president of the CPC was charged with seditious libel, there would be little chance of getting a conviction, largely because American BW experts seconded for the trial "would not be prepared to answer detailed questions on specific operations while under cross-examinations ... [and] would be expected to plead security precautions in order to avoid being led on into divulging classified information."[166] Another deterrent was the concern that American public opinion might turn against Canada if Endicott were acquitted, given the intense anti-Communist crusade being carried out by US Senator Joseph McCarthy.[167]

Instead of criminal action, Foreign Affairs officials decided to adopt another approach – to challenge the scientific validity of Endicott's allegations.[168] As a result, three well-regarded experts were approached: Dr W.H. Brittain, vice-principal of McGill's Macdonald College; Dr A.W. Baker, head of the department of entomology and zoology at the University of Western Ontario; and Dr C.E. Atwood, professor of zoology at the University of Toronto. Their task was to prepare separate scientific critiques of the 1952 Canadian Peace Congress pamphlet, "Documentation on Bacteriological Warfare" (30 April).[169] This strategy proved a great success. After examining all three reports, DEA officials were delighted that the scientists agreed on one key point: the CPC pamphlet was seriously flawed and should not be taken seriously.[170] Dr Brittain's study went even further, dismissing the pamphlet as nothing more than a collection of "biological absurdities," which, he argued was not surprising since "throughout his testimony Dr Endicott has either drawn conclusions he is incompetent to make or has accepted hearsay evidence."[171] These experts did not accept Endicott's claims that he had personally examined some of the BW vectors, given the fact that there were more than 50,000 insect species in this region, making it virtually impossible "that he could ... decide whether or not they were 'germ-laden.'"[172] But their most devastating rebuttal consisted of a basic question: if US military officials had really wanted to use biological weapons for strategic purposes, why would they "have adopted such inept, infantile and altogether stupid methods in a field in which they are supposed to be masters?"[173]

In June 1952 Foreign Affairs prepared a summary report, *Concerning the Charges of the Practice of Biological Warfare by United States Forces in Korea and North-East China.* In short order it was tabled in the House of Commons, and copies were sent to prestigious newspapers in the United States, the United Kingdom, and Western Europe.[174] In addition, Canadian diplomats took advantage of these findings in their presentations at the United Nations General Assembly and at the meetings of the Committee on Disarmament (CD) in Geneva.[175] DEA officials were, however, most gratified by the positive response they received from the US State Department, which praised the Canadian report as an effective antidote in "refuting the communist germ warfare charges recently repeated by Dr. James Endicott … [and] the Dean of Canterbury … carefully calculated to assist the communist effort to discredit the United States and other democratic nations."[176] Another strong endorsement came from the Canadian High Commissioner in India, who arranged to have the report cited in the influential journal *Eastern Economist,* since its readers had "so far received … one-side reporting of the germ warfare allegations."[177]

The second stage in the germ warfare controversy emerged in October 1952 when the International Scientific Commission, chaired by Joseph Needham, released its report *Of the Facts Concerning Bacterial Warfare in Korea and China.*[178] Prepared by seven pro-Communist Western scientists, it was a massive document, containing over 665 pages of text and 100 pages of photographs of 50 specific germ warfare sites (37 in China and 13 in North Korea). Because of the scale of the report, and the credentials of its authors, this document represented a more serious challenge to the US position. As a result, the Needham Report became an immediate target of Allied defence scientists who sought to discredit both its scientific methodology and its ideological bias.[179] For British experts, the commission's report was badly flawed because of its dependence on hearsay evidence, its one-sided approach in investigating the different incidents, and the fact that the commissioners were constantly under the supervision of Chinese and North Korean officials.[180] In the United States there was an even more vigorous reaction, a number of prominent scientists labelling the commission's work as nothing more than a blatant exercise in Communist propaganda. In addition, Defence Research Board specialists rejected the Needham study as a disingenuous document, with "its pseudo-scientific paraphernalia … skillfully designed to impress."[181]

The germ warfare controversy continued to dominate proceedings at the United Nations throughout 1952.[182] However, the situation changed dramatically after the death of Joseph Stalin in January 1953, when, as one American diplomat put it, "a new wind is blowing down from the steppes."[183] A sense of détente was reflected in a series of conciliatory statements by Soviet Prime Minister Georgy Malenkov which, in turn, produced a more cooperative response by Soviet diplomats at the United Nations. Equally important was the willingness of Chinese officials to accept a compromise for the exchange of prisoners of war, which led to the armistice of 27 July 1953, ending the Korean War. By this stage, even the combative Soviet diplomat Yacob Malik reasoned that it was counter-productive for the United States to insist on placing the germ warfare issue on the agenda of the General Assembly, because "after all, the war is over, and the bugs are dead."[184]

Conclusion

The germ warfare controversy coincided with an intensification of security screening of American scientists by the House Committee on Un-American Activities and its Senate counterparts.[185] Similar developments took place in Canada, where some scientists, suspected of left-wing activities, found themselves either excluded from classified research or dismissed from government service. This trend greatly concerned Everitt Murray of McGill University, who had been a key figure in Canada's biological warfare program until February 1952, when he severed his connection with the Defence Research Board with the observation that no "self respecting bacteriologists" should be expected to accept the new security restrictions. Murray also lamented the fact that science was "being used as a political tool, both nationally & internationally ... to the great danger of scientists ... [and] it is best not to have anything to do with any of the classified work."[186]

Yet, significantly, two of Murray's closest friends, Guilford Reed and Charles Mitchell, continued their roles as BW consultants throughout the 1950s because of their belief that Canada's biological warfare program was a vital component of the country's national security. But they now operated in quite a different environment from the heady days of the C-1 Committee, when university scientists had dominated the research and administrative functions. On the other hand, there was still considerable continuity in how Canada's bioweapon program interacted with its counterparts in the United States and the United Kingdom, except that the exchange system was now institutionalized under the 1947 Tripartite CBR Agreement. But like the wartime arrangements, it was also based on the principle of equal partnership and cooperative ventures, despite the fact that the American BW operation, with its large-scale facilities at Fort Detrick, Pine Bluff, and the Dugway Proving Ground, far surpassed those of its allies.

During the postwar years the Defence Research Board, and its Bacteriological Warfare Research Panel, made a number of important contributions on both the offensive and defensive side of the bioweapons equation. For example, Guilford Reed's Kingston laboratory developed innovative anti-personnel weapons such as shellfish toxin and the use of insects as vectors, while the animal BW research at Grosse Ile went far beyond its wartime exploits. But the most important Canadian asset was the Suffield Experimental Station (SES), which carried out a number of important bioweapons trials on behalf of its allies, as both the United Kingdom and the United States sought to enhance their offensive biological warfare capabilities.

During the Cold War, Canadian defence scientists faced another major challenge: protecting Canadian civilians from weapons of mass destruction. Although Canada's civil defence system concentrated on the consequences of a nuclear attack, there were also concerns that Canadian cities might be the target of a Soviet bioweapon assault. In trying to save Canadian lives civil defence officials of the Department of Health and Welfare developed a complex national system, working closely with their counterparts in the United States on cross-border arrangements. This tendency to consider North American solutions remained an integral part of the civil defence system of both countries throughout the Cold War.

The 1952–3 germ warfare controversy had a decided impact on Canada's image at home and abroad. This was not surprising given the close interaction between the Defence Research Board and the US Chemical Warfare Corps in the development of BW munitions and delivery systems. Of particular importance were the attacks by pro-Communist publicists on DRB Director General Omond Solandt, which prompted Ottawa to launch a vigorous campaign to discredit these allegations and to consider the possibility of criminal prosecution. Diplomats of the Department of External Affairs also played a major role in protecting the reputation of both Canada and the United States at the United Nations. The vast majority of scholars who have examined the germ warfare controversy, notably those who have gained access to Russian archival sources, have concluded that the Soviet charges were propaganda, pure and simple.[187]

Another controversial question was whether biological warfare was officially recognized as part of Canada's military policy, or, more specifically, whether members of Parliament had an opportunity of discussing the goals and priorities of the country's BW research program, such as occurred in the United States where there were periodic debates in Congress about these subjects. Basically, the answer is no. There were virtually no discussions of this subject in the annual reports of the Department of National Defence, in the debates of the House of Commons, or in Canadian newspapers. Instead, all of the crucial policy decisions about Canada's biological warfare policies took place behind a veil of secrecy and within a culture of denial.

3 Operational Biological Weapons and Alliance Cooperation, 1955–1969

General Stubbs will arrange a four hour briefing ... for the Science Advisory Committee Panel ... the terms of reference to this Panel may be: (1) provide advisory functions as far as CW explanatory research is concerned including detection and defense; (2) appraise the BW research efforts and potentialities; (3) determine the extent to which the usefulness of BW and CW agents in "limited war" is under study and adequacy thereof; (4) finally determine the extent of interest and effort in the three services to develop and exploit new weapons and defenses in the BW and CW fields.

J.R. Mares to Dr J.R. Killian, US National Science Security Advisor, 1 October 1958, United States Declassified Documents, 163A

Any country intending to incorporate bioweapons into its military arsenal would have a predictable developmental pattern regardless of whether the ultimate goal in using these weapons was deterrence or offensive first use. The six major stages in this weaponization process would be (1) the research, development, and testing of the selected pathogenic agent; (2) the mass production and storage of these agents; (3) the design and field use of suitable munitions (bombs / sprays); (4) the utilization of long-range delivery systems (rockets / high-performance aircraft); (5) the creation of effective civil defence protective measures for friendly troops and civilians (vaccines, antibiotics); and (6) the realization of full operational capacity by integrating offensive and defensive DW measures into existing military strategies. For the United States and its allies, all but the final stage was achieved prior to 1969.[1]

The sixteen years between the end of the Korean War to the end of the American offensive program in November 1969 represented the "golden age" of the biological warfare programs of the Tripartite allies. While most of the activity was carried out in the major US facilities, Canadian and British BW scientists also made important contributions in the development of second- and third-generation biological weapons. This was achieved in several ways. First were the major improvements in the quality of traditional bacterial bio-agents and toxins, along with the weaponization of viral agents such as variola (smallpox), Venezuelan equine encephalomyelitis, and Rift Valley fever. Second, important innovations in the strategic aspects of BW delivery systems were developed, both through more effective cluster projectiles (500 lbs/1000 lbs) and the new techniques of large area coverage (LAC), whereby enemy targets could be attacked by the creation of sizeable aerosol clouds.

While the US Army's Chemical Warfare Corps maintained its administrative control over the country's biowarfare program, the White House and Congress also provided significant input concerning broader strategic questions. In October 1950, for example, the Stevenson Committee advised Secretary of Defence George Marshall to expand the country's biological weapons arsenal and abandon its traditional no-first-use policy. This recommendation was subsequently incorporated into the national defence priorities of the second Eisenhower administration (1956–60), remaining a well-protected secret until the late sixties. Yet despite these policy adjustments, many experts still felt that American CBW capabilities would be insufficient in deterring "the Soviet bloc from initiating chemical and biological warfare, nor to retaliate effectively if the enemy does initiate."[2]

Since Canada had neither a nuclear nor a biological deterrent, its BW research program was fundamentally different from its Tripartite partners. While defence officials paid lip service to a vague plan that Canadian troops might deploy biological weapons if attacked, in reality, this commitment was based on the unlikely scenario that the United States would supply these instruments of war if the Soviets invaded Western Europe. But in many ways it did not matter since Canada's major insurance policy against enemy attack remained the US nuclear umbrella, which was officially extended northward by the 1958 North American Aerospace Agreement (NORAD). While successive Canadian governments endorsed the continental defence system, this dependency was a source of concern for some defence officials who felt that their country should assist the American global containment system where possible. Omond Solandt, for example, consistently argued that the Defence Research Board should assist US military programs in fields such as biological and chemical warfare, where Canada had special expertise and assets. But the most effective manifestation of this bilateral CBW system took place at the ground level, where defence scientists at the Suffield Experimental Station worked on a number of joint projects with their counterparts at Fort Detrick and the Dugway Testing Grounds. This was particularly pronounced after 1963 when the Kennedy administration, under the auspices of Project 112, launched an aggressive and imaginative program to improve US offensive bioweapon capabilities.

Throughout this phase in the Cold War there were fears in Washington, Ottawa, and London that the Soviet Union had surpassed the Western alliance in the strategic use of bioweapons. And there were ample reasons for concern. While initially the Kremlin had viewed germ warfare as a temporary substitute for the Russian atomic bomb, by the late 1950s BW was now integrated into a broader system of weapons of mass destruction to be used against NATO forces in Europe when the inevitable conflict took place. This new appreciation of bioweapons was reflected by the development of an impressive BW infrastructure including secret laboratories, where special attention was devoted to creating more pathogenic and stable strains of anthrax, plague, tularaemia, and smallpox. Once these agents were deemed operational, they were tested at the BW field facilities on Vorzrozdeniye Island located in the Aral Sea, which soon became one of the most contaminated places on earth.[3]

Canadian Defence Science and Nuclear Weapons, 1952–1962

Although Canada did not develop its own nuclear weapons, it actively supported the American and British programs throughout the postwar years.[4] As many scholars have demonstrated, the Canadian government regularly provided the United States with natural uranium and recycled plutonium from its Chalk River reactors that was used for military purposes.[5] Less well known were Ottawa's efforts to assist the British atomic bomb project, including the 1950 offer of a testing site in the Canadian Arctic.[6] The catalyst for this venture was the joint UK-Canadian study entitled *The Technical Feasibility of Establishing an Atomic Weapons Proving Ground in the Churchill Area.*[7] According to the report, this joint undertaking could advance several important goals: expanding Anglo-Canadian nuclear cooperation and gaining the support of the US Atomic Energy Agency for possible joint weapons trials of "one or two per year for several years."[8]

But the costs were daunting. According to British estimates, $5,600,000 would be required for the atomic testing facilities, plus an additional $4,000,000 for building a railway spur line between Fort Churchill and the Broad River proving grounds and millions more for aircraft support and other infrastructure costs.[9] On the other hand, the actual testing site at Broad River, located sixty-four miles southeast of Churchill, had many attractive features: adjacent rivers that would carry contaminants into Hudson Bay; prevailing winds that would transfer radioactive material towards the sea; and, above all, no major population centres within one hundred miles of the site. Another bonus was ready access to the laboratory facilities at the Defence Research Board establishment in Fort Churchill, where "the large influx of scientists to the area ... can be covered as part of the normal operations."[10] Yet despite this promising start, in 1951 the British government decided to concentrate its nuclear testing facilities in the Australian Monte Bellos Islands, since this site could accommodate both atomic bombs and thermonuclear devices.[11]

Another dimension of Canada's involvement with nuclear weapons activities was the challenge of preparing its soldiers for the radioactive / toxic battlefield.[12] One of the earliest training programs took place in October 1952 when units of the Canadian Army participated in the British exercise Medical Rubicon, which proved to be a valuable learning experience in dealing with the threat of tactical atomic bombs.[13] As a result, in March 1954 the Royal Canadian Army Medical Corps (RCAMC) held its own exercise (Medical Broad Front II) at Camp Borden, where troops of the First Canadian Corps were exposed to a simulated enemy attack involving a 20-kiloton atomic bomb. According to the trial organizers, this weapon was sufficiently powerful to produce about 10,000 casualties, from blast, burn, and radiation.[14] Subsequent field tests (Broad Sword and Gold Brick) adopted quite a different perspective, with emphasis now placed on how the RCAMC would deal with an atomic bomb attack on a large Canadian city.[15]

Meanwhile, the March 1954 US Bikini thermonuclear explosions had dramatically changed the civil defence dialogue.[16] Global concerns were further intensified in 1955 when the Soviet Union successfully exploded its own H-bomb, raising the spectre of nuclear war between the two superpowers and mutually assured destruction. The impact of these frightening technological developments on North American civil defence planners was profound, since it was obvious that the hydrogen bomb, with its enormous blast and firestorm, "would make shelters ineffective through oxygen shortage and carbon monoxide poisoning."[17] Initially, however, there was resistance in accepting this harsh reality. Canada's General Worthington, for instance, continued to defend his A-bomb-focused system, denouncing "the abnormal amount of loose thinking ... appearing in the press ... [since] to say that Civil Defence is not warranted by reason of the H bomb is nothing more or less than a venal form of defeatism."[18] Fortunately, this irrational reaction soon passed. And by February 1955 the Canadian Civil Defence Review announced a new set of guidelines because of the threat of thermonuclear weapons: "Areas of high population density MUST be evacuated if lives are to be saved ... [and] Fall-Out of radio-active dust over large areas requires means of protecting life in those areas by taking suitable shelter for approximately 48 hours or by evacuation."[19]

Questions about the adequacy of Canada's civil defence system remained a source of considerable debate throughout the decade.[20] On 20 March 1959, for instance, opposition parties in the House of Commons grilled the Minister of Health and Welfare, the Honourable J.W. Monteith, whether the Soviet Union's intercontinental missile capabilities negated Canada's civil defence policies, which were "based on a possible warning period of 3 to 4 hours."[21] These fears became a "virtual" reality during the Cuban missile crisis of October 1962, when civil defence authorities estimated that if Canada's sixteen major cities were attacked with nuclear weapons, the combined effects of blast, thermal energy, immediate radiation, fallout residual radiation, and subsequent epidemics would result in approximately 4.5 million deaths, or 25 per cent of the Canadian population.[22] There was also speculation that the Soviet WMD attack would include a bioweapons component with "the creation of aerosols of critically sized particles containing aggregates of pathogenic agents." In addition, there were concerns that "biological warfare agents could conceivably be used by subversive methods ... and a saboteur might introduce pathogenic agents into the air of localized but strategically important communities."[23]

Mobilizing Microbiologists for Biological Warfare

While the Western allies successfully mobilized their scientific manpower resources against Nazi Germany and Imperial Japan, this did not occur during the course of the Cold War. On the other hand, during the summer of 1950, with the outbreak of the Korean War, there were intense discussions about the value of drafting medical researchers and practitioners for emergency service, particularly after the 1950 report of the US Interdepartmental Committee on Scientific Research and Development.[24] In Canada the most important booster of medical mobilization was Omond Solandt, Director General of the Defence Research Board. In a 6 January 1951 memorandum he raised this issue with Minister of Defence Brooke Claxton: "In the event of outbreak of major hostilities, the effective employment of the limited number of scientists, engineers and technicians available in Canada will be vital to a successful national war effort. It must be made certain that their knowledge and ability is applied at those points in defence research and development, military duty, war industry and education where it is critically needed."[25]

Significantly, neither Canada nor the United States implemented scientific / medical conscription largely because of their success in recruiting high-profile life scientists as part-time CBW advisers.[26] A similar pattern emerged in the United Kingdom, where the key defence science organization was the Biological Research Advisory Board (BRAB), which included such scientific luminaries as W.W.C. Topley, Howard Florey, Owen Wansborough-Jones, Lord Stamp, and the redoubtable Paul Fildes. And their opinions were not merely window dressing, since both the Ministry of Defence and the Cabinet respected their individual and collective advice, even if it differed from government policy.[27] In fact, on several occasions BRAD members adopted controversial positions, even if they annoyed Washington, and threatened Britain's special BW relationship with the United States. This was certainly the case in April 1952, when Paul Fildes speculated how British defence scientists should respond if the Pentagon used germ warfare in Korea: "If they did, the question would arise whether we concurred. If we did not, and followed our policy of not initiating BW, we should be cut off from scientific liaison. If, on the other hand, we did concur, this would entail a reversal of Government policy, which would upset research workers at M.R.D."[28]

The United States also had an impressive group of civilian biological warfare advisers, many of them prominent members of the American Society of Microbiology (formerly the Society of Microbiology).[29] This was not surprising since during the Second World War an elaborate pattern of consultation between ASM and Camp Detrick had evolved, with prominent civilian scientists such as E.B. Fred, George Merck, and Ira Baldwin being key members of the major biological warfare committees.[30] In an interview, Ira Baldwin described how this relationship worked during the Cold War years:

> I think it is fair to say that the actions of the Committee on Biological Warfare set the pattern for many of the activities ... and we thought it was important that the research and development activities should continue and emphasize both the offensive and defensive aspects of it ... We were concerned about sabotage activities and some kind of effective civil defense program ... In the international aspects of BW we ... did, however, point out that we felt there could be no satisfactory system of inspection and control ... Consequently we felt that the United States should not bind itself to any agreement which would later cause a great deal of difficulty in that field ... [since] we should be prepared to make effective use of BW if it should be necessary.[31]

This commitment to support the Pentagon's biological warfare program was reinforced at the May 1951 conference of the American Society of Microbiologists when delegates addressed the question of how they could "meet their community responsibilities to civil defense."[32] Four years later, the linkages were reinforced when the ASM approved a request from the US Army "that a committee be appointed by the Society to advise the Chemical Corps on microbiological questions." In short order, three high-profile microbiologists were assigned to the liaison Advisory Committee, after the requisite security clearance.[33] This formalized arrangement worked rather well during the next decade, since it involved a strict division of function, with the Society's representatives providing the microbiological knowledge and Detrick's scientific staff focusing on the requirements of specific weapons systems.[34] In all cases, these discussions were regarded as top secret, with rigorous "right to know guidelines.'[35]

Canada's biological warfare advisers also assumed an important role in the country's BW system and within the Tripartite alliance. By the mid-fifties, however, there were major changes in their collective profile, with most of the wartime C-1 Committee group (Murray / Maass / Reed) being replaced by another set of academic and government scientists. This transformation was certainly evident in 1959, when a new organization, the DRB Advisory Committee on Biological Warfare (ACBW), was established with a mandate "to provide research and development information related to Canadian defence against BW ... in cooperation with other government departments and agencies."[36] Under the chairmanship of Professor Armand Frappier, head of the Biological laboratory of the University of Montreal, the ACBW developed a number of innovative policies. One of these was a commitment to make greater use of university and industrial laboratories, rather than rely so heavily on Defence Research Station Suffield (DRES), particularly in dealing with "problems of fundamental research in relation to BW." More specifically, Frappier argued that both his Montreal laboratory and the Toronto-based Connaught Medical Research Laboratory could provide valuable assistance in the development of offensive agents and defensive vaccines, since "production of biological products is a very tricky procedure and requires long experience ... [and] such production has to be kept secret." Closely related were efforts to upgrade the membership of the committee, through the recruitment of Robert Wilson, scientific director of Connaught Laboratories, given his range of contacts among prominent corporate scientists in the United States and Europe, and since he was already an active member of the DRB Infection and Immunity Panel.[37] In contrast, Frappier and his ACBW colleagues showed little interest in the candidacy of Mr J. Gibbard, former head of Ottawa's Laboratory of Hygiene, who was described as a difficult colleague whose "main object ... is to have B.W. Defence transferred to the Dept. of National Health & Welfare."[38]

Tripartite System of Biological Warfare, 1950–1962

The 1947 Tripartite Toxicological Basic Standardization Agreement system was created by the Western allies for the purposes of coordinating research in the fields of biological and chemical warfare, and radiological defence. Although defence scientists dominated the Tripartite deliberations, under its terms of reference there was a commitment to report all recommendations "to Armies' headquarters for initiation of national approval and action." The annual meetings of the organization were based on three major principles: the division of scientific labour in carrying out joint CBW projects; the rapid flow of scientific information between the three countries; and effective coordination between scientific experts and military users. This model worked reasonably well until the early 1960s, when several structural changes were implemented. One of these was the decision to hold bi-annual rather than annual meetings, on the grounds that most of the important CBW work was being carried out by semi-independent technical working groups. A second development was the admission of Australia as a full member in 1964, because of pressure from the United States, which insisted that its Vietnam ally would be a valuable asset.[39] Thereafter, the organization became known as the Quadripartite Conference, under the provisions of the revised Basic Standardization Agreement of 1964.[40] And finally, efforts were made to rationalize the CBW relationship between the older Tripartite / Quadripartite system and the more recent Tripartite Technical Cooperation Program (TTCP), which had been established in October 1957 at the behest of US President Dwight Eisenhower.[41]

The outbreak of the Korean War in June 1950 served as a catalyst in accelerating the work of the Tripartite CBW system, since many allied experts believed that the Kremlin would soon initiate chemical and biological warfare.[42] Faced with this threat, Tripartite planners attempted to correlate the research activities among the three major research centres: Detrick, Suffield, and Porton. One example of this cooperative approach occurred during the British BW sea trials (Cauldron, Hesperus) where "hot" agents such as *Brucella suis, Pasteurella tularensis, Clostridium botulinum* toxin, and *Bacillus anthracis* were used. Another ongoing priority for the organization was the pressure "of getting results out to meet urgent Service requirements." What this meant for Detrick scientists was the development of an effective lethal-type agent which would be dispersed in dry form rather than liquid fill, either within a bomb type munition (E61, E99, F-114) or dispensed by generators. Because of the project's high priority, allied representatives agreed that in the scientific division of labour, "Dugway trials should be primarily Final Engineering Trials and that the British and Canadian facilities could better be used for fundamental studies and for supplementing the Dugway trials where special features made it desirable to do so."[43]

The annual Tripartite conferences dealt with a range of scientific, logistical, and policy issues.[44] And they often provided a forum where national CBW goals could be advanced, as was the case during the 1952 meetings when the Canadian Chiefs of Staff officially endorsed the use of CBW weapons if a major war should occur "since knowledge of our ability to retaliate in kind is a powerful deterrent to their use by a potential enemy."[45] On the other hand, quite a different stance emerged during the Tenth Tripartite (1955), when DRB delegates were forced to admit that "the Government of Canada would be embarrassed if it were known that Canadian Government establishments were engaged in work dealing with the offensive aspects of BW and CW."[46] This request for special treatment did not, however, impress officials of the US Army Chemical Corps, who criticized the proposal "that any and all conference papers in which Canada is included as a participant must be classified."[47] In their opinion, this was an unacceptable procedure, not only because it established a double standard, but also because it would undermine the effectiveness of future Tripartite meetings.[48] Faced with this ultimatum, Ottawa backed down. There was no evidence that DRB scientists were constrained in their research activities during the eleventh Tripartite (1956), given their active participation in a number of CBW operational projects, including the potential battlefield use of the new nerve gas VX, "a quick-acting agent which could circumvent the gas mask."[49] In addition, they were involved in discussions about the military use of psychochemical incapacitating agents such as LSD25 and BZ, which were being developed by the Pentagon and the Central Intelligence Agency.[50]

In many ways the 1957 Conference at Porton Down represented a watershed in the operation of the Tripartite CBW system.[51] On the policy side were the ramifications of the decision by the second Eisenhower administration to adopt a first-use CBW policy "to the extent that the military effectiveness of the armed forces will be enhanced … in general war."[52] At the time, however, there were questions about whether this aggressive stance was justified when Moscow appeared interested in détente. The previous year, for example, the USSR Academy of Sciences had invited a number of outstanding Western microbiologists, physicists, and medical researchers to attend a series of high-profile meetings in the Soviet Union.[53] One of those invited was Everitt Murray, one of Canada's leading BW scientists during the Second World War, now semi-retired at the University of Western Ontario, who welcomed the opportunity of visiting Soviet microbiological laboratories.[54] On the other hand, while the Kremlin was extending the olive branch, it was also determined to assert "Soviet technological superiority over the United States, and hence of the superiority of the socialist over the capitalist system."[55] Russian bragging rights were greatly enhanced on 5 October 1957 with the successful launch of the first earth satellite *Sputnik* 1, an event of enormous international significance.

Another important change, at least for Canada, was Omond Solandt's 1956 decision to step down as Director General of the DRB, after nine years of service.[56] This change in leadership also coincided with the death or retirement of the wartime generation of BW scientists, who were replaced by professional defence scientists such as Archie Pennie, superintendent of the Suffield Experimental Station (1957–1963), A.K. Longair, coordinator of DRB's CBR Special Weapons research, and G.R. Vavasour, Secretary of the Advisory Committee on Biological Warfare.[57] And it was a timely transition, since the DRB faced a challenging task, not only of ensuring its independence within the Defence Department, but also of dealing with the biowarfare priorities of its two allies at a time when the US and UK programs were going in different directions.[58] This trend was clearly revealed in the April 1957 United Kingdom Defence White Paper, which imposed severe cutbacks on the Microbiological Research Department Porton because of the government's new reliance on "the nuclear deterrent and the guided weapon."[59] From the DRB perspective, these British policies were ill-advised since,

> CBR weapons represent the only known alternative to nuclear weapons for achieving large area effects. In fact, in some respects they are more attractive weapons. For example, they may destroy an enemy's will to resist without destroying the economic features of his countryside and cities. In addition, they may be employed covertly ... The present estimate of the threat is low, largely because of the existence and availability of nuclear weapons. There can be little doubt that, if there were an internationally-agreed ban on nuclear weapons, the threat from CBR weapons would suddenly become highly rated, if only because no major power would be prepared to forego entirely the chance of owning some mass-effect weapons.[60]

Porton scientists did, however, continue to make important contributions at the annual Tripartite CBR meetings, particularly in the development of second-generation biological agents and delivery systems.[61] Of particular importance was the movement away from cluster bombs, such as the American M33, which had been used to create BW aerosol clouds through explosions, a process that inevitably resulted in the destruction of many of the bioagents. In contrast the large area coverage system eliminated many of these problems since the pathogenic bacterial and viral agents were sprayed by high-performance aircraft, creating large aerosol clouds that would be carried by prevailing winds over enemy cities and troop concentrations. In turn, this promising new offensive system spurred research in a number of related fields, such as the cause of death of airborne microorganisms and methods of protecting pathogens from environmental conditions.[62] Closely related was research on the pathogenesis, pathology, and immunological and biochemical changes that took place when experimental animals were exposed to BW agents, since this information was essential "in making a reasoned estimate of the infectious doses for man ... a prerequisite to determining the feasibility of volunteer studies."[63] And while human experiments were preferable, this could only happen in unique projects, such as the US Whitecoat trials, when incapacitating agents such as Q-fever and brucellosis were carefully used on religious volunteers.[64]

In February 1958, Dr Hugh Bartlett, the new Director General of the Defence Research Board, launched a review of Canada's CBR policies that was based on the realities of the international situation rather than on "the traditions of the past." One aspect of this exercise was to confirm that most of Canada's biological and chemical warfare projects would be concentrated at the Suffield Experimental Station, where there was a critical group of fifty-one specialists.[65] In a related matter, the DRB was forced to deal with complaints from Washington that Canada's involvement with radiological issues at TTCP meetings, including questions about nuclear weapons, created serious security problems since it would violate the provisions of the US Atomic Energy Commission.[66]

In 1958 Canada hosted the Thirteenth 1958 Tripartite Conference. Overall, the meetings were successful.[67] In terms of BW research, special attention was placed on cooperative research on viruses, notably "the adaptation of an avian virus to monkeys ... [and] this adaptation technique for the selection of novel BW agents was recognized ... [although] the ultimate worth of such adaptations could not be determined without experiments in man." It was also acknowledged that Canadian researchers had made great progress in establishing "the feasibility of developing new diseases by adaptation of viruses from one species to another."[68] On the weapons side, there were further improvements in the large-area offensive, notably an enhanced capability of disseminating potentially lethal pathogens at considerable distances downwind by using additives to prevent aerosol decay.[69] The delegates also decided that the Suffield Experimental Station, with its integrated field and laboratory facilities, "should take primary responsibility for establishing the validity of laboratory predictions of viable decay and give consideration to extending the present work to include field trials with pathogens."[70] The urgency of this work was reinforced by US intelligence reports of the Soviet WMD threat, which predicted that any attack on the West would include "atomic, thermonuclear, biological and chemical weapons ... and that the employment of any one type might well precipitate use of all."[71]

While Superintendent Archie Pennie appreciated this endorsement of Suffield's bioweapons testing program, he was apprehensive about whether his scientific manpower was sufficient to meet these challenges. Not only were the LAC trials labour intensive, but there were also requests from the US Chemical Warfare Corps for assistance in testing non-lethal agents such as *Brucella suis* (Brucellosis), which Detrick had "on the shelf."[72] Another obligation was for the Station to carry out operational research on other weapons systems, including a series of shock and blast field trials calculated to simulate atomic bomb battlefield conditions.[73]

In 1962 Canada staged the Sixteenth Tripartite Conference, which dealt with a number of innovative proposals.[74] On the administrative side, delegates endorsed the merits of the bi-annual meetings largely because most of the essential weapons work was being carried out by seven major working groups, which facilitated "frequent personal contact between the individual laboratory worker and his colleagues in other laboratories."[75] Scientifically, the most exciting developments were in the field of BW basic research. First was the emergence of a number of new agents such as *Staphylococcus enterotoxin,* "a quick acting incapacitating agent." Second was increased us of microbial genetic techniques in enhancing the weapons potential of selected viruses by "inducing multiple antibiotic resistance without loss of virulence."[76] Third, through a combination of freeze drying and reconstitution methods researchers had managed to obtain, for both bacteria and viruses, "a much longer storage life and aerosol recovery; and a much lower aerosol decay."[77] Finally, there was some intriguing research on the linkages between nuclear and biological weapons, based on evidence that animals exposed to radiation became more susceptible to BW agents.[78]

Although priority was given to the offensive side of the BW equation, there was extensive discussion about defensive systems, notably the need to develop effective early warning systems against possible LAC attacks. This was not a new issue, since an earlier Porton study had already warned that "subject to the availability of BW agents, which could not suffer a substantial loss in infectivity during dispersion ... a biological attack, mounted from a ship at sea, against the UK would be feasible."[79] By 1962, however, this theory had become a frightening reality, with clear evidence that "'large area coverage,' by the dispersal of bacteria from aircraft, guided missiles, or sea-borne craft ... [was] the new offensive principle."[80]

Despite these important scientific and technological achievements, Canadian defence scientists had mixed feelings about the Sixteenth Tripartite. Their major grievance was Ottawa's insistence that there should be no formal acknowledgment of Canada's involvement with the offensive dimensions of CBW warfare for public relations purposes. But for defence scientists this policy undermined their status as full partners within the alliance, by compelling "the other two Tripartite partners to adopt phraseology ... consistent with Canadian internal policy."[81] They also deplored the lack of awareness within the Cabinet about CBW issues, particularly the reality that an awareness of the offensive potential of nerve gases and biological agents was essential for Canada's national defence.

Preparing for Biological Warfare, 1962–1968

Canada's close association with the military policies of the United States was highlighted during the Cuban missile crisis of October 1962.[82] It assumed many forms. In both countries there was renewed interest in civil defence that focused on the possibilities of a Soviet assault with weapons of mass destruction.[83] These concerns did not disappear after the Limited Test Ban Treaty of 1963, since a nuclear stalemate might encourage the Warsaw Pact into launching a series of limited wars using biological and chemical weapons. And if such a situation materialized, the crucial question was whether the United States would respond in kind or with its arsenal of thermonuclear weapons

After the October 1962 missile crisis, Canada attempted to redefine its position on *all* weapons of mass destruction, under the watchful eye of the US Department of Defence. Of particular importance was Ottawa's 1963 decision to accept tactical nuclear weapons, in keeping with its NATO and NORAD obligations, a subject that has received considerable scholarly attention. In contrast, little is known about the May 1963 secret policy statement on the use of biological and chemical weapons that was issued by the Canadian Chiefs of the General Staff (CGS). It established three operational principles: "a) Canada will in no instance initiate nuclear, biological or chemical warfare; b) Canadian Armed Forces may be committed to participate in a war in which the use of N, B or C is initiated by an enemy; c) The Canadian Armed Forces will develop the knowledge and the capacity to ensure that protective measures are adequate, and that a capability for retaliation in kind could be quickly instituted if so directed."[84]

In January 1963 G.R. Vavasour, Secretary of the DRB Advisory Committee on Biological Warfare, circulated a memorandum explaining the impact of a bioweapons attack on Canada. Designed as a think piece to stimulate discussion on this relatively unexplored subject, Vavasour's brief explored a number of scenarios. One of these was the possibility that the Soviet Union would launch a BW first strike against the United States "designed solely to kill inhabitants of large cities ... [and] whether this attack would provoke [US] nuclear retaliation." Vavasour answered his own question in the affirmative, pointing out that Washington would never tolerate a situation "in which millions of people ... become mortally sick ... as a result of an undetected, silent, and invisible BW attack." Faced with this predictable American response, he argued, it was highly unlikely that the Kremlin would consider the germ warfare option, even if they had "a complete and effective BW arsenal."[85] Vavasour's second scenario explored the possibilities of an enemy covert attack facilitated by major advances in bioweapons delivery systems. Under these conditions, he warned, North American targets could be attacked "by an aircraft creating an aerosol as it flies a pre-determined course ... [or] a submarine setting up an aerosol as it cruises along the coast ... [which] offers an aggressor the greatest possibilities of surprise and the least chance of being caught in the act." Even worse, Vavasour noted, advances in genetic engineering methods made it possible for scientists to deliberately alter the genetic material of certain microorganisms, creating the possibility "of an exotic microbe being used against North America." Vavasour concluded his report by asking whether the US or Canada could effectively defend themselves against either a covert or overt bioweapons attack. In his opinion, the prospects were grim since neither country had effective BW warning and verification systems, despite ten years of research and development, leaving only an anachronistic laboratory verification system that "would take days to identify the microorganism ... [creating] an undesirable delay in initiating specific therapeutic treatment."[86]

This debate about Canada's biodefence capabilities became even more contentious in February 1963 with the creation of the Sub-Committee on Effects of Covert BW Attacks on Canada.[87] Chaired by Professor C.E. van Rooyen of Dalhousie University, members of the sub-committee were encouraged to submit individual reports about the possibilities of covert BW attacks.[88] Overall, this preparedness exercise was well received by senior officials of the Defence Research Board, with one notable exception – microbiologist J.F. Currie of the Kingston Laboratory, who had serious concerns about the scientific methods and conceptual approach adopted by the authors of the final report.[89] Did these experts, he asked, have sufficient information about the capabilities of the Soviet Union to carrying out a covert BW assault on North America? Or should specialists at Fort Detrick be consulted, since they had studied "in some detail such things as a covert attack on an air field"? Nor did he appreciate the report's dismissal of biowarfare sabotage as merely a military nuisance that should not "be considered to be a serious threat to our national security." Instead, he reminded his colleagues about the challenges of dealing with any form of bioweapons incident, since "there would be no detection possible ... and little or no treatment available especially if a viral agent was used." To reinforce his case, Currie outlined several scenarios in which a covert BW attack on Canada would undoubtedly produce large numbers of casualties: "(a) release in Montreal Forum with 14,000 persons present; (B) release at Malton Airport during a rush hour; (c) release in the Toronto subway; (d) release at railroad stations in large, intermediate and small cities."[90]

By September 1963 the sub-committee had substantially revised its earlier arguments through the joint efforts of G.R. Vavasour and Jack Currie. Instead of trying to cover all the variables associated with a simulated bioattack, they decided to adopt a war game approach.[91] The first stage of this exercise was an LAC line attack by an enemy trawler travelling twenty-five miles off the coast of Nova Scotia "disseminating pathogens during an off-shore wind in January." According to the authors, approximately 25 square miles would be contaminated with the BW agent, meaning that over 115,000 people in the Halifax-Dartmouth region would experience serious illness, with a 40 per cent mortality rate for those left untreated. The report concluded that, despite the presence of suspect foreign vessels, Canadian authorities would probably be unsuccessful in securing conclusive scientific evidence to prove that the Halifax outbreak was caused by an enemy BW attack.[92] Significantly, many of the "lessons" of this exercise were incorporated into the National Survival Plan, commissioned after the 1962 Cuban missile crisis.[93]

Canadian defence specialists were also involved in a number of biological and chemical programs devised by the North Atlantic Treaty Organization (NATO). During the 1950s, the Alliance had established a number of multinational working groups in developing civil defence measures against weapons of mass destruction, with an emphasis on nuclear and chemical warfare.[94] In April 1962, however, biological warfare was added to the WMD list, largely because of a special study carried out by the NATO Military Committee. In the committee's opinion, not only were existing CBW defensive programs ineffective, but the Alliance also lacked a retaliatory capability "due in part, to the sensitive political nature of this type of warfare."[95] This deficiency, it noted, was particularly serious because of growing evidence that the USSR had dramatically increased its arsenal of nerve gases, incapacitating agents, and the means "for spreading biological warfare agents over large areas, thus constituting a new major weapons threat."[96]

The exact nature of the Soviet bioweapons threat was discussed in an April 1965 report by NATO's Standing Group on Science and Technology. It began with the observation that the Red Army appeared to have embraced "the potentialities of biological warfare and have given consideration to its use." But lacking accurate intelligence, the committee was unwilling to go beyond a general warning.[97] By the following year, however, the Standing Group concluded that the Soviet biological warfare threat was real and imminent. This viewpoint was confirmed in its 1966 report, which pointed out that the Soviets had "the necessary background and experience, to develop a complete array of BW munitions systems for a variety of strategic and tactical uses ... [either] individually or in combination with other biological or chemical agents."[98] On 22 September 1967, and facing a crisis situation, NATO's Military Committee issued directive MC 14/3 – *Overall Strategic Concept for the Defence of the North Atlantic Treaty Organization Area.*[99]

> It is not evident to what extent BW or CW capabilities might affect deterrence. However, there is a danger that Soviet leaders might come to believe that their capabilities in these fields would give them a significant military advantage. NATO should rely principally upon its conventional and nuclear forces for deterrence, but should also possess the capability to employ effectively: A. Lethal CW agents in retaliation, on a limited scale; B. Passive defensive measures against CW; C. Passive defensive measures against BW.[100]

Canada's other source of information about the Soviet BW threat came from its Tripartite partners.[101] The Sixteenth Tripartite CBR Conference (1962), for instance, featured wide-ranging discussions about problems in obtaining accurate intelligence about the capabilities of the Soviet BW system and how it compared with the country's nuclear program. While Tripartite scientists were concerned about the possibilities of an enemy attack with integrated nuclear and biological weapons, American and British experts had major differences in opinion about the severity of the threat. These were outlined in the 11 October 1962 report of Suffield's H.J. Fish:[102]

1) The U.K. considers BW as purely a strategic weapon. The U.S. considers it as a tactical weapon, e.g. it could be used by an army to

soften up the enemy prior to a tactical offensive. It is a matter of definition. In any case, the U.S. BW tactical concept seems rather unreal in a nuclear war.

2) The U.K. representative bemoaned the fact … that BW was a dirty word and that too tight security prevented wide instruction of health officers and the population. The U.S. representatives … clearly stating the threat and … they added that education on BW must [be] continuously and vigorously pursued.[103]

Delegates at the 1963 conference also decided to appoint a special Tripartite Ad Hoc Working Group to improve BW detection equipment, a difficult undertaking since any warning device had to take into account "cloud characteristics … particle size distribution … and speed of response." Ironically, this important defensive initiative elicited an angry response from the US Chemical Warfare Corps, which claimed that the research activities of the Ad Hoc Group were interfering with several top-secret US surveillance projects and jeopardizing their security status. Although the situation was quickly defused, G.D. Vavasour used the incident as a reminder for his DRB superiors of the challenges of doing cross-national military research:

> One point that caused a lot of controversy in the early stages of the work of the BW group was the impression of some workers that an attempt was being made to impose standard research methods and procedures on the workers in that field. It is of course a scientific heresy to advocate that all workers in a given field should be limited to the use of identical methods. The problem was solved in the BW field when it was realized that a number of standard methods had to be used by workers in different labs in order to make their results readily comparable … [but] each laboratory was free to use whatever methods and equipment it wished as long as it included in its work the use of the standard reference techniques.[104]

Suffield and the American BW Research and Testing Program, 1962–1968

In 1962 one of the priorities of the newly elected Kennedy administration was the reorganization of the American biological and chemical warfare programs. Under the watchful eye of Secretary of Defence Robert McNamara, the Project 112 Working Group was given responsibility to upgrade essential dimensions of the country's CBW capabilities.

These goals were outlined in McNamara's mission statement to the US Joint Chiefs of Staff, namely the possibility of using bioweapons "as an alternative to nuclear weapons … [and] the development of an adequate biological and chemical deterrent capability."[105] In keeping with this broader mandate, several administrative measures were adopted. First, it was decided that the US Army's Chemical Warfare Corps would be given exclusive responsibility for providing offensive and defensive equipment for all four military services. Second, in order to improve agent and munition development, Fort Detrick's R&D activities and Pine Bluff's production facilities would be placed under the direct control of the Army's Munitions Control Division. Third, a new facility, the Deseret Test Center (Utah), was established with a special mandate to carry out CBR trials outside the continental United States, while the Dugway Proving Grounds would be relegated to low-hazard testing on the domestic front. And finally, to ensure that the Deseret project received the best possible scientific advice, arrangements were made for other Quadripartite countries to participate in the scheduled external trials.[106]

Overall, this new direction in the US biological warfare program was well received by Canadian and British BW specialists.[107] For Suffield scientists there was also a strong sense of continuity since they already had considerable experience in working with their counterparts at Detrick and Dugway with the joint project (WINDSOC) being carried out at the Eniwetok Atoll testing site.[108] These experiments were a valuable learning experience for the SES contingent, not only because they used "hot" agents such as tularaemia, but also because it enhanced knowledge about the impact of long-range travel on biological aerosols.[109] While DRB headquarters recognized the advantages of this kind of cooperative venture, they were also apprehensive about whether the involvement of Suffield scientists with these trials would be "a one-shot or a continuing series of safari's to the South Pacific … [with] the probability of being more or less bound to continue to participate in future years once you start."[110] Their concerns about the need to establish new guidelines were reinforced after the December 1962 meeting of Quadripartite scientists at the Deseret Testing Center (DTC) when the US field trial schedule was outlined: This included

(a) An Arctic (–40oF to 32oF) series in Alaska to collect and evaluate data on the extent of the hazard and persistence of/ these agents and to test contamination and detection procedures under these conditions, GB and VX will be used.

(b) A Tropical series – to collect data on the penetration of and diffusion beneath jungle vegetation of stimulant agents released as an aerial spray.
(c) Sea vulnerability trials – to determine the vulnerability of operational vessels to biological aerosols and to collect data on the effectiveness of various combinations of ship-board collection protection and ventilation systems. Non-pathogens will be used. These trials will be in two series – first a BW generator in a small boat upwind, and second from high performance jets.[111]

From the perspective of DRB headquarters, the Alaskan trials were the most appealing venture since there was a correlation with Suffield's research priorities and they could be linked with Canada's defence of its Arctic regions.[112] On the other hand, Ottawa was reluctant, for reasons of sovereignty and political image, to endorse any arrangement that would be designated as an official Canadian-American "*joint* trials programme."[113] Not surprisingly, these concerns about military and political entanglements were not shared by Suffield's researchers, who applauded the creative role the Deseret Test Center was assuming in mobilizing scientific expertise from "all four branches of the U.S. forces throughout all levels including senior management."[114] They also pointed out that the fifteen trials scheduled for Fort Greeley, Alaska, would greatly assist their research agenda by providing "complementary data." Even more appealing were the prospects of reciprocal trials with certain types of BW agents, since Deseret officials appreciated "SES experience and capability in cold weather field work."[115]

After extensive debate, the Defence Research Board hierarchy endorsed the SES-DTC cooperation option and recognized that if they rejected Washington's offer, it would seriously jeopardize chances "of re-establishing our formerly excellent liaison position in this area of BW-CW."[116] In January 1963 the US Department of Defence was officially notified that the DRB welcomed the opportunity to participate in Deseret trials, since "the tripartite nature of CW and BW investigations has always resulted in an excellent interchange of information and the greater efficiency possible by an integrated programme."[117] But the Pentagon was in no hurry to confirm the agreement; indeed, it was not until August 1964 that a formal letter of invitation was signed by Brigadier General James Hebbeler, Commander of the Deseret Test Center.[118]

By the summer of 1965 Canadian defence scientists were ready to participate in the DTC biological and chemical weapons trials.[119] One of these was Project Copperhead, an Arctic marine trial with BW stimulants scheduled to take place in international waters off the east coast of Newfoundland using an aircraft based at Harmon AFB and a destroyer operating from Argentia. It was not an easy undertaking. First, it was not clear whether the Royal Canadian Navy would participate in the venture, despite having carried out its own CBW hazard ship trials during the early 1960s.[120] Second, there were complex questions about political jurisdiction since the two Newfoundland sites were legally American territory, given their status as US lend-lease bases, which meant that US trials "could undoubtedly be carried out without our [Canadian] knowledge or approval." Third, while Copperhead was primarily a defensive undertaking, it also had an offensive component, including the use "of a cloud of wet biological agent as a method of attacking operational ships." Concerned about unfavourable publicity, DRB officials proposed that this dimension of the exercise be eliminated, despite assurances from Copperhead organizers that these activities would be taking place on the high seas using the invisible and harmless stimulant *Bacillus globigii* (BG).[121]

The Defence Research Board was, however, was most interested in the field trials scheduled for Fort Greely, "a full time CB test area offering Arctic and temperate conditions."[122] Of particular importance was the forthcoming Project West Side, which involved the release of the BG stimulant "on a 12 mile line source from a type Y disseminator carried on a high performance aircraft ... [and] winds will carry the 12 mile line source from the valley ... extending to a distance of some 130 miles downwind."[123] The project had three major goals: to demonstrate the overall effectiveness of the LAC system; to determine the aerosol cloud's penetration of military positions; and to carry out "cold weather sampling ... [of] certain live strain vaccines." Although West Side was an exclusively American undertaking, SES scientists were anxious to help. In November 1964, for example, Field Superintendent A.P.R. Lambert requested permission to negotiate directly with Deseret officials since it would "allow Canadian access to the major part of the [US] CW and BW field programmes ... [since] these are large scale trials which are beyond SES resources, but [advance] our plans and ideas." An additional incentive for this bilateral linkage, Lambert claimed, was DTC's interest in arranging for establishing a joint Large Area Coverage trial at Suffield, using either BG or Phosphorous-32 tracers. Such an undertaking, Lambert stressed, would be in the vanguard of offensive BW planning, since it would establish "an emission line of 130 miles centered across the North boundary of the [SES] range with sampling to points as far South as the U.S. border and including the valley of the South Saskatchewan River, and the City of Medicine Hat." [124]

Predictably, Defence Research Board officials had major concerns about this proposal. First and foremost were the international implications of having a sizeable aerosol cloud crossing from Canada into the United States, despite assurances from SES scientists that the fluorescent particles would "not be noticeable and hence to the ordinary ground observer the releasing aircraft will not be any more noticeable than any other aircraft flying by."[125] Second, there were questions that this exercise could establish a dangerous precedent of having Canadian defence scientists fully integrated into an American military program without any formal agreement between Washington and Ottawa.[126] At the same time, the DRB wanted to maintain its involvement with Project 112. In December 1964 the Department of Defence negotiated a compromise with SES superintendent Archie Pennie: the Suffield Experimental Station could establish direct contact with the Deseret Test Center, with the proviso that DRB Headquarters could veto any SES involvement with trials outside Canada that was deemed inconsistent with Canada's defence priorities.[127]

While most of Canada's biowarfare activities were North American in scope, DRB experts also monitored scientific and operational developments in the United Kingdom. This was evident in January 1965 when the DRB representative in London reported on British urban trials using *Bacillus globigii* (BG):

> As you know, trials with B. coli have been regularly conducted over populated areas in Britain, and these organisms have also been released in the London underground. BG spores are almost always included in these trials as tracer organisms. These trials have been treated as SECRET, and local authorities are not informed in any way. No permission is sought outside of MOD and there is no medical follow-up of any sort on any of the exposed population. Cooperation was of course obtained from the Medical Officer of the London Transport for the Underground trials, and he was specially cleared for the purpose … sampling crews … are instructed to inform curious members of the public that air pollution studies are being made.[128]

This use of BW simulants on urban centres was not unique to the British biodefence program. Indeed, throughout the sixties the US Chemical Warfare Corps carried out an extensive program of high-altitude spraying over select North American cities, including San Francisco, St Louis, Minneapolis, and Winnipeg. These trials became even more elaborate under the direction of the Project 112 Working Group. In February 1965, for example, Deseret officials informed their SES counterparts that they did not require formal permission to release the BG stimulant over populated areas of the United States since "many field trials have demonstrated that BG, thus released, does not constitute a health hazard of any kind to exposed populations."[129] These urban trials became increasingly bold.[130] For example, in June 1966 Detrick's Special Operations Division carried out a simulated bioterrorist attack on the New York subway system, using BG-filled light bulbs, which "proved an easy and effective method for the covert contamination ... [demonstrating] that a large portion of the working population ... would be exposed to disease if one or more pathogenic agents were disseminated ."[131] In addition, under the vigorous leadership of its technical director Riley Housewright, Detrick scientists increased the weapon's potential of many new and existing bacterial and viral agents.[132]

Canada and Tripartite BW Planning, 1966–1970

By the late 1960s, Canada and its Quadripartite allies had developed a formidable biological warfare research and development program. While many of the joint projects were defensive in nature, some of the most imaginative work involved the development of viral BW agents and of more effective delivery systems. Canadian defence scientists, however, faced the ongoing frustrations of being involved with offensive research while their political masters denied such activities were taking place. The challenge of wearing two hats became even more problematic in 1968, when the Canadian Department of Defence carried out a reassessment of its entire biological and chemical warfare operation.

On the positive side, Canada's BW program continued to benefit from its involvement with the Quadripartite Standing Working Group, Biological Warfare.[133] In April 1966, for example, the Working Group held an important meeting at the Suffield Experimental Station which focused on a number of issues associated with the offensive aspects of biological weapon development.[134] The symposium began with a discussion of a DRB position paper on major trends in biological warfare, including how Canada would respond to an enemy bioweapons attack, either "over a large area [or] ... an act of sabotage in the Toronto subway."[135] Another subject of debate was the advantage of using stimulants in BW field trials, compared with non-lethals such as *Coxiella burnetii* (Q-fever), and Venezuelan equine encephalomyelitis (VEE), since the latter group were associated with "unpredictable ecological and epidemiological problems." Alternatively, Suffield scientists proposed that live vaccines could also be used, proceeding from the laboratory and moving towards "an operational trial with human volunteers."[136]

The second section of the Suffield brief proposed a nine-point selection criterion for an effective BW field trial agent.[137] While there was consensus that viruses had advantages over bacterial agents, the challenge was selecting the most suitable viral candidate.[138] In the end, SES scientists recommended influenza A (Myxonvirus category), since it was "a relatively mild disease in man though it may be temporarily debilitating … [and] is dangerous only to the very young and the aged and then only as a result of secondary bacterial infection."[139] Nor were they concerned about the legacy of the 1918 pandemic, which was described as "a freak event and is not relevant to the strains of virus isolated since 1932."[140] Significantly, the American and British scientific representatives were not impressed by these arguments, reasoning that influenza's potential as a BW weapon was undermined by its instability and because its pathogenic features "had already been obtained with candidate agents in the U.S."[141]

The Suffield Symposium also acted as a catalyst for the Defence Research Board to review its own CBR weapons program.[142] High on its list of concerns were: the effectiveness of the biodefence operations at Suffield and Shirley's Bay (Ottawa); the merits of the extramural grants program; and, above all, the long-term viability of the relationship between Suffield and the Deseret Testing Center.[143] In October 1966, for example, DTC officials expressed concern that since their operation was "a well-kept USA secret … the fact that Canada collaborates with the US on DTC projects compounds the sensitivity."[144] Within eight months Deseret announced a new set of restricted guidelines for sharing BW information on the grounds that "the sensitiveness (public and political) of their mission is much greater than that of general research on BW and CW, and [they] would prefer to keep to the essential minimum the number of persons having knowledge of DTC's mission or having access to their plans and reports."[145] Problems of communication between Canadian and American BW scientists further deteriorated when W.D. Vavasour's position as Director of Atomic Research (DRB) was abolished as part of the reorganization of the Canadian Department of National Defence.[146]

Another important development was the October 1968 Third Tripartite Intelligence Conference on CBW in Ottawa.[147] Three major issues were under review: the potential impact of new scientific developments on future bioweapons; the biological and chemical warfare threat from the Warsaw Pact; and the need for greater sharing of intelligence among the three countries. On this occasion, however, Canada was more of an observer than a participant, since the country had "virtually no facilities for the collecting of scientific intelligence."[148] Given this weakness, the Department of National Defence was determined to protect its access to US and UK intelligence sources. Of particular concern was that the Canadian media would provide unfavourable coverage of the conference proceedings, particularly since biological weapons were increasingly a subject of public criticism.

Nor did it help that the operation of the Department of National Defence had become a source of major political disagreement during the 1960s. Despite the passage of time, Paul Hellyer's prolonged campaign of unifying and reorganizing the armed forces continued to divide Canadians, at least until February 1968, when, under the terms of the Canadian Forces Reorganization Act (C-243), the process of change was confirmed.[149] This included: the integration of Defence Headquarters, including changes in the function of the Combined Chiefs of Staff; a reduction in the number of Field Commands from eleven to six; and, above all, the the adoption of the principle that "final responsibility must rest with the civilian head of the department."[150] Another priority for the newly integrated department was to develop policies for the greater utilization of advanced forms of military technology so that Canada's military could work effectively with its NATO allies. Not surprisingly, the Defence Research Board was affected by these developments. Of particular importance was the growing consensus at National Defence Headquarters that "the proportion of the research effort that is devoted to NBC defence is too great in relation to other priority areas and available resources."[151] It did not help that peace groups often targeted Suffield as a place where nefarious biological and chemical weapons experiments were conducted.[152] Within this hostile environment, SES (renamed Defence Research Station Suffield in 1967) experienced a series of budget cuts and reductions in scientific staffing.

In their attempts to keep the station operational, DRES administrators adopted a number of strategies.[153] In January 1966, for example, Superintendent E.J. Bobyn submitted a proposal that his scientists become involved with "field testing pesticides … for future evaluation, research or development concerned with pesticide equipment."[154] This was followed by his 1967 suggestion that the National Research Council use the 700,000-acre testing area for rocket trials since it "had all the necessary safety distances … [and] the NRC team is self-contained and would require little or no technical support from us."[155] But the most promising initiative came not from Suffield, but rather from General Allard, Chief of the Defence Staff, who in March 1968 recommended that the Canadian Forces could improve their battlefield capabilities through annual training exercises at Suffield.[156] Bobyn agreed. To ensure that the proposal became a reality, he assured Allard that DRES would provide logistical support for the Army's operational exercises "when we do not conduct BW, CW or Shock & Blast trials."[157] These negotiations also provided DRES's superintendent with an opportunity to document how Suffield field trials had changed during the previous ten years:

> The operations over previous years entailed the use of non-persistent toxic chemicals and dangerous biological materials which required the full safety distances provided by our range areas in order to allow trials to be carried out with safety to the surrounding civilian population. More recently all chemicals used in field trials have been highly persistent agents, and the main hazard exists in the small contaminated area with much less need for large downwind safety distances. All biologicals used on the range since 1957 have been stimulants which are virtually non-hazardous to humans and animals certainly at distances of a few miles from the source. In the shock and blast programme the main hazard is that associated with effects at long distances … and an area with a radius of 4 miles provides sufficient safety from even a 500 ton TNT charge.[158]

But would Defence Research Establishment Suffield survive another ten years? This was the question that DRB Chairman Robert J. Uffen addressed in his February 1968 memorandum to Minister of Defence Paul Hellyer. In Uffen's opinion the answer was no! And he provided several reasons why DRES should be scrapped. First, because of unification and reduced budgets, all aspect of DRB operations were under review, but particularly "the defensive aspects of nuclear, chemical and biological warfare." Second, for operational efficiency it was recommended that all biological and chemical warfare research be concentrated at Shirley's Bay (DCBRE) because of its proximity to other scientific agencies and because of its central Canadian location. Under

Uffen's proposal, Suffield's 400 scientists, technicians, and support staff would have two options: "either re-locate in Ottawa, or seek employment elsewhere."[159] Finally, once the restructuring plan received ministerial approval, DRES would be phased out within the next twelve months.[160]

Faced with imminent termination, Suffield scientists mounted a vigorous counter-attack. Most of their arguments stressed the fact that the station's field trial and laboratory facilities were essential components of modern warfare, not only in terms of advanced CBW research projects, but also in preparing Canadian Forces for the toxic battlefield.[161] Above all, the DRES team stressed the value of their involvement with the Tripartite CBR system, which provided the Canadian armed forces with "knowledge gained from the very much larger and more comprehensive U.S. and British programs."[162] In the end, Defence Research Station Suffield was preserved – largely because the Canadian Chiefs of Staff were ultimately convinced that the station should remain within the DND family.[163]

Conclusion

Between 1953 and 1969 American, British and Canadian scientists were able to transform biological warfare from a potential future threat into a strategic weapons system. While the most important contributions came from the United States, given that country's sophisticated BW offensive program, Canadian and British defence scientists also made important contributions through the Tripartite (Quadripartite) and TTCP collaborative systems. And the results were impressive. On the research side, the anti-personnel bacterial agents / toxins of the early Cold War years were vastly improved in terms of their lethality or ability to incapacitate; large quantities of this toxic material was freeze-dried and stored in vast vaults at Pine Bluff, Arkansas. By 1969 the US bioweapons stockpile also included a number of major viral agents, notably variola (smallpox), Venezuelan equine encephalomyelitis, and Rift Valley fever. Plans were also being made to utilize DNA and genetic engineering techniques to modify these pathogens, possibly to evade enemy vaccines and antibiotics / antivirals. Equally important were the dramatic improvements in BW delivery systems, as exemplified by the adoption of the large-area-coverage system composed of specially designed pathogens.

Closely related was the 1956 decision of the US Eisenhower administration to adopt a first-use BW offensive policy, even though emphasis was still placed on the retaliatory / deterrent option. Fears of Soviet bioattacks were not unfounded, particularly during the 1960s when numerous Quadripartite and NATO reports stated that germ warfare was an important part of the Kremlin's arsenal. The greatest fear was of an enemy attack, using airplanes and missiles, on North American civilian targets using lethal bioweapons, possibly containing novel BW agents. It was also assumed that the Soviets would deploy well-trained saboteurs, whose bioterrorist attacks on urban centres might initially be undetected, thereby creating the impression that it was a serious natural disease outbreak.

In the United States the challenge of these diverse biological warfare threats created a strong demand for the Pentagon to expand both the quantity and quality of its BW specialists. As a result, by 1969 over 1,500 scientists and technicians worked at Fort Detrick, with thousands more at the Pine Bluff production plant and the Dugway / Deseret testing facilities. One of these experts was Dr Riley Housewright, Technical Director of Research, whose experiences at Detrick reveal many important aspects of the secret world of the bioweaponeer.[164] After obtaining a PhD in microbiology at the University of Chicago in 1949, Housewright spent the next twenty years of his career at Detrick, working on defence-related research projects, which, he claimed, helped advance medical knowledge about many dangerous pathogens, as well as developing effective toxoids and vaccines against " five types of botulism … anthrax, tularemia, Venezuelan equine encephalomyelitis, and Rift Valley fever."[165] Another dimension of Housewright's professional profile emerged in 1966 when he was elected president of the influential American Society of Microbiology, reflecting the positive image Detrick scientists enjoyed within the organization.[166]

For Canada, one of the most influential BW administrators of this period was Archie Pennie, Superintendent of Suffield Research Establishment (1957–64), and then DRB Vice-President of Planning in Ottawa. But Pennie's most memorable experiences took place at DRES, where, as he recalled, American and British BW scientists were "begging for the opportunity to carry out field trials on the 1000 square mile site, or use the large wind-tunnel (The Shed) for smaller controlled experiments."[167] The Station's complement of fifty CBW scientists also worked on a number of joint CBW projects with civil defence officials of the Department of Health and Welfare, including the proper use of gas masks and other defensive equipment. As Pennie quipped, after explaining his own personal exposure to VX during one of these demonstrations, "perhaps that's why I've lived so long."[168]

Remarkably, despite Suffield's isolation and the rigours of the prairie winter, the turnover rate of scientists and technicians at the station was very low. In part this was due to the enlightened policies of the Solandt years (1947–56), when the Defence Research Board provided aspiring young scientists with a stimulating work environment, interesting research projects, steady funding, and the opportunity to interact with some of the world's most outstanding scientists at Tripartite meetings. On the social side, the strong esprit de corps was reinforced by the fact that more than half of the staff lived in the neighbouring hamlet of Ralston, including Superintendent Pennie, who was an active member of the local church, "even giving the occasional sermon."[169]

On 25 November 1969 the scientific world of Housewright and Pennie was dramatically changed when President Richard Nixon abruptly cancelled the American offensive biological warfare program on the grounds that these weapons had "massive, unpredictable and potentially uncontrollable consequences."[170] While Suffield biodefence scientists were shocked by this announcement, they at least kept their jobs; this was not the case at Fort Detrick, where 1,500 scientific positions were lost. Even more devastating was that Nixon's declaration had terminated a major US weapons program that was on the verge of realizing its potential as well as providing the Soviets with a unique opportunity to surge ahead in the biological warfare arms race.[171]

4 Canada and BW Disarmament: National and International Developments, 1968–1975

The Soviet draft convention on BW and toxins is a substantial shift towards our position that (1) BW and CW are separable problems and (2) we should move now towards a separate BW ban ... The Soviet move ... is apparently predicated inter alia on the assumption that the US will be ratifying the Geneva Protocol in the not too distant future ... Because on-site verification could not possibly be effective without also being extraordinarily intrusive, to us as well as the USSR, and since biological weapons have questionable utility, all [US] agencies consider the complaint procedure the only attainable system ... Moreover, it is in our interests to have other countries bound to a policy which we have already adopted.

(Memorandum for President Richard Nixon: From Henry Kissinger, National Security Adviser, 23 April 1971)[1]

In 1968 there was a surge of interest, at home and abroad, in what Canada was doing in the fields of biological and chemical warfare. In part, this was associated with the global debate about weapons of mass destruction and allegations that the United States was violating the 1925 Geneva Protocol through its extensive use of tear gas and herbicides in Vietnam. Unlike the 1952 germ warfare propaganda campaign, opponents of Washington's CBW policies came from respectable peace groups and scientific organizations. As a result, American policy makers, and their Quadripartite allies, were forced to justify why the research, development, and production of CBW weapons was still happening. In the United States, a number of factors undermined military and political support for the offensive BW program. Of central importance was the public's disenchantment with the war in Vietnam, particularly after the Tet offensive of January–February 1968, when predictions about imminent US military success proved spurious. The anti-war movement also criticized the militarization of American science, including claims by the prestigious American Society of Microbiology that socially responsible microbiologists should avoid any involvement with BW-related research. This attack on the Pentagon's bioweapons program was intensified through the appearance of a series of sensationalist articles and books, exemplified by Seymour Hersh's best-selling book *Chemical and Biological Warfare* (1968). The most crucial criticism, however, came from within the US Department of Defence itself, where enthusiasm for bioweapons as a strategic instrument, so pronounced during the days of Project 112, was now replaced by scepticism and hostility.[2] This negative viewpoint was fully expressed on 25 November 1969 when President Richard Nixon declared that the United States would terminate all aspects of its offensive biological warfare program.[3]

Nixon's decision was also strongly influenced by major international developments. One of these was concern about bioweapons proliferation among non-aligned nations, fuelled by evidence that "some of the smaller countries may view biological weapons as the 'poor man's atomic bomb.'"[4] Another factor was the July 1968 initiative of the British Foreign Office in proposing that separate disarmament conventions be established for biological and chemical weapons, given major differences in strategic use and killing potential. This dual approach was subsequently endorsed by the report of the UN Secretary's general panel of scientific experts, which provided useful guidelines on how CBW disarmament measures could be implemented. But would the two superpowers play the game? This was answered in the affirmative in March 1971 when the Kremlin accepted the division of the two weapon systems; in return, the US dropped its demands for an effective BW verification system. After this momentous compromise, the international community moved inexorably towards the establishment of the Biological and Toxins Weapons Convention. It was officially proclaimed on 26 March 1975 when the required number of countries (twenty-two) ratified the 1972 agreement.

Throughout these complex negotiations Canada wore two hats. One of these was the role of trusted Quadripartite partner, with its defence scientists standing firm against possible Soviet CBW aggression. A much more attractive image was the country's role as international conciliator, working with other middle powers at the UN General Assembly to convince member nations to support the Biological and Toxins Weapons Convention. On the domestic front, however, there were serious debates about whether Canada was acting in the name of global peace rather than as a toady of the United States.

Canadian Defence Policy and Biodefence Issues, 1968–1969

During the 1960s Canada's Department of Defence experienced a number of major crises. First, there was a series of acrimonious debates on whether the Canadian Armed Forces should be equipped with tactical nuclear weapons, a controversy that disrupted the administrations of John Diefenbaker (1957–62) and Lester B. Pearson (1963–8) and created difficult relations with Washington. Second was the bitter fight over the unification of the three services, a process that was only completed in 1968 when the country's armed forces were reduced by about 20 per cent. Third, there were important changes in DND leadership when Leo Cadieux, a career politician from Quebec, replaced the colourful and combative Paul Hellyer as Minister of Defence.[5] While Cadieux was not a cabinet heavyweight, he did manage to establish an effective working relationship with the Department of External Affairs (DEA) and its new minister Mitchell Sharp.[6] This sense of common purpose would, however, be undermined after July 1968 when Prime Minister Pierre Elliott Trudeau launched his restructuring of Canada's defence and foreign policies. Unlike previous government reviews, Trudeau was determined that the line departments should justify their respective policies and priorities. The new prime minister was also critical of his predecessors' willingness to accept Washington's interpretation of the Soviet threat, reserving special scorn for those who demonstrated "mindless anti-Communism."[7] Trudeau was particularly critical of the Department of National Defence for its pro-American proclivities and its misguided viewpoint "that Canada had to be involved in everything."[8] In addition, he felt that the Department of External Affairs was not properly focused on the fundamental reality, that Canadian foreign policy "must be in our own national self-interest."[9]

The review of the two departments went through several stages. After initially requesting that National Defence carry out its own internal assessment, Trudeau then commissioned an independent study by Ivan Head, one of his key advisers, who recommended a series of dramatic changes both in terms of DND's operation and its relationship with the North Atlantic Treaty Organization.[10] On 19 September 1969, the new defence policy was announced: Canada would phase out its brigade group and air division in Germany; the tactical nuclear role of Canadian Forces in Europe would be curtailed and the three squadrons of CF-104 aircraft refitted with conventional missiles; the "Honest John" nuclear artillery warheads would be decommissioned. Plans were made to substantially reduce the size of the Canadian Forces and its budget was frozen at $1.8 billion.[11] Similar tactics were used in restructuring the Department of External Affairs. In the first stage, a series of reviews focused on the specific aspects of Canada's present and future relationship with Europe, Latin America, the United States, and the United Nations along with specific discussions about Canada's future role in peacekeeping and international assistance. The July 1969 report, entitled *Foreign Policy for Canadians,* summarized many of these issues and provided a catalyst for further debates about Canada's foreign policy priorities.[12] Meanwhile, behind the scenes, intense discussions about Canada's position on nuclear, chemical and biological disarmament priorities was led by George Ignatieff, DEA's permanent representative to the Conference of the Committee on Disarmament (CCD) in Geneva.[13] Public debate about the threat of biological and chemical warfare was virtually unknown before the late 1960s.[14] But all this changed during the summer of 1968 when numerous media reports denounced the involvement of Western scientists in creating germ warfare weapons.[15] Even more effective was the anti-BW campaign by the Voice of Women, including a series of petitions that were forwarded to the Prime Minister's Office (PMO). But for the most part, PMO officials viewed these submissions as being simplistic and biased towards the Western alliance "while remaining curiously silent on any of the more unpleasant things that the other side has ... been known to be involved in."[16] As part of this dismissive approach, Press Secretary William Morris devised a standardized response towards these "tiresome" critics:

> Mr. Trudeau shares your revulsion against the use of such weapons and would welcome their being outlawed if there were reliable guarantees against their use. However, until such an arrangement becomes an actuality it is the government's responsibility to provide the best possible protection possible for all Canadians, both military and civilians. To abandon such research would, in the present state of world affairs, encourage others to develop newer and more deadly weapons of this type. The deterrent effect of our efforts to develop protective counter-measures would be lost.[17]

By 1969, however, with criticism of Canada's involvement with bioweapons research mounting, this patronizing approach no longer seemed appropriate. Instead, a more nuanced approach was recommended by Brigadier H.E.T. Doucet, Executive Assistant of the Minister of Defence. The first stage would feature a categorical denial that Canada was involved in the manufacture of chemical and biological materials or that Canadian Forces possessed "stocks, or munitions of this type." This would be followed by the assertion that most of the research carried out by the Defence Research Board was "of great value to the medical and civilian community who are daily facing the problems of bacterial infections from natural sources."[18] But despite this polite put-down, Canadian peace groups were not easily deterred. Indeed, throughout the winter of 1969 a series of well-prepared briefs focused on the operation of Defence Research Establishment Suffield (DRES) and posed questions about why 16 per cent of the total budget of the DRB was allocated for projects of the Tripartite Technical Cooperation Program (TTCP). It was alleged that many of the so-called defensive BW projects had offensive applications, a clear demonstration of how "Canadian involvement in Chemical and Biological Warfare adversely affects her influence in international relations."[19]

Meanwhile, the Department of External Affairs (DEA) was carrying out its own analysis of international CBW developments. Perhaps the most contentious question was whether the US use of tear gas and other non-lethal chemicals in Vietnam violated the 1925 Geneva Protocol.[20] Officially, DEA spokesmen defended the military actions of the United States as a necessary response to the brutal tactics of the Viet Cong, stating that tear gas did not "needlessly cause or aggravate human suffering."[21] But at the working level, there were pronounced differences of opinion about the validity of US arguments for the use of CW non-lethals. This was evident in March 1968 when A.W. Robertson of the Disarmament Division sent a memorandum to George Ignatieff, Canada's primary arms control negotiator, claiming that the Geneva Protocol's ban on chemical weapons was outdated, since it was based on the First World War experience, when "gases generally were primitive, painful and crippling." In contrast, Robertson claimed, the present use of chemical incapacitating gases was "increasingly being advocated on humanitarian grounds ... [and] that even the use of nerve gases, though they are lethal, would seem to be an infinitely quicker and less painful way of killing than that provided by ... acceptable weapons." Even more contentious was his insistence that while "the use of some sorts of bacteriological weapons would undoubtedly be most undesirable in light of their effects ... it is at least conceivable that this need not always be the case."[22] Predictably, Robertson's thesis outraged many of his DEA colleagues, with the most biting criticism coming from F. Pillarella, another arms control expert, who pointed out that, whether Robertson liked it or not, using gas or germ weapons was regarded by the international community as a violation of basic moral principles. Pillarella did not accept Robertson's rather benign view of bioweapons, pointing out that "biological warfare in its uses can be more deadly than atomic weapons ... [when] carried away by winds and thus poisoning almost everything they touch: men, animals, plants, cities, villages etc."[23]

Opposition to Biologial Weapons in the UK and US, 1965–1968

The Canadian CBW debate was also influenced by developments in the United Kingdom. Of particular importance were the ongoing discussions about whether Britain should maintain a stockpile of biological and chemical incapacitants "for retaliatory use (and therefore deterrence) in the context of limited war outside Europe."[24] Another concern for British arms control officials was the possible proliferation of biological weapons, since in their opinion these weapons were "cheap and nasty, whereas nuclear weapons are expensive and nasty ... [and] only a few nations can afford it ... [while] any tin pot country can fairly easily run up a line in nerve gas or toxin, and could equally use it without any rational consideration of the consequences."[25] But for the general public the most recognizable symbol of this anti-CBW campaign was Porton's Microbiology Research Enterprise (MRE). This was not surprising, since the British media demonstrated considerable interest in the activities of MRE scientists, particularly after August 1962 when researcher Geoffrey Bacon died as a result of his exposure to a weaponized form of *Yersinia pestis* (plague). But the tone of this coverage became more confrontational during 1968 with such lurid headlines as "Death Mystery at Germ War Plant," "Plague Virus Escapes Security," and "More Germ Safeguards Urged."[26] Porton's image did not improve after the broadcast of the BBC documentary film *A Plague on Your Children*, a devastating critique of the use of biological weapons against helpless civilians.[27] Although Gordon Smith, Director General of MRE, attempted various strategies to combat this negative campaign, including a series of open-access days, these exercises in transparency were unsuccessful.[28] Equally futile was the decision by the Minister of Defence to release classified information about the UK biological warfare programs, including reports that demonstrated "the special vulnerability of the UK to BW attack."[29]

While the British Chiefs of Staff were outraged by these developments, officials of the Foreign Office applauded this change in policy, particularly since they had just completed a controversial reassessment of CBW policies.[30] This working paper on Microbiological Warfare featured three major innovations: it separated biological and chemical warfare; it included a ban against BW offensive research and production; and it minimized the impact of bioweapons verification measures since "there would be no dismantling of existing preparations or defences."[31] In keeping with its alliance responsibilities, the Foreign Office also sent copies of the Microbiological Warfare proposal to Ottawa and Washington, for comment and approval, before being submitted to the Eighteen Nations Committee on Disarmament (ENCD).

American involvement with biological weapons was also intensely debated during the late 1960s as part of a more general criticism of the country's military-industrial-academic complex.[32] By 1968 the anti-war movement was a formidable political force in the country, fuelled by TV images of death and destruction from the Vietnam battle zone, combined with an unwillingness of many young Americans to accept military conscription. Another source of grievance was the militarization of science, notably in the field of microbiology, where prominent scientists such as Harvard's Matthew Meselson and Joshua Lederberg of the Rockefeller Institute, in well-publicized appearances before the US Congress, campaigned against the US biological and chemical warfare programs.[33] And their message was well received. Among the high-profile converts were William Fulbright, Chairman of the Senate Foreign Relations Committee, and representative Melvin Laird, an influential member of the House Defence Appropriations Committee and future Secretary of Defense.[34] Another important aspect of the CBW debate was the publication of many controversial books, the most influential being Seymour Hersh's revelations about Fort Detrick, which used the provocative subtitle *America's Hidden Arsenal*.[35] The pertinent question of the day was: should American microbiologists support the country's biological warfare program in times of national crisis?

Until the late 1960s the leadership of the American Society of Microbiology had answered in the affirmative. By 1967, however, dissident voices within the Society claimed that responsible microbiologists should denounce all aspects of germ warfare. In particular, these critics wanted to terminate the 1953 agreement whereby the ASM provided scientific assistance to Fort Detrick within a classified context. One of the most revealing debates about the pros and cons of this relationship were featured in a series of letters between Dr Alvin Clark, Department of Molecular Biology, University of California, Berkeley, and Dr Riley Housewright, Technical Director of the Fort Detrick Laboratory.[36] The dialogue began with a number of questions posed by Clark: How much basic and specialized knowledge in microbiology is important in biological warfare? Is present biological warfare capability as massive as present atomic warfare capability? Who determines the direction which biological development takes and the grounds for biological weapons use? Does the existence of an ASM advisory committee to the Army Biological Laboratories "imply a moral commitment of the ASM to the precepts of biological warfare or a directive influence of the ASM in the development of weapons?"[37] In his carefully crafted response, Housewright focused on the role of microbiologists in BW research:

> You ask if you were aiding in the development of weapons by being a microbiologist and the answer is no. You also ask who determines the grounds for use of biological weapons. This is a political problem and like all other political problems, it is determined by the elected representatives in government, namely, the President and the Congress – not by the ASM and not by these laboratories. As to your other questions ... the existence of the ASM Advisory Committee to these laboratories implies no commitment whatever, other than an evaluation of scientific people and problems.[38]

Despite Housewright's arguments and his prestige as former president of the ASM, pressure mounted for the immediate cancellation of the Detrick arrangement. This was evident in November 1967 when president-elect Salvador Luria, an outstanding molecular microbiologist, called an emergency meeting of the ASM Advisory Committee to review the subject.[39] On this occasion there was a consensus that Detrick's world of secret research was incompatible with the Society's commitment to the dissemination of scientific knowledge, and that its safety standards in dealing with the expanding field of genetic engineering were inadequate.[40] On the other hand, many of the older and more conservative members of the organization rejected these arguments as being biased and inconsistent with the patriotic obligations of American scientists. These differences in opinion were evident at the May 1968 annual meeting where a groundswell of opposition greeted any attempt to destroy the ASM-Detrick connection.[41] According to Merrill Snyder, a prominent microbiologist from the University of Maryland, the basic premise of the Advisory Committee's policy was incorrect since "the ethical considerations of biological warfare and BW defence research ... never have fallen under the official purview of the American Society for Microbiology."[42] Dr John King of the Cleveland Clinic went even further when he asserted that there was "nothing less moral about killing the enemy by biological warfare than by bombing. Both are essentially bad, but apparently killing of enemies is part of our defense."[43] Nor did the passage of time reduce these tensions within the ASM. At the May 1969 annual convention, for example, the pro-Detrick group vigorously protested that Mathew Meselson, a prominent anti-CBW advocate, had been asked to present his views on the bioweapons controversy. On the other side of the ideological divide, many of the young militants felt that the ASM should become more politicized in opposing the militarization of science, including immediate action to expose "the nature, use and hazards of chemical and biological weapons ... and ultimately to eliminate these weapons from the world's arsenal."[44] Significantly, this debate affected many Canadian members of the Society, notably Robert Murray of the University of Western Ontario, who, as editor of the ASM journal *Bacteriological Review*, often found himself caught between these warring factions.[45]

International Efforts to Control Biological Weapons

In June 1968 the dynamics of disarmament negotiations at the United Nations was transformed. Undoubtedly the most dramatic development was the complex agreement which facilitated the ratification of the Non-Proliferation Treaty on Nuclear Weapons, based on the premise that the five nuclear nations would gradually dismantle their nuclear arsenals.[46] Equally important was the British working paper on Microbiological Warfare, which was submitted to the Eighteen Nation Committee on Disarmament and which sought to ban an entire weapon system.[47]

For arms control specialists at External Affairs the British initiative created serious challenges in terms of its domestic and international implications. As a result, they sought the advice of the Defence Research Board about "the acceptability of the [proposed] Convention generally as they pertain to Canada's research and co-operation with its allies in this field."[48] The real beginning of this interdepartmental dialogue began, however, with the discussions of the DEA memorandum of September 1968, "Some Political Aspects of Agreement on Chemical and Biological Warfare." The first part of the paper reviewed Secretary General U Thant's proposal to establish an international panel of experts to determine the viability of a CBW disarmament system "which would supplement the Geneva protocol and rectify its various defects."[49] The next section examined the complex issue of verification, notably whether developments in the life sciences undermined the technical possibilities of determining compliance in the BW field so that all countries could be confident "that no party is secretly violating the terms of the agreement." But of central importance were three basic issues associated with the British proposal: Would a separate BW convention be more acceptable to the international community than a more broadly based CBW treaty? Was the nuclear model of arms control useful in trying to ban all aspects of biological weapons? And should Canada agree with the British suggestion that most defensive biological warfare should be transparent "to the maximum extent compatible with national security and the protection of industrial and commercial secrets."[50] After considerable discussion DEA and DRB specialists reached a consensus on each of these questions. First, it was agreed that a separate BW convention was a viable option. Second, nuclear and biological weapon systems were viewed as having fundamental differences, given the "unpredictability and uncontrollability of BW ... [which] mitigates heavily against the desirability ... of retaliation in kind." And finally, External Affairs officials did not feel that Canada should abandon its strict "right to know" approach in dealing with biowarfare information because of concerns this would alienate the Pentagon and possibly mean loss of access "to information on developments in both offense and defense not only for BW but in other fields."[51]

During the fall of 1968 A.K. Longair, one of the DRB's leading authorities on biological warfare, carried out his own assessment of the British working paper. Overall he agreed with External Affairs officials that biological and chemical weapons should be considered separately with " a 'no use' protocol for microbiological warfare and 'no first use' protocol of chemical warfare ... [since] intractable problems of BW need not hold up CW discussions."[52] On the other hand, Longair opposed the DEA proposal that biological and chemical incapacitants should be included in any CBW convention since, in his opinion, they "could be the most humane method of warfare available."[53] But his most vigorous rebuttal was directed against the comments by some External Affairs officials that bioweapons had limited strategic value because of their inherent instability and imprecise delivery systems. "If the weapons have been suitably tested," Longair declared, "the effects will be predictable and controllable ... [with] 'on target' bomblets."[54] DEA experts were not convinced. Would it be possible, they asked, to examine scientific reports and field trial results to verify Longair's claims "since it would seem that without actually using pathogens against human beings and on a large scale, any conclusions about their effect must contain an element of conjecture?"[55]

This debate continued in October 1968 when Longair prepared a detailed memorandum outlining the logistics of a biological weapons attack. The first part of his study addressed the relationship between the production capability of BW agents and the scale of the offensive action, pointing out that even a small attack would require "100 and 500 liters ... sufficient to produce casualties over 1000 square miles."[56] Longair then moved to his central thesis: that bioweapons now had strategic value because of improvements in the quality of BW agents / munitions and the military potential of the large area coverage (LAC) system. As he explained, it was now "possible to predict the area where casualties would occur with any given agent ... [with] the margin of error no greater than with high explosive bombs, and since these are area weapons the possible inaccuracies in delivery are not as significant relative to the target area dimensions." Another important issue, Longair pointed out, was the dual-use phenomenon since it was difficult to determine whether research was for medical or military purposes because so many experiments which were carried out for public health reasons "would be valuable to anyone wishing to use a biological weapon."[57]

These discussions between the Department of National Defence and External Affairs were not confined to the field of international disarmament. Indeed, during the fall of 1968 there were sharp differences of opinion about the political ramifications of Exercise Vacuum, a major CBW field exercise scheduled for Defence Research Station Suffield (DRES) involving military and scientific specialists from Canada, the United States, and the United Kingdom.[58] DEA officials were particularly concerned that Suffield's large-scale war game coincided with the sensitive arms control negotiations that were taking place at the United Nations, and they questioned whether Vacuum's value in terms of military knowledge and alliance networking would compensate for "the almost certain adverse publicity."[59] Defence officials strongly disagreed.[60] In a memorandum to Cabinet, Minister Cadieux claimed that the exercise was essential for DND planning purposes since Canadian armed forces in Europe required extensive training in dealing with a toxic battlefield, given US intelligence reports that the Red Army was prepared "to deploy CW weapons ... in their field units." He also emphasized that Vacuum was the culmination of a decade of creative work by DRB researchers so that Canadian Forces could "operate in a war in which CW was used." Cadieux concluded his message by praising Suffield's leadership for arranging such an innovative field exercise that included a large number of American and British CBW specialists.[61]

In the end, Exercise Vacuum took place in October 1968 without incident. This did not, however, end the Defence-External feud over CB weapons. In fact, an even more serious controversy emerged in November 1968 when DEA disarmament officials discovered, as part of their ongoing review of Canada's defence policies, a secret arrangement whereby the United States could call upon Canada "to engage in CW in some unspecified future circumstances." Even more controversial was evidence that experts of the Department of National Defence and the Pentagon were engaged in a joint project to make US chemical munitions compatible with DND weapon systems, "for example whether USA chemical shells can be fired from Canadian field guns." Armed with this evidence, the diplomats confronted their DND colleagues about this secret deal. How would the Canadian public respond, they asked, if these CBW logistical arrangements became common knowledge? And how would these developments, if made public, affect Canada's international image?[62]

Significantly, Canada's military hierarchy appeared unperturbed by these revelations. On the contrary, in December 1968 Brigadier General Henri Tellier launched a vigorous rebuttal of DEA criticism. The first part of his report reviewed Canada's biological and chemical warfare policies at the time of the 1963 DND unification process, reiterating the familiar message that Canada would "in no instance initiate Nuclear, Biological or Chemical Warfare." At the same time, he emphasized that if Canadian Forces found themselves in a situation "where the use of N, B, or C weapons is initiated by the enemy," the Chiefs of Staff would not hesitate to ensure "that a capability for retaliation in kind could be quickly directed if so decided." So far, so good. But the most startling aspect of Tellier's brief was his admission that Canada's CBW retaliatory policies were no longer abstract policy projections; instead, they had evolved into definite operational strategies. This transition, Tellier explained, had taken place in early 1968 when National Defence had decided, in response to the Soviet Union CBW threat in Europe, to adopt a more aggressive military stance. According to the new guidelines, while Canadian Forces had no CBW munitions or delivery systems in Canada or in Europe "our requirements would be held in British or American stockpiles, to be supplied in the event B or CW is employed against NATO Forces."[63]

This secret operational doctrine stunned DEA officials and provided a catalyst for another set of inquiries. They first wanted to know how Canada's new "retaliation if necessary" CBW policy compared with the official NATO position, as set forth in the 22 September 1968 report of the NATO Military Committee. In this document, member nations were encouraged to consider three type of military responses: "(a) Lethal CW agents in retaliation on a limited scale; (b) Passive defensive measures against CW; (c) Passive defensive measures against BW."[64] The second set of questions focused on the actual logistics of how Canadian Forces in Europe would obtain the necessary BW and CW munitions from the United States during a military crisis. In this regard, there was also speculation about whether the Pentagon had negotiated similar arrangements with defence officials of other NATO nations, since there was "not much point having an allied military strategy the implementation of which proceeds unknown to twelve-fifteenths of the alliance."[65] And finally, DEA officials expressed deep disappointment that their Defence colleagues had failed to keep them informed about these major changes in Canada's CBW policies.

These debates over Canada's bioweapon commitments were not restricted to the bureaucratic arena. In October 1968, for example, there were several inquiries in the House of Commons about the activities at Suffield, largely because of the publicity surrounding Exercise Vacuum. The level of public interest in this subject was also intensified with the broadcast of the controversial BBC documentary, *Make a Resolution and Call It Peace.* Not only did this program denounce the "nefarious" bioweapons research being carried out at Porton Down, it also criticized aspects of Anglo-Canadian BW cooperation, based on a revealing interview with DRB Deputy Chairman (Operations) A.M. Pennie.[66] As a result, Leo Cadieux soon faced intense grilling during parliamentary Question Period, with members of New Democratic Party demanding that the Minister of Defence provide more accurate information about Ottawa's relationship with its Quadripartite partners, and whether these countries were carrying out CBW trials at Suffield.[67] While Mitchell Sharp was also an Opposition target, he was well prepared for this ordeal, having received extensive briefing on how to handle questions about bioweapons issues.[68] In addition, the DEA Minister's script included three key policy statements: the usual assurances that Canada's Armed Forces would not initiate this type of warfare and did not have "stockpiles of any agent other than tear gases"; the fact that Cabinet had endorsed the British working paper on a Microbiological Weapons Convention; and Canada's support of the UN Secretary General's proposal for an Expert Panel to examine whether the principles of the 1925 CBW Protocol were being fulfilled.[69]

Canada, the United Nations, and BW Disarmament

Despite Ottawa's commitment to outlaw germ warfare, Canadian diplomats at the United Nations were aware that such policies could alienate the United States, the world's leading bioweapons nation. On the positive side, however, bioweapons had rarely been used strategically, there were formidable logistical and safety problems associated with their use, and they were detested by the international community.[70] Many of these issues were discussed during the October 1968 meeting of the North Atlantic Treaty Organization, when even the US representative acknowledged that "since BW is a potential problem ... we should ... try to reduce its likelihood."[71] Even more important was the ability of Secretary General U Thant to establish a Panel of Experts on CBW weapons with a mandate to explore five major issues.[72] Should biological and chemical weapons be considered separately? What was the relationship between the non-use of bioweapons and non-possession?[73] Could existing verification technology detect BW production facilities and stockpiles? How could the threat of a covert BW attack be prevented? And finally, was there any justification for maintaining national defensive research programs if everyone agreed not to carry out offensive biological weapons research?[74]

Another uncertainty was whether the Communist bloc would resurrect its 1952-style germ warfare propaganda campaign, particularly after the December 1968 allegations by East German officials that the United States was helping the Federal Republic of Germany acquire a biological warfare capability. These charges were based on the testimony of Dr Ehrenfried Petras, who prior to his defection to the German Democratic Republic had been director of the Laboratory for Microbiology at the Institute of Aerobiology at Grafschaft, Sauerland.[75] According to Petras, scientists at his institute were regularly seconded by West German defence officials for a secret project involving "the production of aerosols with B and C poison of prolonged effect ... [and] experiments with highly pathological micro-organisms and breeds of viruses as well as with the bacteriological toxins, especially with the botulinus toxin, the most effective of all poisons."[76] This propaganda campaign, however, was quickly quashed by the powerful rebuttal launched by the United States and other NATO countries, with Canada's veteran diplomat George Ignatieff informing the Eighteen Nation Committee on Disarmament that his country had "no reason whatsoever to believe that the FGR is not adhering strictly to her obligations under the Geneva Protocol and BRU Treaty."[77]

Despite this unpleasant incident, External Affairs officials remained optimistic that meaningful progress in CBW disarmament negotiations would occur. In particular, they strongly supported U Thant's Expert Panel proposal, despite concerns that many of the scientific contingents "would probably act in practice as representatives of national points of view ... since C and B weapons are established in [the] Soviet mind as a distinct category which is also enshrined in Geneva Protocol."[78] Ironically, these kind of issues were not entirely absent from the Canadian selection process. For the Department of Defence, the logical candidate was Dr M.K. McPhail of Defence Research Station Suffield, since it was essential that the country be represented by someone "who has direct and detailed knowledge of the military implications of these subjects and can offer information and competent opinions ... since the Russians and their allies will almost certainly send political advisers." In contrast, DEA officials questioned whether it was appropriate to appoint someone so closely associated with a Defence Department laboratory, rather than a scientist "of some international reputation."[79] To reinforce their case, they pointed out that the United Kingdom had selected the prestigious Sir Solly Zuckerman, while the US choice was Dr Ivan Bennett, a prominent member of the National Academy of Sciences.[80] Yet despite these objections, McPhail was eventually nominated on the grounds that the Secretary General's study was "ultimately a British initiative and any attempt on our part to 'compete' with United States, United Kingdom and Soviet representatives would not only be pretentious but would attract attention to our interest in chemical and biological warfare."[81]

The first meeting of the Secretary General's Group of Consultant Experts took place on 20 January at the Palais des Nations, Geneva.[82] Throughout the next five months this diverse group of fourteen scientists tried to reach a consensus on a number of contentious issues. Of particular importance was whether there should be a separate bioweapons convention, an approach that was vigorously denounced by Soviet representative Dr Reutov, who claimed that the BW and CW weapon systems could not be separated "because of their traditional linkage in the Geneva Convention."[83] Fortunately, an impasse was narrowly avoided by the timely intervention of Sir Solly Zuckerman, who urged delegates to avoid "unduly rigid classification" and instead pursue the broader issues.[84] But reaching a consensus on any subject was a challenging task.[85] Indeed, even attempts to review the historical dimensions of CBW use produced heated debate, particularly when Zuckerman proposed that, given this legacy, the final report should provide "a rational case against their use ... [since] certain political and military leaders clearly believed that they conferred a military advantage and could be used in certain cases." This position was strongly supported by Canada's M.K. McPhail, who stressed that "since these weapons have been used in the past ... a danger exists that they may be used in the future." In contrast, Academician Reutov claimed that the panel's terms of reference "did not require it to enter into the dangerous arguments ... to assess military effectiveness of such weapons ... rather that the use of chemical and bacteriological weapons in any situation was not conceivable and not legitimate."[86]

The second phase of the panel's deliberations occurred in mid-April 1969. By this time, the scientists were operating under a strict deadline since U Thant had assured the UN General Assembly that the final report would be submitted by 1 July 1969. But the prospect of meeting this deadline appeared bleak. Not only was there the difficulty of drafting a multi-authored document, but the experts were also divided along ideological and scientific lines, a situation that was further complicated by the constant threat that the Kremlin would suddenly withdraw from the project. According to one DEA progress report, "the Soviet draft of Section V has high propaganda content ... [and] McPhail is still worried about his ability to exert adequate influence on quality and content of final report ... [fearing] questions will be determined more by political than by scientific considerations."[87] As it turned out this was an unduly pessimistic assessment. The final drafting stage was characterized by a remarkable level of cooperation, with even Academician Reutov recommending that Britain's Zuckerman "undertake coordinating the report as a whole to produce maximum consistency."[88]

The summer of 1969 witnessed several important disarmament initiatives. One of these was the 1 July official release of the Secretary General's CBW report.[89] Nine days later the British government tabled its draft BW Convention with the Geneva-based Eighteen Nation Committee on Disarmament (ENCD). A relatively brief document, the Microbiological Weapons Convention called for a complete ban on the research, development, production, and stockpiling of biological weapons as an essential supplement to the Geneva Protocol of 1925.[90] While there was considerable optimism that these two important briefs would provide clear guidelines for serious BW negotiations, the essential question was whether the two superpowers supported the process. Unfortunately the prospects were not encouraging. Soviet officials refused to countenance a separate BW convention, while the Pentagon and State Department were outraged that much of the CBW discussions at the United Nations quickly degenerated into an exercise of "Yankee bashing." Washington's sense of alienation was reinforced when U Thant publicly declared that "the prohibition contained in the Geneva Protocol applies to the use of all chemical, bacteriological and biological agents including tear gas and other harassing agents."[91] This undiplomatic critique of US Vietnam strategies was quickly denounced by the newly elected Nixon administration, with veiled threats that the Secretary General should understand the limits of his authority. Behind the scenes, however, were signs that the United States was willing to consider a more flexible position on the issue of biological warfare in general and the British Microbiological proposal, in particular.[92]

Throughout the fall of 1969, international support for the biological weapons convention continued to mount. In part, this process was enhanced by the timely appearance of a number of scholarly studies on the subject, notably the four-volume report by the Stockholm International Peace Research Institute (SIPRI), which was widely praised "for its penetrating analysis of CBW … strictly from non-classified sources."[93] Equally important was the World Health Organization's comprehensive analysis of the health implications of biological and chemical weapons prepared by a group of prominent life scientists.[94] But unfortunately the public debate was often dominated by polemical speeches and sensationalist publications. One such incident occurred during the July 1969 meetings of the 22nd World Health Assembly when delegates from the Communist bloc not only denounced the United States for its war crimes in Vietnam, but also claimed that Washington's refusal to ratify the Geneva Convention represented "a threat to world peace."[95] Another anti-American diatribe came in the form of an international petition, prepared by a group of 'distinguished' Soviet scientists, who expressed their concerns about US secret attempts to develop "new pathogenic micro organisms, viruses, and toxins."[96] For many veteran US diplomats, it looked like 1952 all over again.

This anti-American campaign greatly concerned the Canadian government given the close interaction between DRES and Detrick scientists within the Quadripartite system, and through special bilateral arrangements. In the House of Commons, the opposition parties demanded more information about the activities of the Defence Research Establishment Suffield, including questions about whether the Pentagon had tried to negotiate for the transportation, across Canadian territory, "of germs and other materials to be used at biological warfare test centres in Alaska and Greenland."[97] As a result, on 25 June 1969 Prime Minister Trudeau recommended that the Cabinet review the country's biological and chemical warfare policies since "the present review of defence policy affords us a convenient vehicle to become acquainted with these activities and to consider … [our] appropriate role."[98] As part of their crash course on bioweapons, members of the Cabinet Committee on External Affairs and Defence received a number of briefing documents: copies of the 1925 Geneva Convention, the British BW draft convention, and the rival Soviet proposal for a combined CBW convention.[99] On 9 December the committee's final report was submitted under the title *Arms Control and Disarmament: Chemical and Biological Warfare.* It consisted of two closely related policy statements. One of these was an endorsement of the British BW draft convention "as the most logical and immediate step that can be taken to supplement the Geneva Protocol of 1925 by prohibiting the use, production, and development of biological weapons, with destruction of any existing stockpiles."[100] This was followed on 11 December by an official declaration:

> Canada never has had and does not possess nor does it intend to develop, produce, acquire, stockpile or use biological agents of warfare of toxins at any time in the future.
>
> Canada does not possess nor does it intend to develop, produce, acquire, stockpile or use chemical weapons at any time in the future unless these weapons shall be used against the military forces or the civil population of Canada or its allies.[101]

Changes in the American and Canadian BW Program: 1969–1970

The catalyst for BW disarmament was the United States, or more specifically President Richard M. Nixon, whom many commentators regarded as the quintessential Cold War warrior. What these pundits overlooked was the extent to which Nixon and his closest confidant, National Security Adviser Henry Kissinger, were political pragmatists. Nixon and Kissinger cited a number of reasons why the US offensive biological weapons program should be closed down. First was the intense international and national criticism of the Pentagon's use of tear gas, herbicides, and napalm in the Vietnam war. Second, several domestic accidents had occurred involving chemical weapons, the most controversial being the 13 March 1969 aerial release of nerve gas at the Dugway Testing Grounds, killing over 3,500 sheep in adjoining regions. The Pentagon's clumsy attempts to deny responsibility did not enhance its reputation for either competence or honesty.[102] Third, during this period the US Joint Chiefs of Staff faced a difficult choice: Could they retain the use of CW non-lethals in the Vietnam war and develop binary gas munitions while still maintaining their offensive BW program? This debate was also influenced by Secretary of War Melvin Laird's stance that biological weapons were useless in strategic terms, and embarrassing for America's international image.[103]

The future of the US offensive BW program was decided when the report of the National Security Council (NSSM 59) was submitted to the White House in August 1969.[104] The first part of this study discussed the relative merits of biological warfare agents and their CW counterparts: "Compared to lethal C, lethal B is (1) far more toxic, (2) less reliable, (3) relatively uncontrollable ... Lethal B has no effective battlefield uses, being essentially anti-population. (For example, anthrax and pneumonic plague)." The second section of the brief focused on differences of opinion between the generals and the diplomats over whether the US should retain its offensive BW program: "JCS [Joint Chiefs of Staff] favors a lethal B capability as a co-deterrent with nuclears. State favors R &D on lethal B, but no production and stockpiling."[105] Another key development was the October 1969 report of the President's Science Advisory Committee (PSAC), which recommended three basic policies: "renounce all offensive BW; stop completely the procurement for offensive BW; and destroy existing stockpiles of BW agents and maintain no stockpiles in the future."[106] Not only did the authors of this study regard biological weapons as useless in strategic terms, but they also argued that the abolition of this entire weapons system would "be of crucial importance insofar as public reaction, domestic and international, is concerned."[107] What was most surprising about this review process was that the Joint Chiefs of Staff made little effort to defend their long-term bioweapons investment, either by vigorous lobbying or by using the knowledge of experts such as Detrick's Riley Housewright to explain why the American BW program was crucial for national security.[108]

The next stage in the termination process involved a series of congressional briefings, consultations with departments involved with national security, and efforts to ensure that the president's declaration received favourable media coverage at home and abroad.[109] Significantly, despite their close connection with the American BW program, neither Ottawa nor London was notified about these imminent changes in US policy.[110] On the morning of 25 November Nixon made his famous announcement that the US would unilaterally renounce offensive use of biological weapons, support the UK proposal for a BW Convention, and ratify the Geneva Convention.[111] As he explained,

> Biological warfare ... has unpredictable and potentially uncontrollable consequences. It may produce global epidemics and profoundly affect the health of future generations. Therefore, I have decided that the United States of America will renounce the use of any form of deadly biological weapons that either kill or incapacitate. Our biological programs in the future will be confined to research in biological defence on techniques of immunization and on measures of controlling and preventing the spread of disease. I have ordered the Defence Department to make recommendations about the disposal of existing stocks of bacteriological weapons.[112]

Despite Nixon's forceful language, a number of uncertainties about the real meaning of his declaration remained. Of central importance was whether toxins would also be outlawed, a situation that was clarified on 24 February 1970 when the National Security Council recommended that this form of bioweapons be included in the prohibited category.[113] Another challenge was to clarify the relationship between germ and gas warfare. According to Secretary of Defence Laird, the US program was best described as chemical warfare and biological research, since the programs were "so widely different in terms of (a) the strategic concept, (b) the deterrent value, (c) the tactical aspects of retaliation."[114] A final consideration was whether Nixon's desire for rapid Senate ratification of the Geneva Protocol would be realized, thereby completing his BW agenda. This would not occur, largely because of the opposition of Senator William Fulbright, chairman of the Foreign Relations Committee, who was concerned about the US military's continued use of tear gas and herbicides in Vietnam.[115] In general, however, Americans responded favourably to Nixon's announcement. One of the most enthusiastic endorsements came from the Council of the American Society of Microbiology, which at its meeting of 28 April 1970 praised the president's commitment "to end our involvement in the production and use of biological weapons ... [and] we urge that all nations convert existing offensive biological warfare facilities to peaceful uses."[116] Supportive statements also came from other prominent US organizations, notably the National Academy of Sciences, the American Federation of Scientists, and the American Medical Association.

Many of these accolades were repeated by a wide range of national governments, particularly Canada and the United Kingdom. Despite its lack of consultation with Ottawa over termination, the White House was interested in how the Trudeau government was handling bioweapons issues. On 10 November 1969, for example, the US Embassy in Ottawa described an exchange in the House of Commons between members of the Opposition parties and D.W. Gross, parliamentary secretary for the Minister of Defence. At issue was the familiar subject of Canada's CBW research cooperation with its Quadripartite partners, and whether details of these joint programs should be made public. What most impressed the US observers was Gross's ability to avoid divulging any real information, largely by launching into a lengthy account of Canada's wartime development of the rinderpest vaccine, whose threat, he stated, "has been practically eliminated."[117]

On 25 November, what most concerned embassy officials was how the House of Commons would respond to President Nixon's declaration during Question Period.[118] And there was much to report. At 2:40 p.m. Robert Stanfield, leader of the Conservative Party, asked Leo Cadieux the following question: "Does the government of Canada share the view expressed by the President of the United States that all ... biological warfare weapons should be destroyed immediately and research discontinued?" While Cadieux fumbled with his briefing notes, Stanfied continued: "Is it the intention of the minister to continue with the research efforts undertaken by the government of Canada although the President of the United States indicates that research ought not to be continued?" Unprepared for these probing questions, the Minister of Defence tried to divert attention away from the substance of the issue. First, he incorrectly suggested that Nixon's policy was essentially the same as Canadian and British disarmament initiatives since it was still uncertain whether BW offensive research "was to be discontinued in that country." Stunned silence followed. At this juncture David Lewis of the New Democratic Party entered the fray. Did the Minister of Defence, he asked, want to consult with his aides so that they could provide him "with the full statement which does contain a reference to research in that field in the United States as well as to the destruction of present stockpiles of weapons."[119] Another NDP member even offered his personal copy of the *Montreal Gazette* for Cadieux's edification. Once the laughter had subsided, Lewis continued with his interrogation. Would Canada, in light of the US decision to terminate offensive BW research, discontinue its involvement with the Quadripartite arrangement "under which research at

Shirley's Bay and Suffield, Alberta is carried on"?[120] There was no reply. Shortly afterwards the Minister of Defence left the chamber.

While Trudeau obviously did not enjoy Cadieux's performance, he was more concerned about the broader implications of Nixon's declaration.[121] Within the Cabinet, for instance, there was pressure for the release of the government's official statement, entitled *Arms Control And Disarmament: Chemical and Biological Warfare*.[122] There was, however, considerable discussion about whether the Canadian policy position was appropriate for the new BW disarmament context. External Affairs officials, for example, pushed hard for a categorical declaration that all research being carried out by the Defence Research Board was "of a defensive nature only, that it forms only a small portion of the work of the DRB, and that it is being orientated to research problems related to the verification of any agreement on chemical and biological warfare."[123] At the same time, DEA arms control specialists were anxious that the Department of National Defence modify some of its CBW policies, notably those outlined by General Tellier in December 1968 that Canada would obtain operational germ and gas weapons from the United States, if required.[124]

Another DEA priority was to carefully monitor developments in the United States to ensure that the White House carry through with its ban on offensive biological weapons. However, there were some troubling aspects of Nixon's November decree. First, it was not clear whether US biological warfare centres such as Fort Detrick and Pine Bluff would be phased out or whether "production facilities would still be there, and the amounts [agents] needed for a military attack could be produced very quickly, if the relevant strains were kept." A second problem was the scale of the US defensive BW program, since many experts believed such facilities could provide opportunities for future governments "to argue that in order to prepare your defence you must keep studying and testing ... [and] in perfectly good faith the over-ambitious scientist working on defensive problems may argue for the retention of facilities ... which could produce an offensive armoury in a short period."[125]

On 16 February 1970, Minister of External Affairs Mitchell Sharp renewed his request that Prime Minister Trudeau issue an official BW disarmament statement, ideally at the opening session of the Conference of the Committee on Disarmament (CCD) in Geneva. In Sharp's opinion, there would be numerous advantages in having "a firm and forthright statement of Canadian policy ... by openly declaring our capability, intentions and the protective and defensive nature of the work of the

Defence Research Board of Canada ... [and] promote and clarify our position in international negotiations." He did, however, concede that there were some risks associated with this strategy, since "such a statement would invariably be criticized for what it did not contain and invidious comparisons might be made with the statement which President Nixon made on this subject last November." On balance, however, Sharp claimed there were great advantages in adopting a proactive policy, since Trudeau would be able to reiterate, before a major international conference, that "Canada does not now possess any biological weapons and does not intend to develop, produce, acquire, stockpile or use such weapons at any time in the future."[126] Quite a different response came from Ivan Head, Trudeau's foreign policy adviser. In his opinion, such a statement would be inappropriate at this particular place and time. Not only would the opening CCD ceremonies feature the Secretary General, not the Canadian prime minister, but also Canada had no innovative policy to declare since "even the declaration that we do not intend to acquire or even use biological weapons is not really new."[127] Support for this position came from another influential source, British Foreign Secretary Lord Chalfont, who claimed that "a series of unilateral renunciations of this sort might weaken the need that had been felt for a multilateral convention."[128] Consequently, on 24 March 1970 diplomat George Ignatieff, not the prime minister, explained Canada's official position on bioweapons at the Geneva CDD meetings.[129]

Several aspects of Ignatieff's speech attracted national and international attention. One of these was the admission that Canadian defence scientists would continue their involvement with biodefence research and development for the protection of the country's armed forces and civilian population and to assist the international community with problems of BW verification.[130] Not surprisingly, these provisions were roundly applauded by DRES scientists, who could legitimately claim valuable experience in the field of surveillance and detection during their long involvement with the Tripartite / Quadripartite system.[131] They also took advantage of the work of the Stockholm International Peace Research Institute, which discussed the relative advantages of different verification procedures, such as aerial and satellite reconnaissance, remote sensors, and visiting inspection teams, who "would require a high degree of physical intrusion but could provide conclusive evidence."[132] Another useful source was the report of the US Arms Control and Disarmament Agency (ACDA), tabled at the 17 March NATO meeting of CBW experts, which also concluded that only on-site inspection could "be considered reliable."[133] All of these studies were discussed at the 3 June 1970 meeting of the newly formed Interdepartmental Panel on Chemical and Biological Weapons, which included representatives from eight departments and agencies, including Omond Solandt, President of the Science Council of Canada.[134]

Finalizing the Biological and Toxins Weapons Convention 1971–1975

Between 1970 and 1975 the international community gradually moved towards establishing the Biological and Toxins Weapons Convention.[135] But there were numerous obstacles along the way. First, while the UK draft on Microbiological Warfare called for a separate convention, it was not until March 1971 that the Soviet Union finally agreed to accept this approach and drop its insistence that the disarmament model established by the 1925 Geneva Protocol remain intact. Second, as part of this compromise, the Kremlin demanded concessions that substantially weakened the Convention's enforcement machinery: by removing the Secretary General from the complaints mechanism in favour of a Security Council review and by deleting Article I of the UK draft, which had obligated state parties "never, in any circumstances, to use biological methods of warfare."[136] Third, the willingness of the Nixon administration to accept a disarmament agreement without an effective verification mechanism was crucial in advancing the convention process.[137] But for many BW experts this was a disastrous decision both because it prevented any effective way of monitoring the Soviet bioweapons program and because it might provide a catalyst for a covert BW operation. Finally, middle powers such as Canada played an important brokerage role, not only in facilitating the dialogue between the Americans and the Russians, but also in convincing non-aligned nations that the Biological and Toxins Convention would contribute to world peace.[138]

While these complex international negotiations were taking place, the Trudeau government was re-examining its own biodefence situation.[139] Unlike the US experience, where vigorous debates took place in Congress over US biological warfare policies, there was virtually no discussion in the Canadian House of Commons of this subject. This indifference was certainly evident on 5 April 1971 when Minister Sharp's announcement that the USSR had finally accepted the British plan for a separate BW convention was greeted by a collective yawn.[140] On the positive side, consistent support from the Prime Minister and Cabinet for the proposed agreement, along with the creative role of DEA officials in the Committee on Disarmament, helped ensure that the proposed convention became a reality in April 1972. After extensive debate, it was unanimously passed (110 to 0) in the United Nations General Assembly (UNGA Resolution 2826 (XXVI). Three years later, on 26 March 1975, the process was completed when the necessary twenty-two nations ratified the Biological and Toxin Weapons Convention (BWC).

Despite this cooperation between the Western alliance and the Commuist bloc in creating the BWC, Cold War hostilities often surfaced during this period. In June 1970, for example, Defence Research Establishment Suffield (DRES) was the target of an attack by the Russian newspaper *Pravda*, official organ of the USSR Communist Party. In an article entitled "A Devil's Kitchen," submitted by the paper's Ottawa-based correspondent, it was alleged that secret and sinister offensive biological warfare activities were taking place at DRES. These allegations brought an immediate response from Canadian defence officials, who denounced the article as vicious propaganda "replete with falsehoods and deliberate distortions ... [and] published with malicious intent."[141] External Affairs also decided to prepare a detailed rebuttal of the *Pravda* article to safeguard Canada's international image. According to one account,

> DRB Suffield is "fenced" only by three strands of barbed wire to keep cattle off the property. No armed patrols guard perimeter. Air landing strip is small and primitive ... Staff at Suffield is approximately 350 of whom about 50 are scientists and only a small proportion work in chemical and biological research because major research at Suffield is testing high explosives and blast. No chemical and biological weapons are manufactured at Suffield or elsewhere. No testing of chemical and biological weapons is carried out in open air. No CND or other aircraft have ever "sprayed germs of gases." No manoeuvres have been held testing chemical and bacteriological weapons. Lethal chemical and biological weapons are not manufactured at Suffield or anywhere else in Canada and none are passed on to USA or received from that country. Research at Suffield has certainly not been increased and indeed was halved during the late 1950s and two years ago was once again halved. Research is not done on behalf of USA, Britain, or NATO.[142]

Another part of DEA's damage control was direct action in Moscow, where spokesmen for the Canadian Embassy informed their counterparts at the Soviet Foreign Office that the publication of "A Devil's

Kitchen" had caused "considerable harm to CND-SOV relations." While DEA headquarters encouraged this vigorous response, it also warned about revealing any information about the Suffield operation "since this would simply give them [Russia] defence intelligence for nothing, and encourage them to fabricate stories about any other installation they were interested in."[143] Ironically, while this confrontation was taking place, Prime Minister Trudeau was making plans for his first official visit to the USSR, to discuss, among other things, "disarmament, particularly CBW."[144] But the trip was postponed until May 1971, not because of the *Pravda* article, but because of the October crisis in Quebec.[145] Indeed, there was no evidence that the prime minister had any residual hostility towards his Soviet hosts; on the contrary, he apparently amused the Kremlin elite with his satiric comments about Ottawa's periodic differences with Washington.[146] Report of Trudeau's performance was poorly received in the White House, where President Richard Nixon apparently denounced the Canadian prime minister as "a son of a bitch ... I'll never go to that country while he's there."[147]

Conclusion

There is considerable debate about Nixon's 1969 decision to dismantle the US offensive biological weapons program, which had been operating for twenty-seven years, at a total cost of $726 million. For some authors, it was a rational and inevitable decision since bioweapons, with their fundamental flaws, had limited utility in either a strategic or tactical military capacity. Others stress the unique circumstances that prevailed in 1969 when the US military was facing hostile public opinion for its use of tear gas, herbicides, and napalm in Vietnam. As former Secretary of Defence Laird recalled in a 1998 interview, "the service chiefs had confided in him that they supported the BW ban because they believed that most of the U.S. arsenal was obsolete, that it was not as effective as nuclear weapons ... that it was a public relations liability ... [and] that a 'worldwide ban' would ease the burden of competing with the Soviets and reduce the chances for proliferation."[148] This viewpoint was also endorsed by President Nixon, who reportedly said "We'll never use the damn germs, so what good is biological warfare as a deterrent? If someone uses germs on us, we'll nuke 'em."[149]

On the other hand, it could be argued that scientific and technological developments during the previous ten years had transformed the US bioweapons system into a viable strategic option that would complement the country's nuclear retaliatory capabilities. And there is considerable evidence to support this thesis. First, by 1969 the United States and its Canadian and British allies had advanced into the second and third generation of bioagents, through improvements in the lethality and stability of selected bacterial and viral agents. Second, with major innovations in DNA research as the result of the 1954 landmark discovery of Francis Crick and James Watson, microbiologists and geneticists could alter the genetic structure of existing BW pathogens, making them resistant to antibiotics and expanding their host range. Third, there were important developments in the strategic aspects of BW delivery systems through the large area coverage (LAC) system, whereby high performance aircraft, or ships anchored off enemy coasts, could release large aerosol clouds of deadly pathogens which would be carried by prevailing winds over designated targets. By 1969 the operational effectiveness of the LAC system had been confirmed by Project 112, which had directed US armed forces to carry out a series of large-scale trials outside the American mainland, under the direction of the Deseret Test Center. And finally, as contemporary American and Canadian military studies demonstrated, there was the possibility that enemy saboteurs could attack North American urban facilities such as subways, airports, and large office buildings, causing thousands of casualties.

Not surprisingly, the 15,000 defence scientists and technicians at Fort Detrick, and other American BW facilities, felt betrayed by Nixon's declaration. From their perspective, not only had their promising professional careers been disrupted, but the United States had squandered its strategic advantage in using a weapons system that was on the verge of realizing its potential. As Bill Patrick, a veteran Detrick scientist put it, the "stuff is too damn good to go away."[150] Yet surprisingly, the opinions of experts like Patrick and Riley Housewright were not solicited during the 1969 termination process. Nor did members of the US Congress understand the long-term dangers of relying on an unenforceable disarmament convention, given their lack of accurate information and the naive belief that the threat of germ warfare would now disappear.

But the situation was even worse in Canada, where essentially no informed debate about the country's biological warfare policies took place. While members of the Cabinet Committee on External Affairs and Defence were given a brief exposure to the world of bioweapons during the 1969 CBW policy review, this was an isolated event. Instead, most of the crucial decisions about Canada's BW commitments and strategies were carried out by a small circle of civilian and military bureaucrats, with little accountability to either the House of Commons or the Canadian public. Unfortunately, this secretive and self-conscious environment provided many challenges for Canada's biodefence scientists, particularly since their professional responsibilities became more complex. On one hand, they continued to develop BW defensive programs in cooperation with their Quadripartite allies, while at the same time they became international troubleshooters for the Biological Weapons and Toxin Convention.

It was also noteworthy that, in contrast to the establishment of the BWC, the international community failed to advance the cause of a chemical weapons convention in the 1970s, despite Resolution 2827A of the UNGA (1972), which called upon the CCD to reach "an early agreement on effective measures for the prohibition of the development, production and stockpiling of chemical weapons and for their elimination from the arsenals of all States."[151] Moreover, under Article 1X of the BWC, countries were also obligated to assist in negotiations "with a view of reaching early agreement on effective measures for the prohibition ... of chemical agents for weapon purposes."[152] But Cold War rivalries created insurmountable obstacles in reaching a compromise solution, particularly with American insistence that on-site verification was a vital condition of any agreement.[153] Consequently, it was not until 27 April 1997 that the Chemical Weapons Convention became a reality.[154]

5 Triple Threats: Biowarfare, Terrorism, and Pandemics, 1970–1985

It is not too strong to say that access to US defence science and technology is life- and-death to Canadian defence and as a result relations with US heavily influence scientific policy in DND. But this is so for nearly all the NATO Countries, as well as Australia … As regards the benefits of the association, we receive … about 25,000 documents per year from USA … [representing] several hundred million dollars worth of US science … While some "releasing officers" in USA may dislike some of Canada's actions and take a hard line with release of documents … there is no evidence of a national US policy to withhold information from Canada where we have a legitimate need. In one way or another US Department of Defense furnishes DND with lists of nearly all the documents they request. When we make application on the basis of what we believe to be legitimate need, our request is filled in 90% of cases.

(Assessment of Canadian-American BW cooperation by Alex Longair, WMD Division, Defence Research Board, 1972.)[1]

The 1969 American unilateral renunciation of offensive biological warfare had a dramatic impact on the defence priorities and diplomatic policies of all members of the Western military alliance. For the US Department of Defense this transition was particularly difficult, given its sustained efforts to develop operational bioweapons, using both lethal and incapacitating munitions. Now that this program had been terminated, there were challenging questions of what would happen to the Pentagon's three major BW facilities – Fort Detrick, the Pine Bluff Arsenal, and the Dugway / Deseret Testing Grounds. Of these, Detrick's future seemed most assured, largely because of the establishment of a defensive operation called the United States Army Medical Research

Institute for Infectious Diseases (USAMIRIID). In addition, an ambitious plan to transfer many of Detrick's 1,500 scientists and technicians to the National Institutes of Health (NIH) enjoyed wide support within the country's scientific community. As Robert Hungate, president-elect of the American Society of Microbiologists (ASM) explained, the United States could "ill afford the loss of a unique team which has been assembled at Fort Detrick, and it seems folly to close these unusual and uniquely equipped laboratories which are the product of more than 25 years of careful and continuous development."[2]

The early seventies was also a pivotal period for Canadian biological warfare planners. First, with the March 1971 compromise between the Soviet Union and the United States to create the necessary framework for the Biological Weapons and Toxin Convention the possibility of an exclusive defensive biological response was guaranteed. Second, after the abandonment of the Fort Detrick complex there were serious questions about the future of the Quadripartite and TTCP systems of CBW cooperation. Third, despite periodic debate about closing down Defence Research Establishment Suffield (after 1967 DRES), Canada's military hierarchy continued to regard the facility as a valuable asset, particularly with the emergence of a serious bioterrorist threat during the late 1970s. Like many countries, Canada was directly affected by the wave of terrorism that swept through the Western world during the sixties and seventies.[3] Internationally, a sequence of violent incidents was linked to the activities of the Baader-Meinhof group in Germany, the Red Brigade in Italy, the Weathermen in the United States, and, above all, the Palestinian Liberation Organization (PLO), with its daring airplane hijackings, indiscriminate bombings, and its deadly assault on Israeli athletes at the 1972 Munich Summer Olympics. Domestically, the 1960s also witnessed a marked increase in the number of violent acts carried out by the Front de Libération du Québec (FLQ), culminating in the October 1970 kidnappings of British diplomat James Cross and a prominent Quebec politician, Pierre Laporte. Facing possible insurrection, the Trudeau government invoked the War Measures Act, which provided legal justification for the military occupation of Montreal and the arrest of hundreds of suspected terrorists.[4] This traumatic event also encouraged Canadian security experts to establish closer contacts with their counterparts in the United States and United Kingdom, since both these countries already had well-established counterterrorist systems. These connections proved extremely useful in 1976 with reports that international terrorist might attack the Montreal Summer Olympics with biological or chemical weapons.[5]

Another major concern during the Cold War years was the threat of "traditional" and emerging infectious diseases. Many still remembered the frightening legacy of the 1918 influenza A pandemic (swine flu) with its 50 million fatalities. And while medical scientists could argue that this terrible event occurred during the pre-antibiotic era, before viruses had been identified, influenza remained a constant scourge. Closely related were problems with the North American public health system, as was evident during the influenza pandemics of 1957 and 1968, which resulted in many unnecessary deaths because of insufficient supplies of vaccine. Another deadly adversary was the variola virus (smallpox), which killed thousands of people annually, particularly in less developed countries of the world. What was particularly ominous about both influenza and smallpox was the extent to which Western and Soviet defence scientists had assessed their potential as biological warfare agents.

Major Trends in North American Biodefence

Perceptions of American BW Defensive Research after the BWC

In August 1970 Dr Ivan Bennett, chairman of President Nixon's Science Advisory Committee, asked the leadership of the American Society of Microbiologists for their opinions about future linkages between microbiology research and defensive BW development programs, particularly given residual concerns "that these weapons could be employed against our armed forces or our civilian population." In preparing his questionnaire, Bennett selected five key issues, associated with the challenge of finding the correct balance between legitimate defensive research and activities that would undermine the principles of the Biological Weapons Convention.

1. Does the threat posed by possible deliberate artificial dissemination of biological agents or toxins differ qualitatively or quantitatively from that of naturally occurring disease to such an extent that a program of research directed specifically to defend against attack by an aggressor is required above and beyond existing civilian research programs in microbiology, prevention and control of infectious diseases, and epidemiology?
2. To what extent should a research program on biological and toxin defence be carried out within government laboratories and to what extent might the research be better accomplished through the use of grants and contracts in non-governmental laboratories in universities, research institutes, and industry?

3. To what extent do you think university scientists would be willing to engage in the defensive-purposes-only programs in view of the apparent growing distrust and opposition to defence-related research on many campuses?
4. How should a defensive research program include the screening of agents and toxins for their potential as weapons that might be used against us?
5. Do you believe that all elements of defensive research could be conducted openly and still serve the purposes of national security?[6]

This select group of microbiologists, speaking for both American and Canadian scientists, provided a series of thoughtful responses to Bennett's query.[7] One of the most controversial set of arguments came from Robert Hungate, president-elect of the ASM, who strongly endorsed the goals of the Bennett's Committee, since he was "quite out of sympathy with the view that research shall not be done on biological weapons ... [as] I feel that any phase of human experience should be susceptible to research and most of the problems arise in connection with secrecy."[8] But in terms of lethality, Hungate claimed that there was "no question that a deliberate artificial dissemination of biological agents would pose a threat quantitatively different from that of naturally occurring disease ... [which] would best be met through expansion of military and civilian public health and medical organizations." While Professor J.R. Porter, Department of Microbiology, University of Iowa, agreed with this viewpoint, he expressed special concern about the next generation of biological weapons since "with our present knowledge of microbial genetics, I believe it is possible to produce artificially microorganisms with greatly increased virulence, and which are highly resistant to current methods of control."[9] But the most comprehensive response to Bennett's letter came from Riley Housewright, former Technical Director at Fort Detrick, who systematically addressed the five major issues. Housewright began by pointing out that the threat posed by the deliberate dissemination of biological agents would result "in a tremendously large number of casualties," since these engineered pathogens "would apply to unusual routes of inoculation and initiation of infection ... [and] complications of diagnosis and treatment occasioned by the use of combined infections." He was particularly concerned about the use of newly discovered viral diseases, which "would not be recognized by ... even a well trained physician [and] there would be no vaccines to use against them and no satisfactory viral chemotherapy is available." A related problem was that the US public health system was completely inadequate "to meet the demands of such an occasion." Housewright was also pessimistic that US academic microbiologists would be willing to be involved with defensive biological warfare research, which meant "that if work is to be done, then the major portion of it must be done within Government laboratories and the remainder in industry." If this was the case, he strongly advised that a comprehensive biowarfare research agenda be prepared, including "the evaluation of potential agents [viruses] or it will become … in a short time a 'Model-T-Ford' program." Housewright ended his brief with an assessment of the problems associated with transparent biodefence research: "Ordinarily I would say that all work should be open, but I really wonder if it would be wise for us to provide our potential enemies with all the specifications, or deficiencies and defects of our own early warning system. Do we do this with the missile systems" If we do not, then why should we do it for this one?[10]

These ASM luminaries also discussed the possibility that Fort Detrick could become an "open international institute for studies on disease and the environment."[11] Hungate was particularly insistent that the White House and Congress explore this option since Detrick was "an important biological research facility with a large and most capable staff of microbiologists and other scientists … [which] has contributed materially, not only to the security of the United States, but also to progress in public health, to developments in fermentation processes, and to unique concepts in apparatus and laboratory design."[12] A more emotional comment came from the recently retired Housewright, who lambasted the leadership of the US Army for their duplicitous and callous methods in dealing with Detrick's many gifted scientists.[13] This sense of betrayal was also evident in the letter that Robert Lourie Jr, a young Detrick microbiologist, sent to President Nixon in June 1971:

> We have worked diligently within the framework of government to get Fort Detrick converted to peaceful biological research … [as] this is a magnificent opportunity for your Administration to demonstrate to the world that you fully intend to commit the United States to a path of "beating swords into plowshares." … Eighteen months have passed since your announcement … [and] bureaucratic red tape has allowed this unique research complex to deteriorate into a shell of "mothballed" brick and mortar … [and] all of this has been allowed to occur despite the pleading of internationally prominent scientists that Detrick could serve as a bastion in the fight against cancer and other public health problems.[14]

The Fort Detrick saga assumed many forms during the early 1970s. The basic challenge was how to transform a weapons establishment into a civilian research institute.[15] On the positive side, Detrick had a number of assets: 1,230 acres of land, 460 buildings, 1,850,000 square feet of floor space for research, and more than 300 degree-holding scientists (70 PhDs) in bacteriology, virology, immunology, genetics, molecular biology, and cell biology.[16] Yet despite a concerted effort by the National Institutes of Health to absorb Detrick's scientists into its operation, the project was eventually scrapped.[17] While lack of financial resources was a problem, there were also concerns that NIH should avoid any association "with work of a classified nature, or any suspicion that former research activities at the Detrick installation are being continued under a new facade."[18] But this debate ended on 18 October 1970 when President Nixon announced that the Detrick site would be shared by three tenants: the US Army Medical Research Center (later USAMIRID), the US Department of Agriculture, and the Frederick Cancer Research Centre.[19] To assure the world that the former weapons facility had been transformed, the president invited the world's medical community to visit Frederick, Maryland.[20] Ironically, among the first group of visitors were members of a Soviet public health delegation, who were ostensibly exploring the possibilities of American-Russian cooperation in discovering the causes of cancer.[21] Unfortunately, there are no available records of what these scientists reported during their post-visit debriefings sessions in Moscow; but it is noteworthy that a Soviet offensive BW program soon emerged under the guise of a civilian biomedical enterprise.[22]

Disposing of Biological and Toxin Agents

While Detrick's future was being considered, the US Defense Department was busy making arrangements for the destruction of its stockpile of biological agents and munitions. The first stage in this process began with National Security Memorandum 35 and 44 (and Executive Order 11507), whereby the Secretary of Defense was directed "that all stockpiles of biological agents and munitions be destroyed and that the biological research program of the Department of Defense be confined to defensive purposes only."[23] In short order a detailed plan for this formidable task was drafted, followed by an elaborate review assessment by the Office of Science and Technology, the Surgeon General of the US Public Health Service, the Department of Agriculture, and the Federal Water Quality Agency (Department of Interior). It was subsequently determined that three major components of the BW arsenal would be scheduled for destruction: the anti-personnel weapons stored at Pine Bluff Arsenal, Arkansas; the anti-crop agents at the Rocky Mountain Arsenal, Colorado; and small quantities of agents at Fort Detrick and other sites. In total the project was expected to take 107 weeks and cost over $6 million. The operation began with a detailed report from the Directorate of Biological Operations at the Pine Bluff Arsenal describing how the different BW agents would be chemically neutralized and then incinerated. Due to the nature of this material, safety precautions were given the highest priority, with special emphasis on preventing the contamination of "the outside atmosphere in the event of a biological spill."[24] The second stage began with the Army's environmental impact statement of 17 September 1970, which provided assurances that the sterilization process, involving live agents, would be conducted within the existing plant "using the same or even more strict controls than were applied during the manufacturing program." The report then discussed how four different categories of BW material would be destroyed. These included liquid agents stored as frozen pellets, bulk dry materials, bulk stocks of toxins, and biological munitions (filled and unfilled). Of these components, the greatest challenge was in neutralizing the liquid frozen pellets because of the large quantities involved and the complexities of thawing thousands of metal cans, with "approximately one-half gallon liquid [agent] per can." In the end, however, the Army's disposal system was carried out without incident.[25]

In December 1970, the White House announced that the US biological weapons stockpile had vanished. To celebrate this event, plans were made to invite representatives from the other Quadripartite countries "to verify and / or observe the process ... [since] the UK, Canada and Australia are aware ... of the details of our weapons program, and there would be no security risk entailed in their verifying the process."[26] Once the final stages of the project were completed, a detailed inventory documented the scope of the former American BW arsenal:

Liquid agents stored as frozen pellets (bulk):
Incapacitating: VEE- 4,991 gallons; Q Fever, 5,098 gallons
Dry Bulk Agents:
Incapacitating: VEE, 334 pounds
Lethal Agents: Tularemia (804 pounds), Anthrax (220 pounds)
Bulk Toxins: Clotridium botulinum toxin (13 pounds), Staphyloccus aureus enterotoxin (71 pounds)

Agents in filled munitions: Anthrax (167.5 pounds), Tularemia (569.7 pounds), Toxins (Bot Tox, 10.9 pounds).[27]

Significantly, this official declaration did not provide any information about the capabilities of these different biological weapons, either in terms of lethality or their collective strategic impact. Much more serious, however, was the failure to acknowledge the ten biological agents and six toxins retained by the Central Intelligence Agency (CIA) because of a secret arrangement with Detrick's Special Operations Division.[28]

Aspects of Canada's Biodefence Programs during the 1970s

One of the essential aspects of developing an effective biodefence system is the ability to understand the scientific profile and physiological impact of specific biological weapons. Not surprisingly, this was a basic principle of Quadripartite BW research and development programs throughout the Cold War, particularly after 1956 when only the United States had an offensive bioweapons program. As a result, during the next thirteen years BW scientists at Porton Down and Defence Research Establishment Suffield (DRES) regularly requested and received samples of "hot" agents and munitions from Fort Detrick or the Dugway Proving Grounds. But after 25 November 1969 all this changed, and Suffield's pipeline to these "hot" agents dramatically changed. Instead, possession of this BW material was now regarded as a political liability, as was evident in the 1970 exchange between between DRES Director General B.J. Perry and Archie Pennie, Deputy Chairman, Operation, Defence Research Board.

The dialogue began with Perry's admission that his scientists had regularly received small shipments (500 mL) of *F. tularensis* (tularaemia) and *B. suis* (brucellosis) from Fort Detrick throughout the sixties in order to carry out joint testing programs. While this transfer had been curtailed, neither Perry nor Pennie wanted information about these weaponized agents made public, given Ottawa's blanket assurances that DRES scientists were not involved with any offensive BW research. A second major issue was Suffield's large stockpile of blistering CW agents and nerve gases, including "Agent GA, 408 lb; Agent GD, 59 lb; Agent GB, 472 lb; Agent VX, 123 lb; Agent Lewisite, 4190 lb."[29] Even more contentious was Perry's disclosure that his Station had 260 "105 mm. shell charged GB ... [that] represents our only stock of this agent and as such we are reluctant to dispose of it ... [but] they are held under secure guard."[30] While Pennie appreciated Perry's reluctance to relinquish operational CW weapons, since offensive chemical warfare research was still a legitimate Quadripartite activity, he expressed concern about the "touchy subject ... of filled munitions ... [since] many of us, including myself, have made categorical statements that the Canadian Forces and the Department have no stockpiles of chemical munitions." Closely related was the inconvenient truth that "within the last two or three years trials have been carried out at Suffield using American shells filled with chemical agents." Pennie's final advice to his colleague was that he should be careful what he told the media within the prevailing ant-CBW context, and that his statements should conform with the official DRB position since "we want ... to play the same tune to the same people."[31]

Pennie's apprehension about adverse publicity was not unfounded. The summer of 1970 witnessed a number of public demonstrations against Canada's biological and chemical warfare facilities. In May, for example, the Defence Research Establishment Suffield (DRES) was targeted by the Saskatoon Vietnam Mobilization Committee, who gave notice that they intended to organize a peaceful demonstration at the testing site.[32] Yet despite concerns about a possible confrontation, the Royal Canadian Mounted Police, after carrying out a full-scale investigation of the Committee's background and agenda, advised against "deploying troops in the area." Instead, the Mounties recommended that a high-ranking DRES official should meet with the demonstrators, with the understanding that this spokesman would be "carefully briefed ... [having] some psychological questions that he could ... place ... and possibly cool the atmosphere." Accordingly, Director General Bobyn hosted a small delegation at the Station, giving them a guided tour of the laboratory facilities and testing facilities. The mayor of Medicine Hat was not so gracious; he warned protesters to stay away from his city.[33] Within weeks another critique of the Suffield operation surfaced, this time from a member of the governing Liberal Party. On 25 May 1970, Warren Allmand, MP from Montreal-NDG, wrote Prime Minister Trudeau complaining about the contradiction between DRES claims that it was only doing defensive BW research and evidence that it had "an offensive capacity." In order to remove any suspicions that Canada was encouraging the BW arms race, Allmand recommend a number of major changes in Canada's biodefence policies:

> 1. Having this work carried on under non-classified conditions ... If the work is really defensive, then there is really no need for us to keep it classified; 2. Changing this work from the Defence Research Board to the

> Department of Health and Welfare ... this would indicate to Canadians and to the world that this work is not primarily military ... [and] such research could be related to present research against contagious and epidemic diseases ... 3.Withdrawing from the technical cooperation program which we have with the U.S.A., Britain, and Australia which is the basis for our research in CBW. Since we have already decided to withdraw part of our military forces from NATO ... it follows that we should get out of any type of alliance which would promote CBW which is considered by all people as a most abhorrent and distasteful type of warfare.[34]

While Allmand's critique obviously annoyed Trudeau's advisers, they decided that the Minister of Defence should provide the rebuttal. And Leo Cadieux quickly demonstrated that he enjoyed educating his young Liberal colleague about the realities of CBW arms control. First on the agenda was the question of classified research, which, Cadieux claimed, represented only a small portion of the DRES workload. At the same time, he emphasized that advanced research on biodefence was vital to Canadian national security, being the kind of information "which potential enemies would like very much to obtain." The next issue was Allmand's proposal that DRES should be transferred to the Department of Health and Welfare, which Cadieux dismissed as a useless exercise that would "do nothing to calm the unwarranted emotional reaction, which has arisen in a certain section of the public as a result of distortion of the facts by certain self-appointed groups and publicists." Nor did the Defence Minister feel that Canada was being exploited through its involvement with the Quadripartite agreement and Technical Cooperation program (TTCP), pointing out that this kind of alliance cooperation saved Ottawa "millions of dollars through the acquisition of research results made available to us in the various scientific fields." In summary, Cadieux reminded the Liberal backbencher that Canada lived in a dangerous world, and while the government "welcomed the outlawing of these weapons ... if there were reliable guarantees against their use ... until such an arrangement becomes an actuality, this department must ensure that the best protection possible is provided for all Canadians, both military and civilians." Even worse, "to abandon such research would ... encourage others to develop newer and more deadly weapons of this type, and ... our efforts to develop effective counter-measures would be lost."[35]

Obviously, the DRB officials appreciated their minister's robust endorsement of their programs, particularly at a time when new biodefence realities were emerging at the national and international levels.[36]

But the new situation did provide a catalyst for the Board to carry out another review of its CBW operation, with the confirmation that DRES would carry out most of the classified BW defensive projects, while fundamental research project would be farmed out to academic scientists with special qualifications in these fields. In January 1971 Jack Currie, a recent transfer from Defence Research Establishment Ottawa (DREO), provided an overview of the future BW schedule at DRES:[37]

> Priority was given to Project 1 concerning detection of microorganisms for the purpose of physical defence. Fundamental studies will involve the use of the electron microscope for detecting viruses, as well as the continuation of studies on uptake of radioactive phosphorous by microorganisms as a detection mechanism ... Project 2 will be concerned with the evaluation of the hazard of Biological Warfare ... [and] will involve a comprehensive literature review, concept studies, and review of present operational procedures. Under Project 3, an evaluation of defensive systems will be undertaken. Testing of military equipment, as well as determining whether Standard Operating procedures are workable in practice, will constitute investigations in this project.[38]

Another important issue was to assess how military volunteers were being used in defensive equipment trials involving live CBW agents. In particular, reference was made to the recent DRES experiments where soldiers were exposed "to a particulate aerosol of *Bacillus subtilis var niger* spores in a Quonset shed large enough to permit them to carry out physical activities representative of troops preparing a defensive position."[39]

In January 1970 a joint meeting of the Advisory Committee on Defence Against Biological Agents, and the Defence Research Board's Panel on Infection, Immunity & Therapy discussed a range of BW defensive projects.[40] These included "The Hazard of Biological Agents under Arctic Conditions, the "Concept of BW Use in Tactical Situations," and ways in which the DRB could provide medical support "if the enemy did use BW agents against Canadian Forces ... [or] when an unexpected epidemic was detected."[41] In addition, the DRB debated whether Suffield should shift some of its microbiology research "from the current biological defence program to laboratory studies on Public Health problems under military conditions ... that might result from a military occupation of the Canadian North."[42]

Questions were also asked about how Canada's reorganization of its biological warfare programs related to its security system, or more specifically whether DRB's performance in protecting BW secrets met

Pentagon standards. As a result, throughout 1971 there was a thorough review of the Board's classification system, which determined if documents were top secret, secret, confidential, or unclassified.[43] Three goals were adopted in determining the status of information received from its military allies: "(a) to prevent embarrassment ... with the subject matter of the document concerned; (b) to ensure that information of a particular sensitive nature is not disclosed ...; (c) to safeguard certain sources of intelligence." In turn, the documents were given four kinds of stamps: Canadian Eyes Only, Canadian-UK Eyes Only, Canadian-US Eyes Only, and Canadian-UK-US Eyes Only.[44] Not surprisingly, most of the Quadripartite data were placed in this latter group, given the close interaction between CBR researchers in the three countries, with Australia receiving only selected reports.[45] While the system worked rather well, there were occasional glitches when sensitive reports were mistakenly circulated.[46] Of greatest concern, however, was the possibility that sensitive CBW information might be obtained by Communist bloc countries, which routinely requested copies of restricted DRB scientific reports.[47]

Controversies over US Weapons of Mass Destruction

While the 1970–1 destruction of the American bioweapons stockpile attracted little public attention, the disposal of US chemical warfare munitions received quite a different response. In part, this was related to the fact that chemical weapons were integrated into the strategic policies of the US armed forces overseas, including extensive poison gas stockpiles in a number of forward American bases. Until 1970 the Pacific island of Okinawa was in this category, at least until it was scheduled for return to Japan as part of a prolonged negotiated agreement. But before this transfer took place, a serious accident at the US military base involved munitions filled with sarin nerve gas. Facing intense pressure from Japan, the US Department of Defense decided that this material (18,000 tons) should be transferred to a large US Army depot in Oregon.[48] According to the Pentagon's plan, five separate ships would transport the CW munitions to the west coast, with immediate transshipment to the CBW Umatilla Depot.[49] However news of this toxic armada created an immediate political storm. In Oregon both Governor McCall and Senator Hatfield personally requested President Nixon to have the project cancelled for environmental and health reasons. Similar protests also came from Canadian politicians. On 23 January 1970, for example, NDP leader Tommy Douglas (Nanaimo-Cowichan-The Islands) complained in the House of Commons about the public health hazard of having "these five ships pass through Juan de Fuca Strait and be unloaded only 60 miles south of Vancouver Island ... [with] this nerve gas, which has already injured some 24 people who were handling it in Okinawa.[50] Opposition members were not mollified when Minister of External Affairs Sharp claimed that his department had a firm guarantee from US officials that no shipments would occur until all health threats had been eliminated.[51] This did not happen. Instead, in April 1970 spokesmen for the US Department of Defense reaffirmed that the poison gas shipments would proceed since these munitions represented less of a hazard to the ocean environment than "ordinary shipments of high explosives ... [or] normal tanker transportation of petroleum."[52] Most people on the west coast disagreed, and protests continued to escalate. Faced with intense opposition on both sides of the border, the Nixon administration decided that the political costs of this project were too high. Instead, arrangements were made to ship these CW munitions to Johnson Island, a secure US military base in the south Pacific.[53]

For Canadians, the most controversial aspect of American WMD activities was not biological or chemical weapons, but the 1970 underground nuclear tests at Amchitka Island, Alaska. Code named CANIKIV, this project involved a 5-megaton nuclear explosion as part of US attempts to upgrade its nuclear arsenal. But what was most important about this event was the confrontation that developed between US military planners and Canadian peace groups and environmental organizations. On one hand, the Pentagon regarded CANIKIV as a necessary and legitimate undertaking since it did not violate the 1963 Limited Test Ban Treaty and would take place exclusively on American territory.[54] But for CANIKIV opponents, a nuclear explosion of this magnitude "could result in damage to marine life through the accidental escape of radioactivity and, that more important, such explosions could trigger an earthquake that would send a tidal wave crashing down Canada's west coast."[55] Determined to prevent the bomb trials, groups such as the Vancouver-based Committee for Nuclear Responsibility adopted a number of innovative tactics. These included large-scale demonstrations both within the province and across the country. Even more spectacular were the high-seas "sit-ins" near the blast site, involving thirty scientists and reporters, that attracted worldwide attention.[56] Indeed, the 74-ft halibut fishing boat, renamed "The Greenpeace," quickly became a symbol, not only for the anti-nuclear campaign, but also for environmental protection globally.[57] In September 1971 Soviet premier Alexei Kosygin also joined the fray when at the beginning of this Canadian tour he launched a number of well-placed barbs against the US nuclear testing program.

Yet despite this opposition, American nuclear specialists still insisted that CANIKIV was essential for US national security. And President Nixon agreed. Consequently, in November 1970 an underground nuclear device was detonated, causing a seismic shock of 7.0 on the Richter scale. Fortunately, the explosion did not cause a tidal wave or produce a radioactive cloud. But on the political front Amchitka severely tested Washington's relationship with Ottawa and tarnished the image of the United States internationally. Consequently, after a discreet interregnum, the US Atomic Energy Commission announced the end of its nuclear underground testing program in the Aleutian Islands.[58]

On the other hand, debates about American weapons of mass destruction continued. In 1975, for instance, the US Senate and House of Representatives, now under the control of the Democrats, carried out a series of investigations of the Pentagon, the Central Intelligence Agency (CIA), and other national security organizations. In part, these inquiries were associated with the Watergate scandal (1973–5), which forced President Nixon to resign and resulted in the arrest and conviction of his closest aides. While a number of "dirty tricks" were exposed during these years, several of these gained particular notoriety, particularly revelations that between 1955 and 1973 the US Army had operated an elaborate and secret human testing program called Operation Whitecoat. Under this scheme, 2,200 members of the Seventh Day Adventist Church participated in a series of controlled experiments, at Detrick and Dugway, in order to determine how various bioagents affected humans. Most of these agents were incapacitants, with Q fever, tularaemia, typhoid fever, eastern and western Venezuelan equine encephalitis, and Rift Valley fever being the most popular candidates.[59] While US military apologists justified Whitecoat on the basis of informed consent by volunteers, the absence of fatalities, and the long-term contribution to the safety of US servicemen, overall the American public regarded these activities with considerable distaste.

Of even greater importance was the sustained critique of the Central Intelligence Agency carried out by the Senate Committee "To Study Government Operations with Respect to Intelligence Activities." Better known as the Church Committee, this body of senators focused its attention on CIA covert operations, past and present, with special emphasis on two top-secret programs.[60] One of these was related to mind-control experiments carried out under Project MKLUTRA and involving reputable academic scientists such as Dr Ewan Cameron of the Allan Memorial Institute of McGill University. Under Cameron's supervision some unsuspecting patients were given powerful sleep-inducing drugs such as LSD, barbiturates, and temazepam, while others were exposed to disturbing levels of electroshock therapy and psychic-driving.[61] The second controversy was evidence that in 1970 CIA officials, with the assistance of Detrick scientists, had concealed their stockpiles of biological weapons, including the deadly shellfish toxin, in clear violation of the 1969 presidential directive. Although CIA Director William Colby tried to minimize the lethality of this toxin, the committee's experts concluded that the 11 grams of this substance that the agency retained was sufficiently toxic to kill 55,000 people.[62]

This "muck-raking" exercise in Washington was not well received by Canadian defence officials, who were concerned that the testimony of former US defence scientists would reveal embarrassing facts about their long-term bioweapons cooperation with the Pentagon. High on their list of concerns was Guilford Reed's role in helping Detrick develop shellfish poison into a CIA instrument of assassination.[63] A more recent and extensive pattern of cooperation was the involvement of Defence Research Establishment Suffield in the offensive BW work of Project 112. Although these trials had been concealed from public view, this situation changed dramatically during the spring of 1977 when the Senate Subcommittee on Health and Scientific Research carried out an intensive examination of the Pentagon's BW testing program. The catalyst for this investigation was the recently declassified report *US Army Activity in the US Biological Warfare Programs* (vols. 1 and 2), which contained a wealth of information about the logistical and policy dimensions of the American BW program since the Second World War. But the committee was primarily interested in three major questions: How were the 239 secret open-air BW tests organized? What percentage of them deployed disease-causing pathogens? And where had these trials taken place?[64] In this latter case, the 1977 US Army report clearly showed that there had been a number of joint trials between the Deseret Test Centre and Defence Research Station Suffield, with several taking place at DRES in October 1963 (using BG) and in January / March 1965 "to evaluate the area coverage capabilities of a dry agent dissemination system during cold weather."

The American media had a field day with these juicy insights into the secret world of biological weapons research and development. And while most of the criticism was directed towards the US bioweapons program there were no guarantees that this narrow focus would continue. As a result, security officials in Ottawa decided to adopt several damage control strategies. One of these was a collaboration arrangement between DEA officials and CBW specialists in the Department of National Defence "about nature of replies to be made to ensure that they do not contain anything that would embarrass us or compromise our position in CCD." However, this strategy of concealment was tested in March 1977 when an article in the *Ottawa Citizen* described how in 1953 the US Army had conducted secret aerial trials over American cities using BW simulants.[65] But the editors also wanted to know whether any Canadian cities had been included in these toxic experiments. On this occasion DND officials categorically denied that any such experiments had taken place in Canadian airspace, although three years later they were forced to admit that US planes had sprayed aerosols "over a part of Winnipeg."[66]

New Challenges: Changing Canada's Biodefence Priorities

April 1972 marked the twenty-fifth anniversary of the establishment of the Defence Research Board. It was a festive occasion since the organization had enjoyed considerable success in carrying out its responsibilities and employed a sophisticated organizational structure consisting of 1,871 employees, with 500 in the executive / professional / scientific category. Most of the work of the DRB, however, was carried out at six research establishments, which had varying functions and budget allocations: Valcartier (24.9 per cent), Atlantic (9.3 per cent), Defence and Civil Institute of Environmental Medicine, Toronto (6.2 per cent), Pacific (6.0 per cent), Ottawa (13.1 per cent), and Suffield (7.6 per cent).[67] Administratively, the DRB was guided by its Chairman, Dr L.J. L'Heureux, while the Vice-Chairman was Dr Harry Sheffer, formerly Director of Defence Research Station Ottawa. In addition, there were eight appointed members of the Board, three of them ex officio: Sylvain Cloutier, Deputy Minister of DND; General J.A. Dextraze, Chief of the Defence Staff; and W.G. Schneider, President of the National Research Council.[68]

The many achievements of the Defence Research Board were on display at the twenty-fourth annual DRB Symposium (14–16 November), held at the National Arts Centre in Ottawa. It was a celebratory event, highlighted by the keynote address of Professor Hermann Bondi, Chief Scientific Advisor of the British Ministry of Defence, entitled "The Future of Defence Science and Technology." This was followed by a humorous and nostalgic speech by Archie Pennie, DRB Deputy Chairman (Operations), entitled "Defence Research Board – The First Twenty-Five Years."[69]

Unfortunately, the days of the DRB were numbered; by 1974 it was dismantled and its research assets reassigned.[70] There were a number of reasons why the Trudeau government decided that the Board had outlived its usefulness. First was a revival of interest in rationalizing the functions of various scientific organizations within the civil service, a process that had begun during the early 1960s with the critique by the Royal Commission on Government Organization (Glassco Commission) about self-perpetuating scientific bodies that were "allowed to continue for years after they should have been terminated on practical grounds."[71] Another important initiative was the 1964 establishment of a Scientific Secretariat, located in the Privy Council Office, with a mandate to analyse information about the government's diverse scientific programs. Closely related was the work of the Science Council of Canada (1966) "to assess the scientific and technological resources, requirements, and potentialities of Canada."[72] A second major factor in the demise of the Defence Research Board was the Trudeau government's 1969–70 review of Canada's defence policy. Of particular importance was the Management Review Group's negative assessment of the DRB and its call for dissolution:

> The MRG recognized the independence of the DRB and disagreed with it. In their opinion the Board was unresponsive to defence needs of the 1970s and the lack of proper administrative controls. They noted particularly that the DRB continued to work on projects for which there was no immediately discernible need. It was their recommendation that the Board be disbanded ... [Subsequently] the two main Branches of the DRB were taken apart. Operational research was placed under the new ADM (Policy) and research and development was added to the ADM (Materiel) ... Thus no clearly identified scientific head remained in DND ... [and] the adviser to the Minister who once ranked with the Service Chiefs now no longer exists.[73]

But was this termination policy justified? Of course, was the response of the Trudeau Cabinet and the hierarchy of the Department of National Defence. Quite a different reaction, however, came from DRB scientists, who denounced the review process as a flawed exercise carried out by arrogant technocrats. This was certainly the viewpoint of Omond Solandt, who described the destruction of his beloved DRB as "a thoughtless act of mayhem committed in the name of administrative tidiness."[74] Throughout this painful restructuring DRES scientists were apprehensive that their operation might also be placed on the chopping block. And there were good reasons for concern. Not only were all former DRB facilities under scrutiny, but Suffield's status had been dramatically changed in 1971 with the ten-year agreement with the British government to allow UK military forces to use the Military Training Area and the official creation of the Canadian Forces Base Suffield.[75] Even more threatening was the opinion of Minister of Defence James Richardson that DRES should be closed and its chemical and biological programs moved to Defence Research Establishment Ottawa (DREO) on the grounds that it "has been phasing out for several years."[76] Richardson also proposed that Winnipeg, his home town, should become the focal point for all DRB electronics research and development, which meant a shift of personnel from DREO centre (Shirley's Bay).[77]

This exercise in scientific musical chairs did not, however, go unchallenged. Not surprisingly, the most serious criticism came from the DRES scientific community, which pointed out that Suffield's CBW scientific and technological achievements were "legendary," both at home and abroad.[78] In the end, Suffield's defenders prevailed, largely because the Canadian Chiefs of Staff once again decided that its high-quality defensive CBW programs were essential for the Canadian Forces and their international commitments.[79] On 3 May 1977, the Honourable Barney Danson, the new Minister of Defence, confirmed that the Trudeau government had no intention of closing down the DRES facility "any time in the near future."[80] On the other hand, the dismantlement of the DRB complicated Canada's biodefence relationship with the United States. Certainly the profile of Canadian defence scientists was downgraded in the eyes of their US counterparts, particularly in the CBW field. According to Alex Longair, a DRB veteran assigned to the Washington-based Canadian Defence Liaison Staff, there were ongoing problems after 1974 in keeping high-ranking US officials informed about changes in Canadian defence science in general, and the country's biological and chemical warfare capabilities in particular.[81] What made this latter situation particularly serious was evidence that the United States wanted to abandon the Quadripartite CBR system and focus instead on establishing a bilateral biodefence arrangement with the United Kingdom, thereby leaving Canada and Australia on the sidelines.[82]

In 1977 American and British CBW experts signed a Memorandum of Understanding of the Cooperative Program on Research, Development and Acquisition of Chemical and Biological Defense Materiel "to address problems of mutual concern in chemical and biological defense."[83] While initially Canada was not included in this arrangement, the Americans eventually relented and agreed to accommodate their long-time CBW partner. By 1980 this new organization, using the acronym CANUKUS, was involved in the most advanced forms of CBW research and development, placing special emphasis on detection and identification, vaccines and prophylaxis, contamination control, as well as the small-scale production of BW agents for defensive research. In operational terms, the priorities and procedures were determined by the trilateral Steering Group, with more specialized work being carried out by Working Groups and International Task Forces (ITF).[84] While most of the focus was placed on CBW battlefield issues, the 1990s saw a gradual shift towards counterterrorism.[85]

Canada and the Threat of International Terrorism

Throughout the first Trudeau administration (1968–72) there was growing apprehension about the activities of terrorist organizations at home and abroad. In many ways this was quite different from the long-standing apprehension about Soviet aggression, dating back to the 1946 revelations of the Royal Commission on Espionage, when the threat came from spies not saboteurs. Nor was it clear that the counter-espionage skills of the RCMP Security Service and provincial police forces were sufficient in dealing with organizations such as the Front de Libération du Québec (FLQ), which became increasingly radical in its ideology and violent in its tactics.[86] This was exemplified by a number of incidents in Montreal, notably the 1969 bombing of the Montreal Stock Exchange and the October 1970 kidnappings of James Cross and Pierre Laporte. The Trudeau government's decision to invoke the War Measures Act, placing Montreal under virtual military occupation, was both decisive and controversial. Less well known were the sections of Ottawa's 1970 White Paper on Defence, which established procedures on how the Canadian Armed Forces should assist civil authorities in dealing with terrorist bombs and CBW devices.[87]

Although there were sporadic FLQ bombings throughout the sixties, it was only after the October crisis that the Canadian government established comprehensive security guidelines to ensure that its buildings and personnel were safe from terrorist attack. Many of these security priorities were outlined in a top-secret report of 2 December 1970 entitled

"Bomb Threats in Relation to Federal Buildings." Prepared by R.G. Robertson, Chairman of the Security Panel, the document explored three major trends. One was to explore the implications of the marked increase in the number of bomb incidents during the previous two years, particularly in the Montreal region, where 66 real bomb scares and 2,000 hoaxes had occurred. While the number of fatalities was relatively low (five), the Cabinet Committee on Security and Intelligence was concerned about the psychological impact of these incidents and the extent to which these bomb threats disrupted the operations of the federal government.[88] Another priority was to develop a more coordinated response from bomb disposal experts of the Canadian Forces and their counterparts in the RCMP and Ottawa city police; this problem was partly resolved when DND's Emergency Measures Organization was given overall responsibility for the counterterrorism system in the capital region. And finally, the Security Panel report prepared a collective profile of individuals and groups most likely to carry out terrorist acts. It was a strange list. In one category was a bizarre collection of cranks, the mentally deranged, and disgruntled employees "who are simply attempting to obtain time off work." But, as the report emphasized, the most serious threat came from the various separatist organizations, who wanted "to destroy government property, disrupt government business and generally embarrass the government."[89] Significantly, there was little discussion about either the ideological goals of the different terrorist groups or the possibility they might use weapons of mass destruction.[90]

On 23 December 1970 the Cabinet issued a directive requiring all departments and agencies "to create an effective organization to institute adequate security measures and procedures to counteract and minimize the effect of bomb threats and bomb explosions."[91] In the case of the Defence Research Board many of these measures were already in place, given the sensitive nature of its work. Indeed, throughout the 1960s there had been warnings that non-violent protest groups such as the Voice of Women might be hijacked "by extremists to mount open or surreptitious attacks." As a result, all six DRB research establishments had implemented security protocols to screen visitors and examine all packages and briefcases for weapons and explosives. In addition, security guards were given special instructions that if they received threatening telephone calls, efforts should be made to obtain "the exact location where the bomb ... is to be placed ... [and] note such things as male or female voice, accent, peculiarities of speech and grammar, background sounds such as voices, music and traffic."[92]

The counterterrorist capabilities of the DRB were greatly improved during the early 1970s by its enhanced ability to detect concealed explosives "in military operations, in support of civic police during terrorist activity."[93] This expertise was further enhanced through a series of consultations with US defence and security agencies for developing integrated response plans in dealing with terrorist incidents. Of particular importance was the 1972 Memorandum of Understanding on Riot Control Agent Materiel and Technology between DND and the Pentagon for the mutual exchange of information and materiel so that each country could "study and evaluate the results of the other participant's efforts along these lines." This joint venture concentrated on four shared programs: protective equipment and clothing; non-lethal tracer material; decontamination procedures; and the development "of various types of chemical riot control agents effective through the respiratory route or ... on the skin."[94]

Canadian authorities were also very interested in the security strategies adopted by the newly established US Cabinet Committee on Terrorism.[95] This multi-agency organization was created on 25 September 1972, under the chairmanship of Secretary of State William Rogers, with a mandate for "bringing the full resources of all appropriate United States agencies to bear effectively on the task of eliminating terrorism wherever it occurs."[96] Initially, President Nixon had high expectations that all members of the United Nations would be willing to follow the lead of the United States in holding back "the perimeters of lawlessness."[97] Unfortunately for Washington, the international response was disappointing. In fact, a solid bloc of African and Asian members of the General Assembly were able to defeat a 1972 Canadian resolution that would have moved the organizations towards a UN Convention on Terrorism.[98] However, as an alternative, Ottawa and Washington negotiated an agreement "for dealing with international terrorist acts against aircraft," which was later reinforced by the October 1972 Montreal Sabotage Convention.[99]

The US Cabinet Committee on Terrorism enjoyed its greatest success at the domestic level.[100] These included rigorous enforcement of visa requirements to identify possible terrorists; the installation of equipment in the nation's post offices to detect letter bombs; and special safeguards "for protecting nuclear raw materials which, if captured by terrorists, can be made into crude atomic bombs or exploded to cause contamination."[101] In 1973 additional counterterrorist measures were adopted, notably an arrangement whereby the Department of Defence was given permission to assist the Federal Bureau of Investigation "in the event of a terrorist emergency." And the Bureau continued to work on its own emergency response system. In November, for example, in a simulated terrorist attack on Chicago's O'Hare airport, FBI operatives were able to set up a command post in 15 minutes, and have "160 agents in place in 40 minutes."[102] After 1974, the committee experienced a rapid decline, largely because of President Nixon's preoccupation with the Watergate controversy.[103]

But despite its relatively short life span the organization had established a number of important security measures which strengthened Washington's ability to deal with present and future groups of terrorists.[104] American security officials also appreciated the assistance of their northern neighbour, as was evident in a January 1973 committee report which praised Canada's role "in the crusade against terrorism ... at Brussels, at UNGA, at Montreal, and in bi-lateral plans for discussing ... trans-border anti-sky-jacking procedures."[105] Security specialists with the Royal Canadian Mounted Police and the Department of Defence also profited from this crash course in counterterrorist techniques, which would prove very useful as Montreal prepared to host the 1976 Summer Olympics.

On 29 January 1976, J.H. Meek, Director of Scientific and Technical Intelligence of the Department of National Defence (DND), circulated a top-secret memorandum, "Plausibility of Use of Nuclear, Biological or Chemical Agents for Terrorist Coercion."[106] It began with the observation that the basic strategies of international terrorist organizations had changed dramatically in recent years, notably by rejecting the doctrine "that the massive terror inherent in the use of super-lethal agents would be counter-productive ... [and] would alienate the public sympathy." The report also identified three categories of terrorists who threatened Canadian national security: individual psychopaths operating independently, persons involved with "separatist / terrorist" organizations, and groups "operating on behalf of some external terrorist organization."[107] On the positive side, Meek claimed that there was no evidence that any terrorist group in Canada was "known to be actively considering utilizing or threatening to utilize nuclear, biological or chemical agents / devices for political or criminal purposes." But, as he quickly noted, there was no guarantee that this situation would continue. As a safeguard, Meek recommended that all Canadian security personnel be briefed about the scientific and operational dimensions of nuclear / radiological, chemical, and biological weapons. In his opinion this would not be a taxing experience since the operational principles were relatively straightforward. For nuclear devices, he noted, the greatest threat came from radiological, or "dirty bombs," which would be dispersed "in a mass of explosion detonated in free air … [so] that an aerosol … would contaminate the atmosphere to a large distance." For poison gas, he predicted that while terrorists had a range of choices, they would most likely use nerve gas agents (sarin / soman / VX). But the weapon system Meek feared most was germ warfare.

There were a number of reasons why bioterrorism was regarded as the greatest threat. First, BW agents could easily "penetrate buildings and their virulence is maintained with viability over long distances. " Second, this vulnerability was intensified by Canada's lack of an early warning and detection system, which meant that "a biological attack could be carried out by an individual, or group with the complete certainty that it could not be detected until it was too late to afford protection." Third was the possibility that terrorists would select agents of high lethality since they "would have a great psychological impact." And finally, it was assumed that a bioterrorist attack would target Canadian cities, where even primitive forms of dissemination, such as light bulbs filled with anthrax, could be placed in a busy subway station. Most serious of all, Meek conceded, would be more sophisticated BW attacks, with "dispersal over a targeted populous area from helicopter or light aircraft."[108]

Tellingly, the Meek report was immediately followed by a second study prepared by the DND Director of Scientific and Technical Department.[109] Again the focus was on possible bioterrorist attacks on Canadian cities, with three different scenarios being outlined. One of these was an attempt to contaminate the water supply of a major urban centre (Montreal was cited), using 100 litres of typhoid cultures, continuously injected over 24 hours. Such an attack, the report claimed, "would result in 35,000 casualties of which 200 would die." Another possible scenario involved terrorists attacking a local food supply by injecting "a biological agent into plastic milk containers, thereby causing wide-spread infection." But the most deadly form of bioterrorism, the experts argued, would be the dissemination of a biowarfare agent in the crowded downtown areas of Canada's largest cities, such as Place Ville Marie, Montreal, where "a ten second emission from some devices will produce infection in humans 800 m downwind."[110] The DND briefing paper ended with a brief discussion of whether international terrorists might use either biological or chemical weapons at the 1976 Summer Olympics. Significantly, the authors concluded that this was a very remote possibility "so long as other means – hostage, explosives etc. serve their purpose."[111]

In March 1976 the Department of National Defence established a special counterterrorist organization for the Montreal Olympics. The major features of this operation were outlined in the top-secret report "Security Measures for Gamescan," prepared by Major General H. McLachlan, Deputy Chief of the Defence Staff. It was a strange document, full of qualifications, confirming the fact that DND officials remained unconvinced that a CBR threat existed "in the context of Gamescan." Another complication was that under federal security guidelines for the games, the Ministry of the Solicitor General had been given responsibility for preparing a WMD threat assessment report. But it was not expected until April 1976. Given the urgency of the situation, DND officials decided to establish their own preliminary CBR incident response procedures "pending receipt of the definitive threat assessment."[112] Under this tentative plan six mobile counterterrorism response teams were established, with most operating in the Montreal region, including "a platoon (CBR), trained and equipped ... [with] its own assigned vehicles, pre-loaded with all necessary CBR equipment and protective gear." Operationally, if a biological, chemical, or radiological incident occurred, these specially trained teams would take immediate action "to isolate, contain, assess and eventually dispose of the hazard ... [and] to perform large scale decontamination or clean-up operations." The most elaborate set of procedures was associated with a biological weapons attack:

> The principal initial requirement ... would be the taking of samples for rapid laboratory analysis ... [which] would then indicate the nature and magnitude of the hazard and would dictate the measures necessary. These might include quarantining specific buildings or facilities; isolation of persons known or suspected to have been exposed; immunization programs, and chemical decontamination of specific buildings or facilities. The majority of such actions would of course be the responsibility of the provincial health authorities; nevertheless the CBR platoon may be called upon to assist ... [while] the FMC Command surgeon should be prepared to assist in coordinating any DND assistance in this area with the provincial medical authorities.[113]

The biodefence expertise of the CBR response team was considerably enhanced when Jack Currie, one of the leading BW experts at Defence Research Station Suffield, was seconded to the Gamescan operation, on the grounds that he had "for some time been engaged in research into some aspects of this potential area of terrorist activity."[114] One of Currie's major duties was to identify the specific agent used in the bioterrorist incident and assist medical first responders in treating anyone "exposed to the potentially deadly pathogen.[115] By the summer of 1976, fears about a possible CBW terrorist attack on Montreal were further intensified by an article in the German news magazine *Der Spiegel* claiming that a clandestine organization "was planning to produce nerve agent for use by underworld clients ... concerning the possible terrorist employment of chemical, bacteriological or radiological agents during the 1976 Olympics."[116]

In May 1976, the Ministry of the Solicitor General finally provided the necessary legal documentation for the so-called "CBR Incidence Response."[117] Under this system the Department of National Defence was now authorized to protect both Olympic facilities and the greater Montreal area. In operational terms, a special CBR platoon was assigned to the Longue Pointe suburb, ready to provide immediate assistance on a 24/7 basis, with additional help, if needed, coming from DRES experts, who were on full alert.[118] Behind the scenes, Ottawa security officials also benefited from the expertise of American BW experts through the auspices of the US Military Attache in Ottawa. Fortunately this assistance was not required, since there was no terrorist attack on the Montreal Olympics. But the task of preparing for such an exigency had been a draining experience for CBR specialists of the Canadian Armed Forces and counterterrorist specialists of the RCMP Security Service. Regretably, this learning experience proved of short duration. Indeed, it was not until the devastating 1995 sarin attack on the Tokyo subway by the radical religious sect Aum Shirinko, that Canada once again took CBR terrorism seriously.[119]

Infectious Diseases and Biological Weapons

The symbiotic relationship between natural disease outbreaks and biological warfare was evident throughout the Cold War. In 1964, for example, the Canadian BW Advisory Committee prepared a report on the military implications of a large-scale outbreak of typhoid fever in Aberdeen, Scotland, caused by contaminated food cans from South America.[120] What was most impressive about this epidemic was that it demonstrated "that the public health resources of a present-day city were taxed to the limit, despite the fact that only 0.2% of the population were infected, the disease was well-known, the therapy was known, the causalities were spread over 4 weeks, and the death rate was very low."

What would happen, they speculated, if Aberdeen had been the target of a BW attack which worked on the basis "of infecting 25-30% of the people of a city all at once"?[121] Nor was this an isolated opinion. At the November 1964 meeting of the Advisory Committee on BW Research, virologist C.E. van Rooyen of Dalhousie University pointed out why the Aberdeen typhoid incident could be regarded as "an effective demonstration of the potential hazard associated with BW even in its present imperfect state of development, since methods of attack, both overt and covert, could cause infection of much greater numbers of people with much less tractable diseases over a shorter period of time."[122]

Other similarities involving these two disease systems were evident during a series of postwar influenza pandemics. The first of these occurred in 1957–8, when a novel influenza virus (A/H2N2) killed 2 million people globally, including about 69,000 in the United States and 7,000 in Canada.[123] The Asian flu pandemic provides an interesting case study of the relationship between North American biodefence and health security policies. Above all, scientists still faced a number of problems understanding the causative influenza virus and developing the most effective strategies in containing the pandemic. On the other hand, forty years after the 1918 Spanish flu outbreak, which claimed over 50 million lives, medical researchers now understood the basic characteristics of the influenza virus and were able to use antibiotics against secondary infection by opportunistic bacteria.[124] Of crucial importance was knowledge that the influenza virus was divided into A, B, and C types, and that sub-types of A/influenza regularly changed due to the activity of two important surface glycoprotein receptors: the 16 haemagglutinin (HA 1-16) and the nine neuraminidase (NA 1-9).

According to WHO specialists, the 1957 appearance of the H2N2 virus (Asian flu) represented a major antigenic drift "in which 2–3 avian gene segments were re-assorted with the then-circulating human-adapted virus."[125] Like its infamous 1918 predecessor, the Asian flu pandemic went through three distinct stages: a spring-summer smattering of cases, a major surge of infections during the fall of 1957, and the deadly wave of January–April 1958, which accounted for almost 40 per cent of the total fatalities.[126] Another characteristic of the 1957 pandemic was the unprecedented level of international cooperation, largely through channels established by the World Health Organization.[127] In May 1957, for example, the WHO Committee on Influenza warned that because of antigenic shift existing vaccines would "not give protection against the new Far East strain."[128] As a result, the committee warned that there was "the possibility of a repetition of the 1918 situation ... [and] that the first sign of such a change would probably be the occurrence of deaths due to proven virus infection in unusual age groups, particularly young adults."[129] But while the logistical support and guidance of the WHO was important, the real task of coping with this health crisis occurred at the national level, where officials faced the enormous challenge of developing an effective H2N2 vaccine, assisting hospitals in dealing with the surge of infected patients, and coordinating laboratory and clinical resource facilities across the country.[130]

In Canada, most of this work was carried out by the special Advisory Committee on Influenza, which worked closely with the federal Department of National Health and Welfare ((DNHW) and the ten provincial ministries of health.[131] Of particular importance was the need to establish a national vaccine policy, concentrating on the production of monovalent influenza vaccine by the country's two leading laboratories: Connaught Medical Research Laboratories of Toronto and the Montreal-based Institute of Microbiology and Hygiene.[132] In addition, there was a debate whether vaccine should be imported from the United States if shortages should occur, with the Quebec provincial government being one of the most vocal advocates of this policy.[133] In contrast, defenders of a Canada-first policy emphasized the advantages of having a guaranteed and safe vaccine supply, particularly given reports in the US media that the American H2N2 influenza vaccine had caused many serious health complications.[134]

Throughout the 1957–8 public health crisis a number of bilateral influenza prevention strategies emerged.[135] One of these was the high level of cooperation between Canada's Department of Health and Welfare and the US Public Health Service, which was evident in the development of new vaccines and innovative medical research, and in the exchange of epidemiological information about the incidence of outbreaks in selected North American cities.[136] The civil defence dimension of the influenza outbreak was another area of shared concern. In August 1957, for example, Canadian health officials announced a new federal initiative "to establish a 'pilot' system of laboratory reporting which could be expanded in event of national or local emergency such as epidemics, natural disaster or war ... so far as defence problems are concerned, much more stress is being laid on the probability of biological warfare attack than formerly."[137] In a subsequent report, Deputy Minister G.D.W. Cameron was even more direct about the national security value of monitoring the Asian flu pandemic: "We have been interested in these effects, as from the standpoint of defence against B.W., since the dislocation caused by a 'natural' epidemic may give some hint as to what may be expected from a 'man-made' epidemic."[138] In Washington, a similar approach was adopted by civil defence officials, who stressed that "should the United States be hit by an Asian flu epidemic of the magnitude predicted by public health authorities, valuable data pertaining to the vulnerability of this nation to large numbers of casualties can be collected ... [which] could be used to ascertain the effects of causalities of any sort on the war potential of the nation, through correlation of the medical records and current data on 'target' effects."[139]

The 1968 influenza (A/H3N2) pandemic also demonstrated important linkages between infectious disease outbreaks and biodefence strategies. First, officials needed to anticipate the specific form of the disease threat – whether it was a mutating influenza virus or a bacterial or viral BW agent.[140] A second challenge for both groups of scientists was early detection and diagnosis, particularly given the similarities in symptoms between many different types of infectious diseases, whether naturally occurring or deliberately spread.[141] Third, while the Hong Kong flu (A/ H3N2) was not as severe as the 1957 outbreak, it still represented a serious public health threat, overwhelming the resources of many Canadian and American hospitals.[142] And finally, public health and military officials were concerned about the difficulty of securing regular supplies of effective vaccines. In 1965, for example, the US Department of Defence decided to address its recurring problems of securing regular supplies of vaccines for a number of bacterial and viral BW agents when it signed a lucrative long-term contract with the Richardson-Merill (RMI) company of Swiftwater, Pennsylvania on the grounds that there was "no commercial market for biological warfare vaccines in the United States ... [and] vaccine producers lack the facilities needed to produce them."[143] Significantly, the Swiftwater facility was also one of the leading producers of influenza vaccine in North America, with close connections with the Toronto-based Connaught Medical Research Laboratories.[144]

Fears about a large-scale public health disaster, either from natural causes or bioterrorism, also emerged during the 1978 swine flu pandemic. Of particular importance was the 24 March 1976 decision of US President Gerald Ford to request congressional support for the immunization of 230 million Americans, based on the fears that the country was facing a serious public health crisis. According to medical experts at the Centers for Disease Control and Prevention (CDC), isolates obtained from infected patients revealed that this A/H1N1 virus were "antigenically related to the influenza virus which has been implicated in the cause of the 1918–19 pandemic which killed ... more than 400 out of every 100,000 Americans."[145] The Trudeau government also heeded this warning. On 30 March, Marc Lalonde, Minister of Health and Welfare, informed the House of Commons that his department would be implementing a program of selective immunization for over 12 million Canadians and that contracts would be signed for the Connaught Medical Research Laboratory (CMRL), and the Frappier Institute in Montreal for the delivery of the necessary quantities of H1N1 vaccine.[146] As it turned out, however, neither country would require these medical counter-measures. By August 1976, the Ford administration placed its mass immunization program on hold, both because the anticipated number of swine flu cases had not materialized and because some of those inoculated showed signs of Guillain-Barré syndrome and accompanying paralysis.[147] While Canada experienced only a few of these complications, it was sufficient to convince Ottawa and the provinces to abandon the large-scale vaccination effort.

The swine flu crisis left a powerful legacy. It resulted in serious political ramifications, particularly in the United States, where the reputations of President Ford and the Centers for Disease Control were discredited because of the false alarm. A second factor was the surge of liability suits launched against major North American pharmaceutical companies such as Richardson-Merill (RMI) and the Connaught Medical Research Laboratory because of adverse reactions to the H1N1 vaccine.[148] And finally, a spate of conspiracy theories surfaced, including allegations "that the American government was carrying on biological warfare experiments at Fort Dix and unwittingly infected the 12 soldiers [and] the hue and cry about swine flu was a cover-up."[149]

Another dual-threat pathogen was *Variola major*, the causative agent of smallpox. Not only was this one of the world's most feared natural diseases, its potential as a biological weapon had been assessed by American and Soviet scientists during the 1960s because of its high infectivity, its stability, and its relatively high fatality level (30 per cent). But despite variola's effective BW qualities, after 1965 debates arose on whether it was appropriate to weaponize a pathogen that the World Health Organization had targeted for global eradiction. The WHO campaign was based on two major factors: a consensus among its members that smallpox eradication was a priority and the willingness of the US government to provide initial funding. The Johnson administration, however, was insistent that the director of the project should be D.A. Henderson, a prominent public health physician with the CDC. It was an excellent choice. During the next ten years, Henderson skilfully directed the global program from his WHO headquarters in Geneva, drawing on the talents of the world's foremost medical experts, including his friend Robert Wilson of Connaught laboratories.

While the eradication of smallpox was an enormous struggle, it was aided by a number of important factors. First, important technological innovations facilitated large-scale vaccination, such as the availability of freeze-dried vaccines, which were administered either by the bifurcated needle or by jet injector.[150] Second, Henderson and his team developed the so-called ring system of vaccination to contain a smallpox outbreak; once this was achieved, mass vaccination was carried out in the endemic regions in order to achieve an 80 per cent immunity status.[151] Third, under WHO supervision the quality of smallpox vaccines in developing countries was greatly improved, with Connaught scientists providing valuable assistance in Latin America.[152] Above all was the remarkable level of international cooperation, despite the ongoing Cold War; indeed, the Soviet Union eventually donated 75 million doses of dried vaccine.[153] In 1976 WHO officials were confident that the world was now free of smallpox.[154] But despite this achievement, major questions still had to be resolved.[155] Above all was the question of what should be done with the stocks of *Variola major* in laboratories across the world. This important issue was addressed by the WHO Global Commission for the Certification of Smallpox Eradication, which recommended that the number of laboratories holding the virus should be restricted to "WHO collaborating centres with maximum containment facilities."[156] In 1979 an elaborate list of fourteen biosafety guidelines were established for all laboratory experiments involving variola, with the proviso that research should only take place in high-containment facilities with "access to the laboratory ... under strict control."[157] These measures, however, were only regarded as having temporary relevance since it was anticipated that all remained stocks of the variola virus would soon be destroyed. Unfortunately, this has not happened since after the terrorist attacks of 2001 the United States has determined that the smallpox samples at the CDC should be available for its biodefence scientists.

Conclusion

Many of the debates about bioweapons and biodefence during the 1970s have resurfaced during the first decade of the twenty-first century. Of particular importance are the major questions raised by Dr Ivan Bennett, chairman of President Nixon's Science Advisory Committee: How does the impact of natural occurring disease outbreaks compare with bioattacks? Is it possible to differentiate between research on BW agents for defensive and offensive purposes? Can the specialized knowledge of academic scientists be mobilized for biodefence purposes? What are the relative merits of secret and open research? Fortunately, the responses of the select ASM scientists provided insights into these challenging questions, particularly about the dangers of new viral BW agents and genetically altered pathogens. Equally important, on the policy side, is Riley Housewright's view that the United States should not expose its biodefence vulnerabilities because of some naive attachment to open research. Canadian and British defence scientists encountered similar challenges.

In terms of biodefence preparation the two 1976 National Defence discussion papers about possible bioterrorist attacks provided a number of insights that have relevance today. Above all was an awareness of the vulnerability of urban targets, the range of BW agents and delivery systems, and the degree to which "modern" terrorists wanted to maximize casualties and create widespread psychological disruption. In operational terms, the deployment of the special DND units and their DRES advisers during the Montreal Olympics was an impressive beginning for Canada's CBW counterterrorist system. But, of course, it was an exercise not a real event. Recently, the connection between global health security at the national, regional, and global levels has become a subject of increased importance for medical scientists, public health specialists, academic scholars, and government administrators. Indeed, the challenge of coping with the dual threats of natural epidemics and man-made disease outbreaks has increasingly brought these different fields together, with scientists adopting innovative forms of analysis which encompass "threats from both biological weapons and naturally occurring infectious diseases … integrating two policy realities previously separate from one another – security and public health."[158]

This was not a new phenomenon. Using the model of the 1957 and 1968 influenza pandemics, biodefence planners developed many useful comparisons between the public health response in dealing with natural diseases outbreaks and the defensive strategies associated with a biological weapons attack. First, in each situation medical first responders are forced to deal with an unknown pathogen of unpredictable consequences. In the case of the influenza virus, for example, there was always the possibility that through mutation it could become deadly, as was the case during the pandemics of 1918–19 and 1957–8. It was not clear that the specially designed vaccine for this novel influenza virus would be effective. A second similarity is the enormous burden that pandemics and bioattacks placed on the public health system of even advanced countries, with hospitals and their workers being severely tested to provide essential services over a sustained period of time. A final comparison was the challenge of retaining public confidence during a pandemic or bioattack in order to avoid panic and social disorder.

On the other hand, several important factors distinguish biological weapons from natural disease outbreaks. One of these is the reality that neither Canada nor its allies has ever experienced an external biological weapons attack – all of the BW war games were simulated situations rather than real events. Closely related is the grim fact that even the 1918 Spanish flu, the worst pandemic of the twentieth century, had a fatality rate under 3 per cent, while most lethal bioweapons are based on a 25–35 per cent killing level.[159] A second consideration is that in the past, most biodefence strategies were controlled by military experts, who used battlefield assessments of the effectiveness of biological weapons as their framework of analysis. This was certainly the case during the 1976 Montreal Summer Olympics when the CBR defensive strategies were constructed by scientists from Defence Research Establishment Suffield, not public health experts from the Department of National Health and Welfare. This situation would, however, change dramatically after the 2001 anthrax letter bomb attacks in the United States, when the civilian side of the biodefence equation emerged as the dominant partner.

6 Preventing Germ Warfare in the Age of the Biotechnology Revolution

Each State Party to this Convention undertakes never in any circumstances to develop, produce, stockpile or otherwise acquire or retain:

(1) Microbial or other biological agents, or toxins whatever their origin or method of production, of types and in quantities that have no justification for prophylactic, protective or other peaceful purposes; (2) Weapons, equipment or means of delivery designed to use such agents or toxins for hostile purposes or in armed conflict.

(Article I – The Biological and Toxin Weapons Convention, 26 March 1975)[1]

The ratification of the Biological and Toxin Weapons Convention in March 1975 was a remarkable event since it outlawed an entire weapon system. Most countries of the world pledged that they would renounce the research, development, stockpiling, or use of germ weapons. Unfortunately, this did not occur. Indeed, even before the ink was dry on the original disarmament agreement, the Soviet Union intensified its efforts to develop the world's most formidable biological weapons arsenal. While many questions about the offensive dimensions of the Soviet program remain unanswered, available evidence indicates that many of the world's most frightening pathogens were weaponized for possible military use, either in a major conflict with the West or in covert operations.[2]

Throughout the late 1970s, intelligence agencies in the United States suspected that Russian scientists were carrying out clandestine bioweapons research and development. These fears were confirmed during the spring of 1979 when a deadly accident occurred at the secret anthrax weapons facility at Sverdlovsk, killing at least sixty-six people.[3] The following year saw reports of another Soviet BW violation, this time in south-east Asia involving so-called "yellow rain" mycotoxins. Unlike Sverdlovsk, where international inspection was not possible, the United Nations was able to commission on-site investigations in Laos and Kampuchea to determine the validity of American allegations. In both these situations, Soviet officials vigorously denied the US charges, claiming that the Pentagon was carrying out its own nefarious biowarfare activities and alleging that it was responsible for the global HIV/AIDS epidemic.[4]

Unfortunately, the Kremlin's violation of the Biological Weapons Convention (BWC) encouraged other countries to develop biowarfare capabilities. Indeed, by the late 1980s Western intelligence estimated that at least six countries were pursuing the BW option, with several others waiting in the wings.[5] International arms control organizations such as the BWC and the Australia Group could not prevent the proliferation of bioweapons if these "rogue" countries had the necessary scientific talent, industrial infrastructure and financial resources.[6] This pattern of WMD proliferation was clearly evident in the experiences of Iraq, South Africa, and North Korea, where biological weapons were developed simultaneously as nuclear devices.[7] It was also apparent that while all of these countries were members of the BWC, this did not deter their activities for either moral or diplomatic reasons.

Canada was affected by these international developments in a variety of ways. On the one hand, DFAIT officials were determined to uphold the integrity of the Biological Weapons Convention, working hard at the periodic review conferences (normally every five years) to improve transparency and accountability. At the same time, Canadian defence scientists were actively involved in developing a comprehensive Canadian biodefence strategy for the Canadian Armed Forces, consistent with the strategic priorities of its NATO and Quadripartite allies. Even more challenging was their involvement with top-secret CANUKUS biodefence research projects, designed to deal with three major threats: the CBW potential of the Communist bloc, the activities of rogue states, and the emergence of international terrorist organizations. Canada also participated in the UN-sanctioned intervention during the First Gulf War (1990–1) by supporting the work of the United Nations Special Commission (UNSCOM) in locating and destroying Iraq's BW weapons.[8]

The last two decades of the twentieth century witnessed major changes in the fields of molecular biology and genetic engineering, as exemplified by the Human Genome Project. While many countries were involved with the challenging task of mapping and sequencing the human genome (a complete set of chromosomes and genes), many of the most successful projects were located in the United States, where life scientists from academic institutions, research foundations, biotechnology / pharmaceutical companies, and government agencies developed a number of competing and complementary strategies. These efforts were greatly facilitated by millions of dollars in funding, the emergence of new computer technologies (bioinformatics), and medical techniques such as polymerase chain reaction (amplifying short structures of DNA). And success was soon forthcoming. In June 2000, at a special ceremony at the White House, two teams of US researchers announced that the first survey of the human genome had been completed, with compelling scientific evidence that it consisted of approximately three billion genes. At the same time, they realized that "the task of identifying all the genes, and determining their function, and the relationship to the traits and diseases of interest had really just begun."[9]

Upholding the Biological Warfare Convention

Alliance Biodefence Programs after the BWC

By 1975 the United States biowarfare program was a shell of its former self. Yet despite the loss of resources and prestige, American biodefence scientists continued to warn their military and political masters of a possible Soviet BW attack. Of particular importance were a series of reports prepared by the Materiel Test Directorate, Dugway Proving Grounds, designed to convince the Pentagon of the severity of these threats. For example, in the November 1976 study "An Evaluation of Biological Treaties and Some Relationship to Defense Planning," it was pointed out that the Kremlin regarded Nixon's 1969 decision to abandon offensive biological warfare "as weakness ... [and] a 'winning' event for the USSR." In order to remedy this unfortunate situation, the report recommended that Washington insist on a comprehensive on-site verification system, even if it meant the dismantlement of the BWC, since this "would afford the US an opportunity to re-establish a deterrent-level BW capability."[10]

During the early 1980s these Dugway studies had become even more pessimistic. According to one report, if the Soviets used a BW warhead on their intercontinental ballistic missiles, such an attack on urban America would cause enormous casualties "both in numbers and percentages ...

[of] the population."[11] A related document examined the possibility that the Russians were exploring the development of deadly new biological agents through recombinant DNA research techniques for use "against US military or civilian populations." In analysing this scenario, Dugway BW specialists developed a developmental model consisting of five different criteria: survival in the environment, mechanism for penetrating the body, multiplication within the host, systemic spread through the host, and the ability to overcome host defence mechanisms. At this stage, however, the report dismissed fears that the Soviets were using DNA recombinant techniques to produce new biological weapons, since existing biotechnology techniques did not "offer a reliable approach for the production of bacterial agents that are either novel or superior to those already available."[12]

That American defence scientists should be concerned about the potential benefits and abuses of recombinant DNA research is not surprising. Indeed, throughout the seventies this subject was the focus of intense debate at the national and international levels. Of particular importance was the February 1975 meeting at Asilomar, California where 150 of the world's leading experts on genetic engineering addressed a number of crucial questions about the future of recombinant DNA (rDNA) research. First, did certain types of risky experiments outweigh their possible medical benefits, specifically with "new kinds of hybrid plasmids or viruses with biological activity of unpredictable nature"?[13] Second, did concerns over high-risk experiments justify a national moratorium on this kind of scientific activity? And finally, if such experiments were permitted, what laboratory containment level was necessary to safeguard laboratory personnel and local communities?

But devising effective guidelines was a daunting task, given the novelty and diversity of recombinant DNA research. It was also difficult to obtain a consensus because of the conceptual and methodological differences among the five major disciplines involved in this field – molecular biology, cell biology, genetics, biochemistry, virology. Yet in the end, the Asilomar process worked. After sustained debate, a consensus emerged that the rDNA research moratorium should be lifted, with the understanding that "some forms of experiments should proceed; some not."[14] A more difficult task, however, was to establish an oversight system that was acceptable to the different stakeholders: academic research scientists, private sector biotechnology companies, government health officials, and the general public. By July 1976 this goal was largely achieved with the publication of the *National Institutes of Health Guidelines for Recombinant DNA Research.* Although still a work in progress, the document warned about certain kinds of dangerous experiments such as the cloning of recombinant DNA from certain high-risk pathogens and the biosynthesis "of toxins of high toxicity." The NIH report also advised against the release of DNA molecules until after controlled tests were completed, using only test pathogens that could not survive outside the laboratory and eliminating any drug-resistant organism.[15] Despite a few unfortunate incidents, the NIH strategies enjoyed widespread support in the United States and elsewhere. According to one media report, "Not everyone is happy with the guidelines. Some scientists feel it is idiocy to attempt the curbing of scientific pursuits. Others, believed that all experiments that lead to genetic manipulation should be prohibited. The guidelines steer a narrow, but apparently safe ground between these extremes."[16]

Another aspect of the NIH recombinant DNA program was the establishment of maximum containment laboratories (BSL-4).[17] Ironically, some of the laboratories selected for these high-risk experiments had formerly been part of the offensive biological warfare system at Fort Detrick, which had "a proven performance capability for handling hazardous microorganisms."[18] The campaign to secure these facilities began in 1976 when officials of the National Institutes of Health submitted a request for the use of fifteen laboratories and four pilot plants owned and operated by the Frederick Cancer Research Centre on the grounds that there was mounting pressure "to support extramural research involving recombinant DNA molecules which require P4 level physical containment." Another key argument was that involvement with this BSL-4 experience would help train American life scientists, "thereby improving laboratory biosafety standards throughout the United States."[19] Although there was vigorous opposition from some environmental groups, by March 1978 the new Detrick high-containment laboratory was ready for business.[20] An article in the *Washington Post* vividly described this convergence between Detrick's BW past, and its recombinant DNA future:

> Just as the first man-made atomic chain reaction in an old squash court beneath the University of Chicago's stadium grandstand in 1942 led inexorably to Los Alamos, so the discovery in 1953 of the DNA molecule – life's blueprint – has led to Room 215 in building 550 at Fort Detrick here, former centre of the Army's chemical and biological warfare program … [where] Army researchers were fooling around with deadly diseases like

> anthrax in the same room, assessing the "safety" of further experiments with them as agents of biological warfare. The anthrax is gone now and within two weeks it will be replaced in Room 215 with experiments in genetic engineering.[21]

These developments had a decided impact on Canadian life scientists because their own research activities often included collaboration with their counterparts in the United States, as well as large-scale involvement in the American Society of Microbiologists. In addition, there was considerable pressure on Canada, most notably from the federal Department of Health and Welfare, to develop its own BSL-4 system. But until the late 1990s, the country's only high-containment facility (BSL-3) was at Defence Research Establishment Suffield, where biodefence scientists monitored debates about how developments in the rDNA field could affect the future role of biological weapons. Indeed, by the early 1980s US defence officials now viewed germ warfare as "a militarily significant weapon," largely because genetically altered BW agents could circumvent enemy BW countermeasures.[22] On the other hand, the Pentagon's renewed interest in BW research was not well received by many civilian US scientists. In June 1982, for example, the National Institute of Health's recombinant DNA Advisory Committee denounced the US military's increasing interest "in expanding its biological warfare research into gene-splicing techniques," particularly since the BWC prohibited "any use of molecular cloning for weapons development."[23]

Biological Warfare Incidents and the Western Alliance

Most of the Soviet offensive program was operated by the All-Union Science-Production Association Biopreparat, established in April 1974 as a civilian biotechnology front organization to conceal the country's accelerated BW operation. By 1990 Biopreparat was responsible for over twenty research and production facilities, employing over 45,000 scientists and technicians; the Ministry of Defence operated five of its own secret bioweapons centres. In strategic terms, the bioweapons system was based on a number of factors. The first priority was to have bioweapons production plants on a state of readiness, or standby capacity – rather than stockpiling germ warfare munitions. Second, a range of delivery systems were developed: multiple warheads for intercontinental and short-range missiles; spray tanks and aerial bombs; and a number of covert weapons for attacking US cities. Third, Soviet scientists developed a number of unique bioagents capable of evading conventional vaccines and antibiotics, notably special forms of *Bacillus anthracis*, and *Yestina pestis*. Even more fearsome were selected "BW cocktails" such as the one that combined "smallpox with a hemorrhagic disease virus to make it more lethal and an encephalitis virus to make it more transmissible from person to person."[24] And finally, all this research and weaponization took place behind a pervasive system of secrecy, based on rigid compartmentalization and strict "right to know" principles. To reinforce their sense of importance, Biopreparat scientists were repeatedly told that their work was essential for the safety and well-being of the USSR, since the United States was well ahead in the BW arms race. As former weaponeer Ken Alibek explained, "We didn't believe a word of Nixon's announcement. Even though the massive U.S. biological munitions stockpile was ordered to be destroyed, and some twenty-two hundred researchers and technicians lost their jobs, we thought the Americans were only wrapping a thicker coat around their activities."[25]

On 18 March 1980, the American embassy in Ottawa forwarded an urgent message: "An outbreak of disease in Sverdlovsk, USSR ... raises a serious concern as to whether country possesses biological warfare agents in quantities exceeding those justified for prophylactic, protective or other peaceful purposes, which would be a violation of the Biological Weapons Convention." In short order the Canadian government also received copies of a special report on Soviet BW activities prepared by the US Defence Intelligence Agency which castigated Russia's violation of BWC principles: "In recent years," the report stated, "we have seen significant acquisition of technology and equipment, building of large-scale biological fermentation facilities, and ... storage / bunker areas with identical configurations at the different sites ... [which] appear to be constructed for storage of explosives, or explosive components, suggesting weapons activity ... intelligence information makes a strong circumstantial case that the Soviets are pursuing the development and probable production of biological weapons." In justifying these charges US Defence Intelligence officials had an impressive array of sources, including the testimony of a Soviet surgeon who provided vivid details about the disease symptoms demonstrated by exposed workers from the Sverdlovsk weapons facility, including severe pulmonary edema and toxaemia. According to American medical specialists, the rapid death of these men from pulmonary anthrax not only provided conclusive proof that the accident had occurred at a sophisticated weapons facility, but it also refuted Soviet claims that this was "vaccine research or

BW defence research."[26] Yet despite this damning evidence, the Kremlin continued the charade that the Sverdlovsk outbreak was caused by anthrax-tainted beef.[27]

This image of Soviet CBW duplicity was reinforced by a series of reports in 1980 that the USSR had supplied North Vietnam with experimental bioweapons. Dubbed "yellow rain," because of its peculiar physical characteristics, American CBW experts claimed that this material was not "conventional chemical warfare agents but rather biological toxin, mycotoxins [Trichothecene Mycotoxins] possibly representing a third generation of Soviet biological / toxin weapons." Among the sources for these US charges were interviews with victims of the attacks, notably Hmong tribesmen in northern Laos; the analysis of "yellow rain" samples, carried out by qualified US laboratories; and intelligence reports that "Soviet research on mycotoxins has taken place in several restricted biological institutes."[28] On 13 September 1981 the situation escalated into a major international crisis when US Secretary of State Alexander Haig, during a highly publicized speech in Berlin, accused the Soviet Union of violating the BWC. In response, Russian officials denounced the United States for its war mongering, as symbolized by its development of the "neutron bomb and ... binary [CW] weapons."[29]

Because of concerns that this superpower confrontation was a threat to world peace, UN Secretary General Kurt Waldheim requested assistance from member states to determine whether the yellow rain allegations had any scientific validity. Ottawa was quick to respond, in part because of its strong stance against CBW proliferation and in part because Washington requested its assistance. In short order two teams of Canadian scientists were mobilized for on-site investigations: a DFAIT mission led by Dr Bruno Schiefer of the University of Saskatchewan and a contingent from the DRES and the DND Directorate of Preventative Medicine.[30] After extensive fieldwork, the Schiefer team submitted its preliminary report, which concluded that biological warfare agents had been used in south-east Asia since "most of the features described with 'yellow rain' attacks are consistent with trichothecene mycotoxicosis ... chemicals produced by living organisms."[31] The military group agreed, pointing out that in certain areas of Cambodia, Laos, and Thailand mysterious substances "possibly mycotoxins ... are being used against unprotected troops and civilians with inadequate medical care ... to instil fear or ... being used as an experimental (CW/BW) laboratory."[32] Reactions to the two reports varied. For some observers, these were solid scientific studies that demonstrated Canada's strong commitment to uphold the

Biological and Toxin Weapons Convention. But for others, the Canadian submissions were deeply flawed because of their faulty scientific methodology and Ottawa's slavish support of the Reagan administration's hawkish position on bioweapons.[33]

These suspicions were carefully cultivated by the Soviet Union in its various propaganda initiatives throughout the 1980s. Of special importance were the attempts to convince world opinion that the HIV/AIDS virus had been artificially created by Detrick scientists, proving that the United States would stop at nothing to advance its imperialistic goals. The first stage of this smear campaign began in July 1983 when an anonymous letter was published in the Indian newspaper *Patriot,* claiming that the appearance of AIDS was the result of the Pentagon's attempts to develop new and more deadly biological weapons. Two years later, these allegations became even more strident when the Soviet weekly *Literaturnaya Gazeta,* official organ of the USSR's Writer's Union, published an article entitled "Panic in the West or What Is Hidden behind the Sensation about AIDS," which charged that the prestigious US Centers for Disease Control had been involved with the HIV/AIDS biological weapon.[34] Equally contentious was the June 1987 story that appeared in the Soviet newspaper *Sovyetskaya Mololdezh* entitled "AIDS: Its Nature and Origin," claiming that the HIV virus had been "engineered by the United States through the artificial synthesis of VISNA virus (a retrovirus causing a complex disease syndrome in sheep) and HTLV-I (a human retrovirus causing a rare leukemia)."[35]

From Washington's perspective, the Kremlin's propaganda offensive was even worse than the 1952 germ warfare controversy. Not only did it identify the US biodefence program with AIDS/HIV, one of the world's most destructive pandemics, but it also linked CDC researchers with these activities. Given the high stakes involved, US diplomats and military officials decided to launch a multi-faceted counterattack. First, they registered an official complaint with the editors of selected Russian newspapers and with high-ranking members of the Gorbachev administration. In addition, American public health officials warned their Soviet counterparts that future collaboration on AIDS research "would be impossible as long as the disinformation campaign continues."[36] A second approach was to convince prominent Soviet scientists such as Academician Viktor Zhdanov, the country's most outstanding expert on HIV/AIDS, to challenge the scientific validity of the allegations. This goal was achieved in June 1986, at the Second International Conference on AIDS, when Zhdanov categorically rejected claims that American scientists could have artificially synthesized the AIDS virus.[37] A third strategy was to convince members of the international media that the Soviet explanations about the origins of AIDS were nothing more than vicious propaganda.

Another factor in reducing tension between the two countries was the positive working relationship between President Reagan and General Secretary Mikhail Gorbachev.[38] This was particularly the case after their October 1986 summit meeting in Reykjavík, Iceland, where plans for the important Intermediate-Range Nuclear Forces Treaty were successfully negotiated. While the issue of bioweapons was not discussed, this situation would quickly change as vital details about the Soviet offensive bioweapons program became available. The most valuable source of information came from high-placed defectors such as Vladimir Pasechnik, former director of the Institute for Ultra-Pure Biological Preparations in Leningrad, and a key member of the Biopreparat scientific elite. In his October 1989 debriefing by British MI6 and American CIA officials, Pasechnik provided essential data about the types of biological agents being researched at his institute and how recombinant DNA techniques had been used in creating fourth-generation BW munitions.[39] Although Soviet hardliners such as General Kallin, head of Biopreparat, urged Gorbachev to maintain the veil of secrecy surrounding the country's bioweapons system, Moscow's new leadership was more interested in accommodation, not confrontation, with the West.

In January 1991, a team of thirteen British and American biodefence experts were allowed to visit four of the major Biopreparat research institutes.[40] It was an illuminating experience. Not only did it confirm general suspicions about the Kremlin's biological warfare program, but it also provided evidence that Soviet scientists had weaponized smallpox, despite their international commitment to eradicate this terrible disease. But despite this betrayal of trust, Western experts were anxious to begin meaningful BW arms-control negotiations with their Soviet counterparts. On 14 September 1992 the three countries issued a joint statement that "reiterated their commitment to the BWC and agreed to host reciprocal visits at selected facilities in order to enhance confidence in treaty compliance."[41] Unfortunately this trilateral system of mutual inspections was short-lived. Within a year, the Russian military hierarchy demonstrated that they would not provide any information about their top-secret bioweapons laboratories or discuss its previous BW programs.

Although Canada was not directly involved with these Tripartite negotiations, DRES scientists were kept informed about the revelations concerning the Soviet/Russian BW program through their involvement with the MOU / CANUKUS system. A February 1990 report, for instance, discussed the activities of the Scientific Research Institute of Military Medicine in Leningrad, which, it was noted, "has long probed the possibilities of mass immunisation against agents of biological warfare ... [including] the use of aerosols of live, freeze-dried vaccines." In addition, Canadian defence scientists, after examining many Soviet-era scientific publications, were impressed with the number of studies involving viral agents such as Ebola, Marburg, and, above all variola, particularly since smallpox immunization in North America had not been authorized since 1970. But of greatest concern was that the Soviets had developed a new generation of biological weapons. These fears were reinforced in December 1990 when Canadian BW experts were invited to top-secret meetings in Washington, DC, where they were briefed about how "progress in genetic engineering and in other fields of bioscience has created new incentives for the development and use of biological weapons ... for tactical purposes behind the front lines, as strategic weapons, or for sabotage."[42]

Developing Canada's Future CBW Defensive Policies

During the late 1980s three official pronouncements helped establish the priorities of Canada's defensive programs in biological and chemical weapons research, development, and training. One of these was the 1985 National Defence Headquarters Policy Directive P3/85 on Nuclear, Biological, and Chemical Defence, which predicted that Canadian Forces deployed outside Canada would encounter weapons of mass destruction for the foreseeable future, thereby necessitating "tactical exercises ... and biological and chemical safety systems."[43] The December 1988 Barton Report reinforced this message when it called upon Ottawa to ensure that all Canadian troops receive adequate training and equipment "to protect themselves against chemical and biological warfare ... consistent with the international obligations undertaken by the government."[44] The report expressed special concern about the BW threat because of possibilities that developments in biotechnology could create "new agents ... for which existing protective equipment and detection capacity would be ineffective."[45]

Canada's CBW strategies were also influenced by the policies of its major allies. Of particular importance was the 1986 Quadripartite Working Group report, *The ABCA Armies Operational Concept, 1986–95*. Several key priorities were outlined in this document. One of these stressed that innovations in the field of NBC weapons included not only battlefield devices but also "the sabotage use of biological agents, in aerosol form or by contamination of food and drink." A related topic involved the different BW delivery systems that Warsaw Pact countries might deploy: "1. high speed drones; 2. guided missiles and free light rockets; 3. bomblets filled with agents; 4. bombs, artillery shells and mines; 5. sprays from manned aircraft." The report concluded with the observation that ABCA countries were at a serious disadvantage in deterring the Warsaw Pact's bioweapons threat since their national policies "preclude any use of biological agents and limit the use of chemical weapons for retaliation only."[46] Another source of useful information came from the special studies that were part of the BWC review process. Of particular value was the September 1986 American submission, which claimed that major changes in biotechnology had undermined existing verification measures since countries seeking to evade the strictures of the Convention could conceal BW manufacturing plants, particularly for biological chemicals such as toxins and peptides whereby "large quantities ... can be produced in small facilities ... very quickly."[47]

But the most important aspect of Canada's biodefence planning was associated with the 1980 Memorandum of Understanding on "Research, Development and Acquisition of Chemical and Biological Defense Materiel."[48] Under this agreement, the broader operational functions of the CANUKUS system were carried out by the trilateral steering committee, with specific projects being allocated to different working groups and task forces. Not surprisingly, the United States provided the most advanced information and devices, meaning it also determined the level of security classification. While the author had only limited access to the records of this important organization, available records of CANUKUS meetings provide many valuable insights into the priorities and procedures of the MOU system during the late 1980s. For example, the 5 June 1986 meeting of the Steering Group in Ottawa reached a consensus that the differences between biological and chemical weapons "were expected to disappear as biotechnology is applied to the production of toxic agents ... [and] it is unlikely that the nature of all CB threat agents available ... will be known prior to actual deployment."[49] Another important subject for discussion was Canada's evolving counterterrorist system, including the breakdown of federal and provincial responsibilities, and the functions of the newly created role of the Special Threat Assessment Group (STAG).[50] The meeting concluded with an analysis of the exchange among the three participants under MOU guidelines, whereby "each country will give classified material the same degree of protection as it would to its own material."[51] Canadian officials, however, expressed concern that the quality of intelligence they received from their allies was often quite disappointing, particularly when it involved an assessment of "Soviet directions in Biological and Chemical Warfare."[52]

It was a justified complaint, since the fundamental goal of the CANUKUS system was to develop new technology for defensive CBW purposes.[53] This priority was highlighted in the 1986 report of the Tripartite task force (IFT-4) "to coordinate R and D work on the detection, identification, monitoring and reconnaissance for new chemical and biological (CB) agents ... for a variety of operation scenarios."[54] While Canadian defence scientists endorsed these guidelines, they also acknowledged the specific challenges of developing technologically advanced biological and chemical warfare detection systems:

> There is likely to be a requirement for a device to detect the presence of toxic aerosols at a given location, to provide an alarm signifying that an attack is taking place ... The device should be compact and respond near instantaneously to all toxic aerosols with a low false alarm rate. It should respond to challenge levels as low as 1 agent particle / litre of air, although it is unlikely to be achievable in a first generation aerosol detector. Furthermore it is unlikely in the case of CW aerosols that this level of sensitivity is required, though it would be for BW. Because of the likelihood of materials of high toxicity being used ... it is vital that the response time of the device should be a few seconds. The particle diameters of interest are from 0.1 to 15 cm, which includes the respirable range. There is currently no system which meets these requirements.[55]

Another major issue was whether there should be a uniform CANUKUS counterterrorist strategy.[56] In January 1986, for example, the Steering Committee noted that "the requirement or desirability for direct involvement of the MOU in this area [counterterrorism] has not been established ... [since] agencies already exist in each country that address the problem." This emphasis on national procedures was Ottawa's first preference. As DND's Chief of Research and Development [CRAD] pointed out, his department had already negotiated an agreement with "the Solicitor General's STAG committee which responds in this area."[57] On the other hand, MOU experts from the US and the UK disagreed with the narrow national approach and recommended that a study be carried out to determine whether trilateral collaboration should be institutionalized by establishing "a technical evaluation of the potential hazard and to indicate the possible technical responses to such a hazard."[58] As a result, in November 1986 arrangements were made to hold a joint Canadian / American NBC anti-terrorist exercise in Ottawa for senior-level security officials.[59] In order to facilitate discussion, a detailed breakdown was provided of the Canadian agencies involved with counterterrorism planning:

> Health and Welfare Canada (HWC)
> Responsible for threats to Public Health; Enforces Legislation on Importation and Use; Liaison with other countries for NC threat assessment
>
> Solicitor General of Canada (SGC)
> Coordinates Federal response to acts of Terrorism; Uses RCMP for investigations; Participates in development and assessment of NBC threat exercises; Liaison with other countries for NBC threat, Intelligence information
>
> Special Threat Assessment Group (STAG)
> To advise and assist government and law enforcement officers in responding to the threat or use of NBC agents by terrorist, criminal, or psychopathic elements. To assess credibility and feasibility of a threat; To assess magnitude and consequences of execution of the threat; To identify medical and physical resources to cope if threat is executed.[60]

Criticism of Canada's Biodefence Programs

Throughout the late 1980s certain aspects of Canada's biological warfare policies received adverse publicity. One of the most serious charges was that the country was not upholding the non-proliferation sections of the Biological Weapons Convention. In 1986, for instance, the *New York Times* reported that Iranian military officials had attempted to obtain samples of "T-2 mycotoxin producing fungi" from Dr Bruno Schiefer, Director of the Toxicology Research Centre at the University of Saskatchewan. This article attracted immediate attention from Foreign Affairs officials, both because Schiefer had been involved in the 1981 "yellow rain" controversy and because of intelligence reports that Iran was developing an active BW research program, since "unlike nuclear and chemical weapons, there are no international agreements to stem the spread of technology needed to make weapons from germs, microbes or toxins."[61] Ironically, while DFAIT was investigating the Schiefer case, it received a formal complaint from Dr Erhard Geissler, director of the East Berlin Central Institute of Molecular Biology, that Canada had violated BWC guidelines on several occasions by failing to submit reports about serious disease situations. One of these incidents was the 1987 outbreak of shellfish poisoning in Atlantic Canada; another case involved a Canadian citizen who had contacted Lassa fever, a traditional BW agent. In their rebuttal, Foreign Affairs officials made a concerted effort to explain why these two incidents did not fall within the Convention's confidence-building system. They also reminded Geissler that unlike the East German government, Canada was committed to the principles of transparency and accountability as a dedicated BWC member.[62]

Even more controversial were claims that Canada was secretly assisting the Pentagon in developing a new generation of biological weapons. In part, this was a spin-off from a series of acrimonious debates in the United States, where influential groups such as the Federation of American Scientists and the American Society of Microbiology questioned why financial support for the US Biological Defense Research Program had increased by 262 percent between 1980 and 1985.[63] Of even greater concern was the 1984 announcement by the US Department of Defence that it intended to construct a high-containment laboratory (BSL-4) at the Dugway Proving Grounds, capable of handling the world's most dangerous pathogens. Predictably, there was a groundswell of opposition in Utah, given memories of the 1968 Skull Valley nerve gas disaster, reinforced by scientific reports that the Dugway site was still contaminated with anthrax spores from earlier BW field trials. The campaign against this new laboratory soon gained momentum, fuelled by emotional warnings that the Reagan government would stop at nothing to develop its own genetically engineered bioweapons. The situation did not improve when US Secretary of Defense Caspar Weinberger tried to explain that US biodefence efforts were driven by the Soviet threat: "To ensure that our protective systems work, we must challenge them with known or suspected Soviet agents."[64] This vintage Cold War rhetoric impressed neither the US Congress nor the American public. Within a few months, facing intense legal and political opposition, the Pentagon decided to abandon its Utah BSL-4 venture.

But what if the US Army were to locate its high-containment laboratory in Canada? Or more specifically at Defence Research Establishment

Suffield? These questions, while entirely speculative, periodically surfaced throughout the mid-1980s, when organizations such as the Toronto-based Science for Peace pursued a number of CBW conspiracy theories.[65] This sense of Canadian apprehension about a bioweapons arms race was reinforced by the appearance of many new books, articles, and documentaries on this subject, often adopting a sensationalist and worst-case scenario approach. For some commentators, Ottawa's subservience to the goals of the American military-industrial-academic complex was part of a recurring pattern in which Canadian interests were sacrificed in order to satisfy the demands of its more powerful allies.[66] This approach found full expression in John Bryden's 1989 book *Deadly Allies: Canada's Secret War, 1937–1947,* with its claims that during the Second World War Ottawa's chemical and biological weapons policies were perverted in order to satisfy British defence priorities. As Bryden explained, "after 1942 Porton arranged to have all its human experiments with mustard gas done in Canada, Australia and India rather than use U.K. troops. It also successfully pressed the Canadians to do large-scale aircraft trials with persistent bacterial agents like anthrax and brucellosis even though it had its own anthrax disaster at Gruinard. Large tracts of windblown land at Suffield must still be contaminated to this day."[67] In reality, these accusations were based more on Bryden's fertile imagination than on historical evidence. But unfortunately, it was not until 2008, with the availability of relevant Canadian British and American historical sources, that the myth about large-scale open-air anthrax trials at Suffield was finally put to rest.[68]

Criticism of Suffield's past and present operation did, however, force the Department of Defence to carry out a full investigation of its biological warfare legacy. In July 1988, William Barton, a former DEA arms control diplomat, was commissioned to examine all contemporary and historical records, to interview defence scientists, and to recommend changes in the country's CBW policies. The investigation had mixed results. While Barton exonerated DND of violating any national or international laws, he criticized the department for its obsession with secrecy and its arrogance in dealing with public requests for information. To overcome these flaws, Barton recommended the establishment of an independent advisory committee of senior / high-profile chemical and biological scientists who would conduct an annual review of Canada's CW / BW defensive research projects and facilities.[69] This goal was achieved with the 1989 establishment of the Biological and Chemical Defence Review Committee, which during the past twenty-two years has provided a useful liaison between the defence science community and the Canadian public.[70]

Preventing BW Proliferation in the 1990s

Canada, Biological Weapons, and the Gulf War

Throughout the 1980s Iraq's dictator Saddam Hussein embarked on an ambitious chemical and biological warfare program as part of his long and bloody war with neighbouring Iran (1981–8).[71] Unfortunately, despite Iraq's documented use of mustard and nerve gas munitions, the international community demonstrated a lack of resolve in punishing this blatant violation of the Geneva Convention. This failure to impose sanctions encouraged Saddam Hussein not only to launch ever more deadly gas attacks, but also to commence an aggressive biological weapons program.[72]

Indeed, as early as 1987 US intelligence agencies reported that Iraq had acquired a modest biological warfare capability with "agents anthrax and botulinum toxin ... [delivered] in an aerosol made with dust-like carriers ... which could penetrate mask filters and permeable protective suits."[73] At this stage, however, Washington did not appreciate either the scale or sophistication of Baghdad's germ warfare operations, which included substantial amounts of growth media, large-scale fermenters, and weaponized strains of B. anthracis and botulinum toxin.[74] Indeed, by the outbreak of the first Gulf War in December 1990, Iraq had sufficient BW munitions for 166 aerial bombs and 25 warheads for Al Hussein missiles.[75] The question was whether the Iraqi dictator would authorize a first strike with his BW arsenal or be deterred by the threat of nuclear retaliation and regime change.[76] Faced with this ominous situation, in January 1991 the US military decided to immunize about 150,000 of its invasion force with a special anthrax vaccine.[77]

Concern over Iraq's bioweapons capabilities during the 1991 Gulf War had a number of ramifications for Canadian defence scientists. First, it reinforced the previous arguments of the Chief of Scientific Research and Development (CRAD) that the CBW research programs should be given high priority since "the proliferation of chemical and biological weapons into the third world is increasing, not diminishing, and our forces must continue to be equipped to counter this threat under all contingencies from high to low intensity warfare and peacekeeping operations."[78] A second consideration was to ensure that effective detection equipment and protective devices were available for all Canadian personnel involved with Operation Friction (Canada's Gulf War contribution). One of the most important developments was the deployment of the Canadian Integrated Bio / Chemical Agent Detection System (CIBADS), which could detect and collect suspected aerosol samples for analysis.

Once Iraq was defeated, a number of Canadian biological and chemical specialists were seconded for service with the United Nations Special Commission (UNSCOM). As numerous studies have demonstrated, the search for Iraq's hidden CBW weapons was both demanding and time-consuming. Indeed, despite his pledge of compliance, Saddam Hussein and his subordinates consistently denied that they had an active biological warfare program, at least until July 1995 when crucial information was provided by a high-ranking defector. But even with this insider information, it was still difficult to find stockpiles of agents / munitions, production equipment, and BW delivery systems. Consequently, most of the fourteen Canadian UNSCOM inspectors were required to serve multiple tours of duty, both in Iraq and at UNSCOM headquarters in New York City. One of these was Dr Ken Johnson, Director of Science and Technology at DRDC Headquarters, who participated in nine separate UNSCOM missions between June 1991 and July 1998.[79] While this was often a draining experience, Johnson and his colleagues also appreciated the professional rewards in working so closely with outstanding biodefence scientists such as Britain's David Kelly and David Franz of the United States.[80] Another useful spin-off was the opportunity to refine some of the techniques that were associated with the proposed BWC Verification Protocol. These included unrestricted interviews with the target laboratory's scientific staff; the right to conduct biological sampling; and the use of innovative scientific methods "based on polymerase chain reaction technology that made it possible to detect the traces of BW agents."[81]

The Gulf War also had a powerful impact on the CANUKUS system.[82] Above all was the awareness that if Iraq had used its germ and gas weapons, this would have "significantly reduced the operational effectiveness of coalition forces." As part of the learning process, MOU planners strongly recommended that the alliance should develop new biodefence strategies in terms "of doctrine, the procurement of equipment and the definition of R andD shortfalls." Attention was also drawn to the importance of accurate and timely intelligence, effective detection and identification, and reliable medical countermeasures that would "enable

CANUKUS forces to fight effectively in a BW environment."[83] Many of these issues were subsequently investigated by the special working group IFT 32 (Impact of BW on Operations), which recommended several operational strategies "with the aim of destroying the enemy's ability to wage biological warfare."[84] Another high priority was the need to ensure that trilateral vaccine development was "able to meet the key performance parameters for ... Variola Virus Vaccine PA ... Plague Vaccine PA ... Next Generation Anthrax Vaccine PA ... [and] Brucellosis Vaccine PA."[85] To facilitate this work, a special Orthopox Working Group was established with the goal of extending "collaborative detection and diagnosis testing using antibodies and gene probes against orthopox viruses ... [working] with WHO Group of Experts, where appropriate."[86]

Meanwhile, the controversy over the "Gulf War syndrome" was gaining momentum in the United States, the United Kingdom, and Canada. Once the conflict ended in 1991, there were numerous reports about veterans experiencing serious health problems, many of them claiming that their postwar illness was directly related to the anthrax vaccine they received during Desert Storm.[87] This debate became even more intense in 1997–8 when the US and Canadian governments both decreed that compulsory anthrax immunization would be enforced for their respective armed forces.[88] Almost immediately, legal challenges were initiated on both sides of the border. In Canada, the most celebrated case was the court martial proceeding launched against ex-sergeant Michael R. Kipling, one of the most vocal critics of the Canadian military's compulsory vaccination program, who was supported by veterans groups and civil liberties organizations.[89] For its part, the Department of Defence mobilized an impressive list of expert witnesses, including Dr A.M. Friedlander, Chief of the Bacteriology Division at the US Army Medical Research Institute of Infectious Diseases (USAMIID), who claimed that the anthrax vaccine in question was completely safe and had been used for over twenty years "without any noticeable negative long-term effects."[90] This viewpoint was challenged by Kipling's defence attorney, who noted disturbing reports about the safety and purity of certain batches of the anthrax vaccine produced by the Michigan Biologic Products Institute, the main supplier of the US Army. In addition, the vaccine in question was not licensed in Canada. After a lengthy trial, the defence arguments ultimately prevailed. On 5 May 2000 Lt Colonel G.L. Brais, Chief Judge, Canadian Forces, rejected DND's case against Kipling on the grounds that "by imposing vaccination against anthrax ... the military authorities have effectively infringed the accused's right to life, liberty and security of person by not affording him the right to informed consent, his right not to be subjected to cruel and unusual treatment and his right to the equal protection and equal benefit [under the Charter of Rights] of the law without discrimination.[91]

Other aspects of the Canadian Forces biodefence program were less contentious. For instance, Defence Research Establishment Suffield continued to make improvements in its CIBADS II detection system, as was evident during joint field trials at Dugway in 1997, when it identified "all 23 blind challenges of biological agent simulant in real time."[92] Other key defensive equipment included a portable kit for the Sampling and Identification of Biological and Chemical Agents (SIBCA), a special Reactive Skin Decontamination Lotion (RSDL) for CBW exposure, and a novel water-based foam that decontaminated surfaces or materials "contaminated with nerve agents, vesicant agents, organo-phosphorous pesticides and biological agents.[93] "On the training side, military volunteers were carefully monitored during live-agent outdoor exercises "using standard chemical agent detectors, collecting samples for confirmatory analysis, and conducting decontamination of equipment which has been identified as containing agent."[94] Scientists at DRES also worked on a number of projects to help civil authorities deal with bioterrorism. One of the most innovative experiments demonstrated that anthrax spores could be disseminated by ordinary packages / letters laced with this pathogen. As the final report stated, "if anthrax spores were finely powdered, a letter could release thousands of lethal doses of the bacteria within minutes of being opened ... [and] large amounts of material leaked out of sealed envelopes even before they were opened."[95] Significantly, as part of the MOU exchange arrangements, this information was relayed to biodefence experts at the US Centers of Disease Control, approximately two months before the terrorist anthrax letter bomb attack. Unfortunately, it was not accessed until after the immediate crisis had passed.[96]

Canada and the Biological and Toxin Weapons Convention

Throughout the 1990s Canadian arms control officials at National Defence and Foreign Affairs consistently supported international attempts to establish an effective compliance and verification system under Article V of the BWC. This process was facilitated by the evolving détente between the United States and the USSR / Russia and the consensus that the Convention drastically needed improvements. The final report of the Second Review Conference (1986), for example, called upon all member countries to adopt key confidence-building procedures: by providing information about their BSL-4 / BSL-3 laboratories; by reporting unusual outbreaks of disease; by circulating publications of all biological research; and by encouraging the exchange of scientists with other states parties. During the next five years there was considerable optimism in Ottawa that the BWC would be transformed into an effective disarmament system. And the stakes were high, since there was evidence that at least twelve states parties were trying to develop bioweapons capabilities.[97]

The Third Review Conference of 1991 was a landmark event. As part of a general agreement about the need for more accountability and transparency, it was decided that all states parties should describe their past BW work under Confidence Building Measure E/F.[98] Not surprisingly, there was considerable scepticism about the historical accuracy of these declarations. The US declaration, for example, contained a complete account of the different stages of its offensive program between 1942 and 1969. This included specific details about the weaponization of different lethal and incapacitating agents, along with major changes in its BW policies, including the secret 1956 decision to authorize bioweapons first use "at the discretion of the President for the enhancement of military effectiveness in a general war."[99] In justifying its post-1969 defensive program, the US report outlined two areas of major concern:

> One is evidence that other nations are developing and maintaining offensive BW capabilities. That threat suggests prudence in maintaining a level of preparedness for defence against potential BW agents. The second factor is realization that new methods in molecular biology and genetic engineering can be applied for the creation of novel biological agents or to the production of specific agents ... In 1984, the Army established its Biological Defense Research Program (BDRP) to encompass all efforts underway in biological defence. All work conducted under the BDRP has been unclassified.[100]

In contrast, the Soviet / Russian declaration was characterized by evasion and falsehood. First, while it acknowledged that offensive bioweapons work had taken place during previous decades, this was offset by the spurious claim "that munition development never proceeded past the prototype stage and that no biological weapons were produced or stockpiled."[101] Second, the Soviet declaration only listed 20 per cent of the

Biopreparat operation, with even those facilities lacking information about the size of their laboratories or scientific workforce. Even worse, key military BW sites were not even mentioned. Third, the Soviet report contained no discussion about the role that key institutions, such as the Academy of Sciences, the Ministry of Health, and the KGB had played in the Soviet / Russian BW system. Finally, and most surprising, was the failure to mention Soviet President Boris Yeltsin's famous March 1992 declaration that the Soviet Union had maintained a secret offensive BW program and that the 1979 Sverdlovsk anthrax outbreak was directly linked to a secret BW weapons plant.[102] Moscow's cynical response towards this confidence-building exercise was roundly criticized. As the *Washington Times* noted, since the Russian declaration only contained information about four of the twenty known facilities, it was obvious that "the Russian military [was] lying about the extent of the biological weapons program."[103]

In April 1992 Canada provided members of the BWC with an official account of its involvement with biological weapons. Overall, it was a disappointing document, repeating many of the half-truths that had often characterized Ottawa's attempts to rationalize its BW activities. Predictably, it contained the familiar mantra that "the Canadian Armed Forces have never used biological or toxin weapons ... does not possess them and has no intention of acquiring them." While this was technically correct, the statement ignored the important role Canada had assumed in assisting the US and the UK in their postwar bioweapons programs. Equally suspect was the assertion that Canada's direct involvement with BW agents and munitions lasted only ten years (1946–56) and never went beyond fundamental research. "There was no large-scale production, stockpiling or weaponization of BW agents," the report claimed, "and no work was aimed at determining the suitability of specific agents as weapons."[104] But if this was the case, why did the BW Panel of the Defence Research Board spend so much time selecting the most lethal biological warfare agents and endorsing Guilford Reed's proposal for the establishment of a BW pilot production plant at Kingston, Ontario?

Even more misleading were claims that after 1956 the Canadian program focused only on defensive research. The exception to this pattern, the report claimed, was on those rare occasions when specific bioweapons had to be examined, since it was "only through a thorough understanding of the properties and behavior of potential BW agents that the potential threat can be appreciated, and work on suitable defensive measures can be undertaken." Clearly this was not the case. Throughout the

1960s Canadian defence scientists were actively involved with the Tripartite / Quadripartite cooperative research system, with its emphasis on both offensive and defensive measures. Of particular importance was the commitment by Canada and the United Kingdom to assist the United States in developing an effective retaliatory bioweapon capability to deter Soviet BW adventurism. Likewise, the Canadian submission makes no mention of the 1960s bilateral testing arrangement between Defence Research Establishment Suffield and the Deseret Testing Center, which substantially advanced the strategic potential of the US offensive bioweapons system.[105]

Ottawa's confidence-building declaration to the BWC was more accurate when dealing with the period after 18 September 1972, when the country officially endorsed the Biological Weapons Convention. Nor is there any historical evidence that Canada has been involved with any offensive BW research during the past forty years; on the contrary, the country has steadfastly upheld the goals of CBW non-proliferation through its involvement with the BWC and the Australian Group. Another useful aspect of the contemporary section of the Canadian brief was its inventory of the different research facilities in Canada where work on dangerous pathogens was being carried out. These included Canadian Forces Base Suffield, notably "laboratory areas in Building 1 by containment level ... BL2-262 m, BL3-148 m."[106] In addition, reference was made to the linkages between DRES and other important Canadian research and clinical facilities such as the Toronto-based Connaught Medical Research Laboratories, with its advanced work on bacterial and viral vaccines.[107]

Encouraged by the end of the Cold War, throughout the 1990s Canadian scientists and arms control experts were actively involved with the campaign for a verification protocol to strengthen the Biological Weapons Convention.[108] This was evident in their participation in the work of the special verification study group (VEREX), which eventually produced a workable strategy of how suspected bioweapons activities could be detected and perpetrators exposed to international censure. Canada was also actively involved with the 1994 Special Conference on BW Verification, attended by seventy-nine interested states parties, which created the specialized Ad Hoc Group of Experts with a mandate to develop proposals for a legally binding verification protocol in time for the Fourth Review Conference in 1996.[109]

The governments of Brian Mulroney (1984–93) and Jean Chrétien (1993–2003) supported these initiatives for a number of reasons. First, it was consistent with the country's traditional work in the arms control and disarmament fields. Second, Canadian diplomats had already worked hard to preserve the BWC during the turbulent 1980s, when controversies over Sverdlovsk and yellow rain threatened to destroy the organization. In addition, both Canada's allies and respresentatives of the United Nations praised the work of the fourteen DND scientists who had been involved with the UNSCOM mission in Iraq. And finally, there were encouraging signs that the United States might abandon its traditional opposition towards a BWC verification system, in part because of the lobbying efforts of groups such as the Federation of American Scientists and the American Society for Microbiology, who called for a more creative approach to "solving the difficulties that impede removing the threat of biological weapons from our world community."[110] Even more encouraging was US President Bill Clinton's strong endorsement of the campaign to implement "a compliance and transparency protocol" in his January 1998 State of the Union address.[111]

Another encouraging development was the 1993 international agreement for the Convention on the Prohibition of the Development, Production, Stockpiling and Use of Chemical Weapons and on Their Destruction (CWC). Although it took another four years for the CWC to become operational, it was widely applauded for being "the first multilaterally negotiated treaty which provides for the ban of an entire category of weapons of mass destruction under strict and universally applied international control."[112] For supporters of the BWC verification protocol many aspects of the chemical convention appeared to support their agenda. One of these was the establishment of the Organization for the Prohibition of Chemical Weapons (OPCW), with its mandate to ensure compliance control, facilitate cooperation between states parties, and supervise the destruction of CW weapons and facilities.[113] The protocol boosters also noted that the US Chemical Manufacturers Association, representing the country's top fifty chemical companies, had assumed a critical role in the CWC ratification campaign.[114] Would it be possible, they asked, for the powerful Pharmaceutical Research and Manufacturers of America (PhRMA) to adopt an equally creative stance towards the BWC? Or, on the contrary, were PhRMA members unalterably opposed to the proposed inspection system, because of their concerns about corporate proprietary information and patent rights?[115] These questions would remain unanswered until the December 2001 BWC Review Conference.

The Threat of Bioterrorism

Biodefence Lessons From the United States

In its long involvement with biological warfare, Canada has been strongly influenced by major paradigm shifts in the American BW program. This was certainly the case in 1969 when President Nixon cancelled the US offensive biowarfare operation and supported the establishment of the BWC. Another example of this linkage occurred during the 1990s when the Clinton administration decided to confront the growing threat of bioterrorism. This trend was directly related to a series of terrorist incidents, notably the 20 March 1995 sarin attack on the Tokyo subway by the Aum Shinrikyo cult that killed six and sent thousands to the hospital.[116] This incident also reinforced fears in the United States about the possibility of terrorist attacks on American soil with either conventional explosives or biological or chemical weapons. These dire predictions were confirmed by the 26 February 1993 al Qaeda bombing of the World Trade Center in New York City and the deadly assault of 19 April 1995 by American right-wing extremists on the federal Alfred P. Murrah Building in Oklahoma City. Less well known was the 1998 arrest of American white supremacist Larry Wayne Harris, whose threat to release weapons-grade anthrax against Las Vegas turned him "into a poster boy for the emerging threat of bioterrorism."[117]

The possibility that terrorists would use biological weapons, including genetically engineered organisms, was reinforced by numerous intelligence reports, congressional hearings, and media accounts.[118] At the popular level, Richard Preston's blockbuster 1997 novel *The Cobra* featured a deranged scientist who develops a genetically engineered virus that combines "the worst attributes of smallpox, the common cold, a rapidly proliferating insect virus, and the gene for Lesch-Nyhan syndrome, a neurological disorder that causes victims to mutilate themselves." Among its many readers was US President William Clinton, who became an early convert to the apocalyptic view that the United States was facing an imminent threat from biological terrorism, where "the image of a cloud of anthrax killing millions ... gained the same kind of symbolic strength as the mushroom cloud of a nuclear explosion."[119] An early manifestation of Washington's new fixation on bioterrorism was the March 1998 White House–sponsored exercise, which involved forty cabinet members and senior officials and was structured around a simulated BW biological attack on the United States by terrorists using a genetically altered hybrid of smallpox and Marburg hemorrhagic fever.[120]

In May 2000 the first of the congressionally mandated training exercises, named TOPOFF One (Top Officials), took place in three different parts of the country. An essential goal of this mock CBRN crisis was to determine how specialists of thirty-five government departments and agencies could deal with three simultaneous WMD attacks. In Portsmouth, New Hampshire, the terrorist device was CW nerve gas; in Washington DC, it was a radiological "dirty bomb"; and in Denver, Colorado, the city's population was exposed to the bacterium *Yersinia pestis* (plague).[121]

The Denver simulated pneumonic plague attack required the most innovative response from high-level planners and first responders.[122] First, as the exercise evolved there was considerable confusion over whether the local or federal authorities were in charge, a situation that was further complicated by "turf wars" between officials of the Federal Bureau of Investigation and the Centers for Disease Control. This administrative confusion not only impeded the flow of vital information, but also eroded public confidence in the ability of public officials to deal with the crisis. A second problem was the nature of the bioagent. Since pneumonic plague is highly infectious, a surge of patients overloaded Denver's health care facilities, representing "10 times the usual caseload ... [and] hospitals that participated in the exercise were beyond capacity in less than 24 hours of the epidemic." Trying to contain the outbreak within the state also proved difficult, despite the enactment of restrictive travel measures, including the closing of the city's international airport. As one public health official observed, "How do you keep 1 million people in their homes?"[123] Finally, TOPOFF One demonstrated that a bioterrorist incident of this magnitude would produce serious psychological problems among medical first responders, who experienced rapid "burn out" due to job fatigue, combined with concerns about bringing the infection home to their families. Equally disturbing was the gradual erosion of public confidence, and even localized riots, because of rumours that food, gasoline, and antibiotics were no longer available.[124]

In order to deal with this type of situation, the Clinton government implemented a number of important biodefence measures. These included the passage of the Antitrerorism and Effective Death Penalty Act of 1996 (Title V), the National Security's publication entitled *National Security Strategy for the New Century*, and the 1998 Domestic Terrorist Program, which sought to enhance first-responder capabilities in 120 US cities. In addition, the level of funding for counterterrorism increased dramatically from $5.7 billion in 1995 to $11.1 billion in 2000.[125] Yet despite these initiatives serious questions remained about the ability of the United States to cope with a major bioterrorist attack, because of serious problems of coordination between security specialists in the fields of defence, law enforcement, intelligence, health policy, and foreign affairs.[126]

Another dimension of US security concerns was the possibility that a terrorist attack might be launched from Canada, particularly after the December 1999 apprehension of Ahmed Ressam, the so-called "millennium bomber."[127] Ressam's arrest at the British Columbia–Washington border not only prevented the bombing of the Los Angeles international airport, it also revealed an elaborate North American terrorist network. According to Richard Clarke, Special US Adviser on Counter-Terrorism, "the strings from Ressam, the man on the ferry, led to a sleeper cell of Algerian mujahedeen in Montreal. How the Canadians had missed the cell was difficult to understand ... but now they were cooperating ... [and] the leads the Royal Canadian Mounted Police provided went to what looked like cells in Boston and New York."[128] CIA Director George Tenet had similar observations: "In looking back, much more should have been made about the significance of this event [with] ... our northern border vulnerable, the United States did not have a comprehensive and integrated system of homeland security in place ... [with] borders, visas, airline cockpits, watch-lists ... managed haphazardly."[129]

Building Canada's Counterterrorist System

After the Montreal Olympics, Canada's level of counterterrorist preparedness rapidly deteriorated. In 1985, however, two incidents forced Ottawa to become more proactive in dealing with the international terrorist threat. One of these was the 12 March seizure of the Turkish embassy in Ottawa by members of the so-called Armenian Revolutionary Army.[130] More serious was the 23 June bombing of Air India flight 182, resulting in the deaths of 329 passengers and crew.[131] Carried out by ultra-nationalist Sikhs living in British Columbia, this ruthless act of political violence stunned the nation. It also forced federal officials to reassess the country's airport security and counterterrorist response systems, particularly since it was obvious that neither the RCMP nor the Canadian Security and Intelligence Service (CSIS) was prepared for this type of terrorist attack given their focus on airplane hijacking, while "communication within and between security, law enforcement and transport agencies was often flawed or non-existent."[132]

Facing serious criticism about the performance of its civilian security agencies, the Mulroney government decided to implement a number of important changes. One of these was the 1988 National Counter-Terrorism Plan (NCTP), devised by the Office of the Solicitor General. It consisted of three functional measures: the Counter-Terrorism Consequence Management Arrangement, the Government Emergency Operations Coordination Centre, and the National Infrastructure Assurance Program. Another dimension of the NCTP were cooperative arrangements with the United States in dealing with civil emergencies by providing cross-border mutual assistance, coordinating transportation and communication policies, and facilitating the exchange of information about terrorist groups. Administratively, these tasks were carried out by a joint bi-national Consultative Group, "operating under the guidelines of Canadian and American legal principles, and respecting existing defence procedures negotiated under the North American Aerospace Defence Agreement (NORAD)."[133] The first real test of this North American security plan came during the First Gulf War (1990–1) amid concerns that pro-Iraq groups might attack US and Canadian targets using either conventional explosives or weapons of mass destruction. In 1993 the al Qaeda bombing of New York's World Trade Center, symbol of American corporate capitalism, took place. But it was the 1995 sarin attack on the Tokyo subway system that provided the major catalyst for Washington and Ottawa to revise existing methods of dealing with CBR terrorists.

According to a DFAIT report, the most important dimension of the Tokyo incident was that the Aum Shinrikyo cult had experimented with both chemical and biological agents, the former being the most successful, with "7 dead and 270 requiring medical attention in Matsumoto, and 12 dead and 5,500 requiring medical attention in Tokyo."[134] CSIS was equally impressed, as is evident in Ron Purver's analysis of August 1995, which pointed out that certain BW agents, notably anthrax and botulinum toxin, could approximate "the lethality of a nuclear explosion in terms of the potential number of casualties caused." But unfortunately, Purver claimed, the Canadian government was still unprepared for a bioterrorist attack, despite the ongoing work of the Special Threat Assessment Group (STAG), consisting of medical professionals and scientists working to prevent, contain, or mitigate "terrorist threats and incidents involving the use of nuclear, biological, or chemical agents."[135]

According to a subsequent CSIS report these resources might be urgently needed since Canada "remains as vulnerable as any of the other Western industrialized states to the kind of nightmarish, mass-casualty CBRN terrorist attack that until recently was confined to fiction."[136]

In March 1998, Canada's counterterrorist system was further enhanced with the formation of the Special Senate Standing Committee on Security and Intelligence. After carrying out a series of national hearings, the Senate Committee developed a systematic action plan.[137] It began with the basic premise that Canada was vulnerable to a possible bioterrorist attack since "most of the major international terrorist organizations have a presence in Canada ... [and] geographical location also makes Canada a favourite conduit for terrorists wishing to enter the United States, which remains the principal target for terrorist attacks world-wide." Another concern was the negative impact of government underfunding on CSIS, whose staff had declined by about one-third during the previous eight years, despite having an expanded workload.[138] In addition, there were questions about whether Ottawa's threat analysis capabilities had been severely weakened when the Department of Foreign Affairs decided to curtail its foreign intelligence assessments, with the consequence "that current lists of terrorist organizations were no longer readily available."[139] On the positive side, the Committee approved of recent changes in the National Counter-Terrorism Plan whereby "responsibility for armed response to a serious terrorist threat [shifted] from the Royal Canadian Mounted Police to the Department of National Defence."[140]

The Senate Committee's report also examined the elusive subject of terrorist motivations and tactics. Was there any validity, it asked, in the traditional paradigm that terrorist groups would be unwilling to use weapons of mass destruction "lest they inspire public revulsion and antipathy to the terrorist organization and its cause"? The committee decided that this was merely an outdated myth according to recent US and Canadian security studies. For instance, the FBI reported that in 1997 it had investigated "over 100 threats or incidents involving nuclear, chemical or biological materials." Reference was also made to the fact that the RCMP had recently arrested an American white supremacist "trying to smuggle 130 grams of Ricin from Canada into the United States."[141]

In December 1999, Solicitor General Lawrence MacAulay tabled a report in the House of Commons calculated to protect Canadians against a terrorist CBRN attack.[142] In his presentation, MacAulay noted that the timing of his bill coincided with evidence that "ad-hoc extremist

Islamists … [were] preparing terrorism acts from Canadian territory … to cause the maximum amount of damage."[143] However, it was not an easy task to prevent these attacks from taking place. According to the Solicitor General's report, it was possible to obtain "deadly pathogens … through the mail from scientific supply houses … [or] steal agents from civilian research facilities or military stockpiles."[144] In addition, Canadian first responders had technical problems in detecting and identifying biowarfare agents at the time of release, a situation that was aggravated by the country's shortage of high-containment laboratories.

On the positive side, however, several cooperative defensive programs were already in place, notably the RCMP / Canadian Forces Joint Biological, Chemical, Radiological Response Team (JCBRNRT), with its responsibility "to help isolate, contain, assess and eventually dispose of a limited NBC hazard."[145] Also in place were four major federal emergency plans: the National Counter-Terrorism Plan (Solicitor General), the Federal Nuclear Emergency Plan (Health Canada), the Food Agriculture Emergency Response System (CFIA), and the National Support Plan (Office of Critical Infrastructure Protection and Emergency Preparedness). Equally important was the valuable assistance provided by the Canadian Science Centre for Human and Animal Health in Winnipeg, one of the world's leading high-containment laboratories, while the interdepartmental Special Threat Assessment Group (STAG) had already demonstrated its ability to coordinate Canada's biodefence responses.[146]

By 2001 responsibility for Canada's CBRN counterterrorist system was shared by three government agencies. First in line were scientists at Defence Research Establishment Suffield, who had a long-established role in CBW research and testing for both the Canadian Forces and for civil defence purposes. These skills had been further enhanced through their participation in the advanced biodefence projects being carried out by CANUSUK/MOU and by their involvement with the UNSCOM teams in Iraq.[147] Moreover, despite the impact of the austerity cuts of the mid-1990s, there was now a concerted effort to upgrade the CBW facilities at DRES, notably through a major upgrade of the BSL-3 laboratory, with useful assistance being provided by experts at USAMRIID. This enhanced ability to work with deadly pathogens was evident in the special 2001 report on the ability of weaponized *Bacillus anthracis* to be widely disseminated, even through unopened letters.

The Office of the Solicitor General (OSG) also assumed an important role in dealing with the growing threat of bioterrorism, largely because it was responsible for Canada's two leading security agencies: the RCMP and CSIS.[148] The OSG had actively participated in biodefence planning during the 1976 Montreal Olympics, and in 1988 it had assumed a lead position in administering the 1988 National Counter-Terrorism Plan.[149] Another important development was the 1996 establishment of the Solicitor General's new Counter-Terrorist Division, which soon became an active player in the Ottawa security environment, working closely with the US Federal Emergency Administration. According to Mike Thielmann, one of the Division's key administrators, "it was a fast learning curve."[150]

Health Security and Biodefence

The transformation of Health Canada's role in the country's counterterrorist system occurred during the late 1990s. In part, this was related to the expanded role of medical and public health professionals in dealing with the bioterrorist threat and, in part, to the influence of Dr Ronald St John.[151] After an active career with the US Centers for Disease Control, in 1992 St John became director of the newly established Office of Special Health Initiatives. This office had a mandate to deal with a number of serious health and administrative problems, including drug-resistant tuberculosis, hospital infections, and reorganization of quarantine procedures at the country's international airports. Canada's vulnerability to global disease outbreaks was demonstrated in September 1994 when it was reported that an Air Canada flight from Surat (India), the site of a pneumonic plague outbreak, contained many passengers infected by this dangerous pathogen. Fortunately, St John and his Health Canada associates were able to convince Toronto airport staff that there was no danger in contacting this disease, because the plane's occupants had been thoroughly screened prior to departure and because Pearson airport was well stocked with antibiotics.[152]

But if Canada had problems coping with a natural infectious disease, what would happen if there was a major bioterrorist attack? This question was foremost in the mind of Ron St John when he attended a top-secret briefing at CDC headquarters in 1998 that attempted to reorganize US biodefence policies, including ways of controlling pathogens most likely to be used in a terrorist attack.[153] On his return to Ottawa, St John was able to convince his superiors that Health Canada should develop its own bioterrorist strategies through the creation of the Centre for Emergency Preparedness and Response (CEPR).[154] Under CEPR an expansive set of programs was established. These included the upgrading of the National Emergency Stockpile (NESS) for vaccines and antibiotics / antivirals and the development of the National Capital Response Plan, a cooperative venture between Health Canada's emergency response specialists and their counterparts in the Ottawa police, fire department, and local hospitals.[155] According to Dr Marc-André Beaulieu, one of its key organizers, this initiative enjoyed considerable success in reconciling the different occupational cultures of first responders and producing effective protocols for emergency situations.[156] The Office of Laboratory Safety also complemented the work of CEPR by monitoring all Health Canada's laboratories, by assessing permit applications for the importation of Risk Group 3 and 4 human pathogens, by assisting the Canadian Food Inspection Agency in the development of containment standards for veterinary research, and by fulfilling its obligations as a WHO biosafety collaborating centre.[157]

Health Canada had another major asset in carrying out its biodefence work – the Winnipeg-based Canadian Science Centre for Human and Animal Health (CSCHAH).[158] This large-scale facility, shared by Health Canada and the Canadian Food Inspection Agency, provided the unique opportunity for scientists to combine human and animal research within the same facility, particularly since so many zoonotic (animal) diseases were among the most deadly pathogens affecting humans.[159] Over 30,000 square metres in area and consisting of five major laboratories, the CSCHAH was officially opened in 1999, after ten years of design and construction.[160] The human pathogen part of the operation, concentrated in the National Microbiology Laboratory (NML), had four BSL-4 containment laboratories "equipped to deal with the most serious of agents and diseases that may ... have the potential for aerosol transmission ... [and] have maximum containment equipment to completely seal in infectious agents."[161] The NML's stature was further enhanced in 2000 when Dr Frank Plummer, an internationally recognized expert on HIV/AIDS, became director general.[162] Shortly afterwards, a major stage in the laboratory's evolution occurred when the NML received its first shipment of six deadly pathogens from the Centers for Disease Control: small vials containing samples of Lassa fever, Junin virus, and three strains of the Ebola virus.[163] Because of its elevated status as one of only fifteen BSL-4 laboratories in the world, the Winnipeg laboratory became an increasingly important component in Canada's biodefence system, particularly during the 2001 anthrax letter attacks, the 2003 SARS crisis, and the 2009 H1N1 (swine flu) influenza pandemic.[164]

Conclusion

During the last two decades of the twentieth century, the United States continued to define the parameters of biodefence responses, both unilaterally and through the two major cooperative programs. Of these, the 1980 MOU arrangement was the most important, given its high-priority research agenda in analysing biological weapons agents, in developing advanced detection and identification technology, and in discovering new vaccines and antibiotics / antivirals. While many of these projects were successfully tested during the first Gulf War (1991–2), serious questions arose during the early 1990s about whether the level of MOU preparedness was sufficient to cope with CBW challenges from rogue states or international terrorist organizations.

Another important debate centred on whether the Russian government, after President Boris Yeltsin's famous 1992 pronouncement, had really curtailed its offensive BW program. While some US officials were prepared to give the Kremlin the benefit of a doubt, others feared that the Russian military was still engaged in secret BW research. In preparing for such an exigency, the US authorized several secret biodefence projects. In one of these, code-named Clear Vision, the Central Intelligence Agency sought to develop a model of the Red Army's anthrax bomblet munition "that measured ... its dissemination characteristics and how it would perform in different atmospheric conditions ... after release from a warhead."[165] Another venture, Project Bacchus, carried out by the Defence Advanced Research Projects Agency (DARPA), focused more on the terrorist threat, demonstrating that it was possible to build a factory capable of making biological weapons with readily available commercial materials and that detection of this facility would be extremely difficult. Not surprisingly, once these two projects were exposed, there were accusations, at home and abroad, that the US government had violated the spirit, if not the law, of the Biological Weapons Convention.[166]

Overall, these controversies had little impact on Canadian defence scientists, since there is no evidence that they participated in either the DARPA or CIA projects. But they were certainly aware of advanced bioweapons research through their involvement with MOU/CANUKUS, in terms of both battlefield preparations and the threat of bioterrorism. In addition, there was growing counterterrorism cooperation between Ottawa and Washington because of shared concerns that international terrorist groups were prepared to use biological and chemical weapons against North American targets.[167] These bilateral arrangements not only included the well-established liaison between military scientists, but also increased interaction between health security specialists at Health Canada and their counterparts at the US Centers for Disease Control in dealing with bioterrorism and natural disease outbreaks.[168]

Indeed, during these years eight major disease threats emerged, including AIDS/HIV, avian influenza (H5N1), and drug-resistant tuberculosis.[169] In dealing with these broader issues of health security, officials adopted an all-hazards approach that allowed first responders to cope with natural disease outbreaks and bioterrorist incidents. As these methods gained ascendancy, emergency planners began to include other natural disasters such as floods, ice storms, and hurricanes in their mandate.[170]

Were Canada and the United States successful in breaking down the traditional boundaries between different groups of biodefence stakeholders – medical scientists, public health experts, law enforcement personnel, military CBW specialists, and arms control diplomats? While there had been some progress in overcoming departmental and agency "turf wars" in the two countries during the 1990s, largely because of the pressure of dealing with terrorist organizations on both sides of the border, serious residual problems still remained. Indeed, the events of 9/11 revealed that the cooperative biodefence model remained more a goal than a reality.[171]

Frederick Banting prepares for his scientific mission to the United Kingdom, 14 February 1941.
F.C. Banting Papers, Thomas Fisher Rare Book Library, University of Toronto

Otto Maass of McGill University, chief of Canada's biological and chemical warfare operations during the Second World War.
Library Archives Canada (LAC): PA-171607

Professor Everitt Murray of McGill, coordinator of Canada's biological warfare program in the Second World War.
LAC: PA198207

Minister of Defence Colonel Ralston visits Suffield Experimental Station, 30 May 1942.
Defence Research Development Canada Suffield Collection

Aerial view of Experimental Station Suffield in May 1945.
DRDC Suffield Collection

Approaching the loading dock at Grosse Ile, Quebec, site of BW research between 1942 and 1955.
Donald Avery Collection

Omond Solandt (left), Director General of the Defence Research Board, greets Sir Henry Tizard during the 1947 Commonwealth defence science meetings in Ottawa.
University of Toronto Archives: Omond Solandt Collection, B94-0020/003 (07)

Guilford Reed receives an honorary degree from the University of Saskatchewan, 8 May 1953, for his achievements in microbiology research.
Queen's University Archives, Guilford Reed Collection

Aerial view of Queen's University medical laboratories and animal house used for secret BW research, 1942–53.

Queen's University Archives, Queen's Picture Collection (B-28)

Guilford Reed's BW laboratory: Top floor of the New Medical Sciences Building.
Queen's University Archives, Queen's Picture Collection (B-28)

Everitt Murray captivates his defence science colleagues
at the Suffield officers' canteen (1947).
LAC, E.G.D. Murray Collection, E008406495

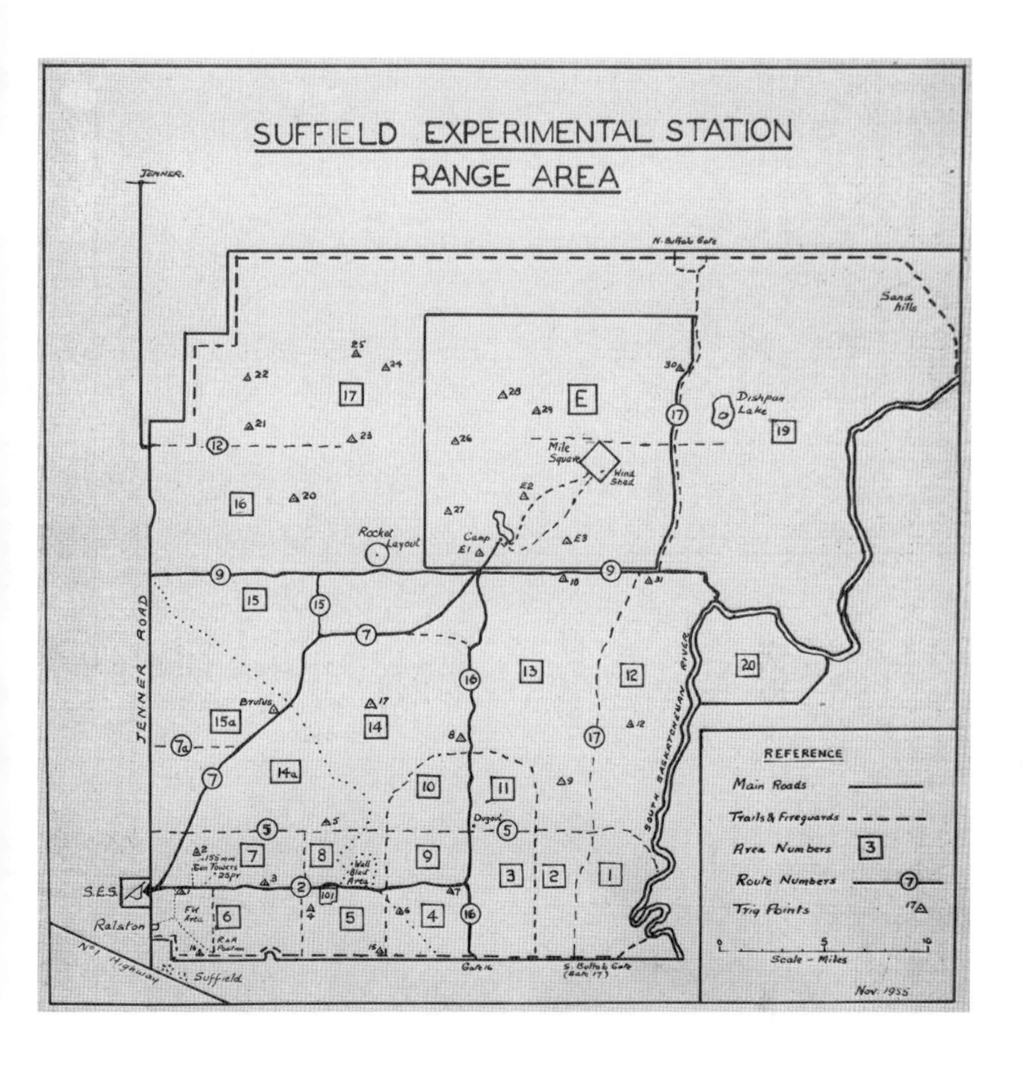

Map of Suffield Range showing Area E bioweapons testing site.
DRDC Suffield

Wind shed used for controlled tests with BW agents in July 1952.
DRDC Suffield Collection

Superintendent Archie Pennie (centre) greets Prime Minister John Diefenbaker at SES on 12 May 1962.
DRDC Suffield Collection

The 15th Joint Dugway and Suffield Conference, 8 September 1965.
DRDC Suffield Collection

U.S. Honest John Rocket with CBW cluster bomb (1960s).
Wikimedia Commons, and U.S. Library of Congress Prints and Photographs Division (HAER COLO, 1-COMCL, 53)

Anti-war demonstration outside SES entrance, 6 August 1964.
DRDC Suffield Collection

State-of-the-art aerosol chamber for CBW detection at DRDC Suffield.
DRDC Suffield Collection

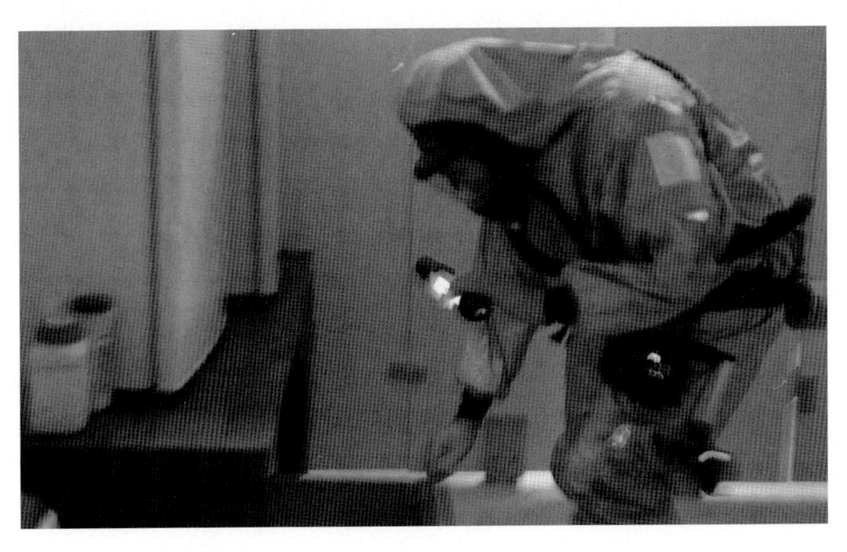

Suffield CBW first-responder training program.
DRDC Suffield Collection

High-performance mannequin for simulated BW attack at Suffield.
DRDC Suffield Collection

Cartoon showing panic during 2009 swine flu pandemic.
Stock Photo: 9314999 (Rights purchased)

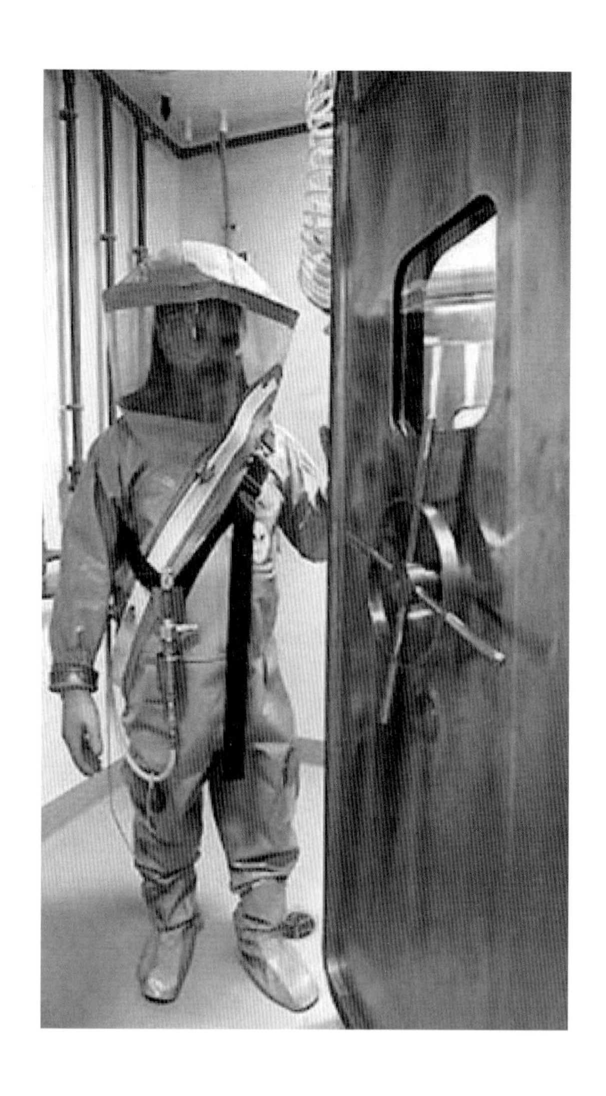

Researcher at the Winnipeg National Microbiology
Laboratory entering high containment section.
Public Health Agency of Canada

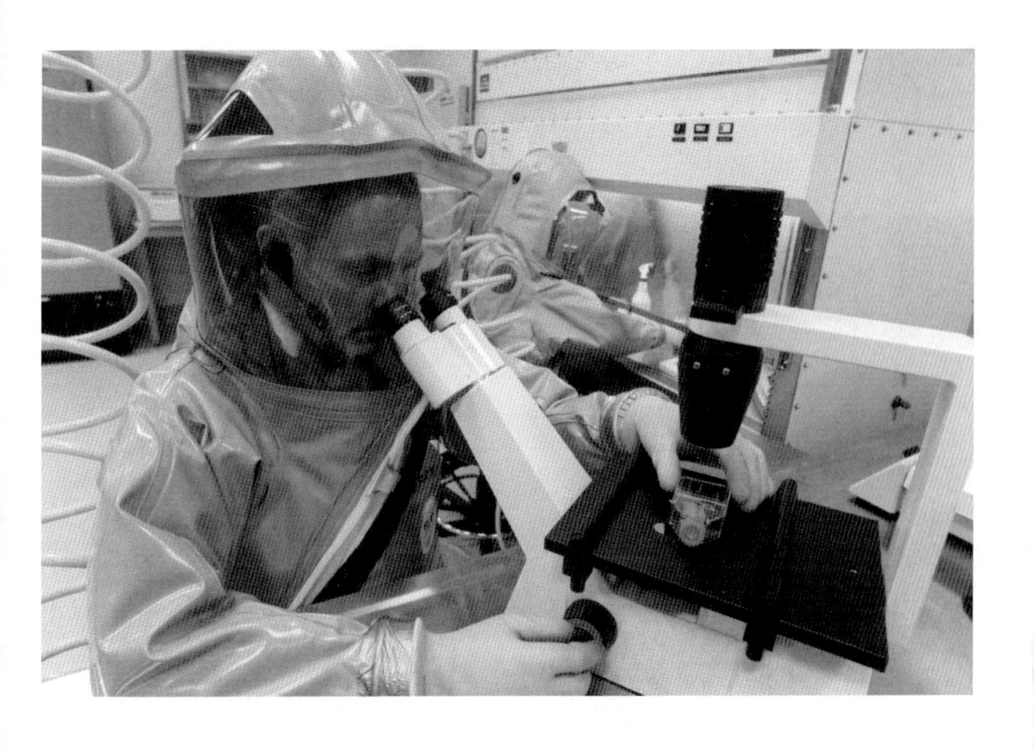

Analysing some of the world's most deadly pathogens at the Winnipeg NML.
Public Health Agency of Canada

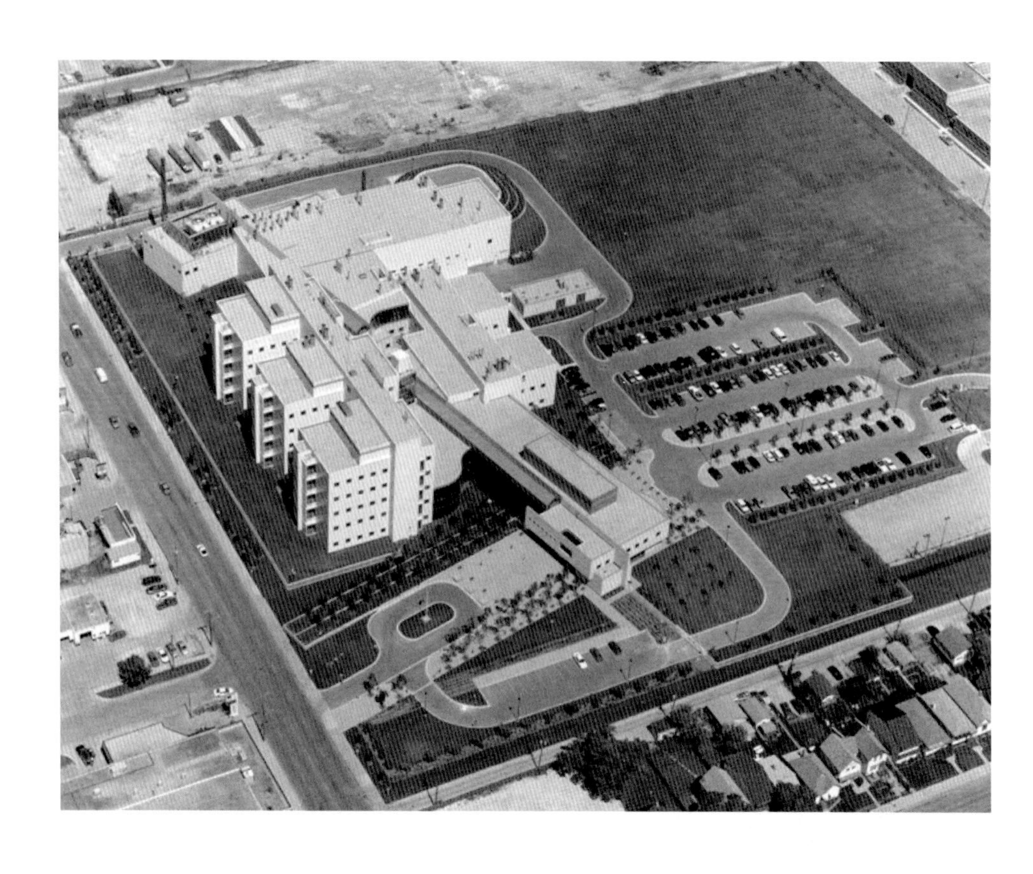

Aerial view of the Canadian Science Centre for Human and Animal Diseases.
Public Health Agency of Canada

Frank Plummer, Scientific Director of the National Microbiology Laboratory.
Public Health Agency of Canada

Ron St John, Director General of the Centre for Emergency Preparedness and Response.
Ron St. John Collection

Director General Camille Boulet of DRDC Suffield.
DRDC Collection

7 Biodefence after 9/11: Old Problems and New Directions

Official U.S. statements continued to cite around a dozen countries that are believed to have or to be pursuing a biological weapons capability. In addition to the efforts by terrorists or states with malevolent intent, we must be concerned about the grave harm that may result from the misuse of the life sciences and related technologies by individuals or groups that are simply careless or irresponsible. The continuing threat of bioterrorism, coupled with the global spread of expertise and information in biotechnology and biological manufacturing processes, has raised concerns about how advancing technological prowess could enable the creation of new threats of biological origin possessing unique and dangerous but largely unpredictable characteristics.[1]

The sense of foreboding about a possible bioterrorist attack on Canada was evident in the 1999 report of DRDC Suffield entitled "Hogtown Disaster: A BW Terrorist Attack on a Major Canadian City."[2] In this simulated study, metropolitan Toronto's 4.5 million people were the target of a group of domestic terrorists using the deadly pathogen *Bacillus anthracis*, which caused "casualties in at least 20% of the exposed population."[3] A year later, Health Canada's Centre for Preparedness and Response (CEPR) prepared its own BW disaster model, based on those developed by the US Centers for Disease Control and Prevention.[4] Again the projections were ominous: "Under certain conditions, an anthrax attack on 100,000 Canadians would result in 50,000 anthrax cases, 32,875 deaths, 332,500 hospitalization days, and at a cost of $6.5 million."[5]

A number of scholars have argued that while the 9/11 al Qaeda airplane attacks on New York and Washington were traumatic events for most Americans, there was a certain familiarity with the terrorist's weapon of choice. In contrast, the anthrax letters delivered between 18 September and 19 October were an unfamiliar and frightening type of assault on the United States, "creating the impression that the social order itself was genuinely threatened by Islamic radicalism."[6] To make things worse, a wave of copy-cat BW hoax letters ensued, demonstrating the psychological impact that could be achieved "with nothing more than a powdery substance and a threatening note."[7] This sense of vulnerability was intensified by sensationalist media reports, by the Bush administration's lurid system of colour-code threat levels, and by serious problems of cooperation among different groups of first responders, with some of the most bitter confrontations taking place between forensic science experts from the FBI and disease control specialists from the Centers for Disease Control.[8] And, above all, the public expressed widespread dismay over Washington's inability to prevent further attacks or to arrest the perpetrator(s).[9]

Canadians also experienced a heightened sense of vulnerability after 9/11, reinforced by home-grown anthrax hoaxes.[10] One of these was the 10 October incident at the Montreal offices of Globe International, which forced the local Hazmat teams to evacuate the entire downtown area of the city, causing widespread panic.[11] Yet surprisingly, debate in the House of Commons about the threat of bioterrorism remained cursory and superficial.[12] Nor did the Canadian media rise to the challenge; most of the coverage consisted of regurgitated US reports that did not attempt to explain whether Canada was facing its own bioweapons threat.[13] On the positive side, Ottawa's counterterrorist response managed, for the most part, to avoid the jurisdictional disputes that disrupted the American anthrax investigation. For example, immediately after 9/11 federal authorities attempted to coordinate the work of the major biodefence stakeholders – the Defence Research Centres, Health Canada's Center for Emergency Preparedness and Response, the Office of the Solicitor General, and the Department of Foreign Affairs and International Trade (DFAIT). All of these agencies also participated in the establishment of the National Advisory Committee on Chemical, Biological, Radiological Safety, and the CBRNE Research and Technology Initiative (CRTI), with its mandate to develop partnerships between government agencies, industry, and academia for relevant counterterrorist projects.[14] Additional cooperative measures were also arranged with provincial security and health authorities to cope with the CBRN threat. This network also proved useful during the 2003 outbreak of severe acute respiratory syndrome (SARS).[15] Indeed, this crisis provided a powerful incentive to restructure Health Canada and create the Public Health Agency of Canada.

Of major importance was Ottawa's official statement on counterterrorism, entitled *Securing an Open Society: Canada's National Security Policy* (2004). A brief but focused document, it addressed three core security issues: protecting Canadians at home and abroad; ensuring that Canada was not a base for terrorist activity; and expanding Canada's contribution to international security projects. In addition, several new security agencies were created: the Integrated Threat Assessment Center, to co-ordinate cooperation among the RCMP, CSIS, and National Defence, and the Department of Public Safety and Emergency Preparedness "to co-ordinate government-wide responses to emergencies and to manage national security and intelligence issues."[16]

This chapter will address five major questions: (1) How did the terrorist attacks of 2001 affect Canada's security relationship with the United States? (2) What role did the Global Health Security Initiative play in Canada's evolving system of health security at the national and international levels? (3) To what extent did Canadian participation in major American WMD exercises encourage greater cooperation among health security officials of the two countries? (4) In what ways has Canada responded to the challenge of dual-use biotechnology at the international and national level? (5) Were Canadian attempts to deal with domestic terrorists effective in maintaining the balance between national security and individual liberties?

Canada's Response to the Terrorist Attacks of 9/11

During the fall of 2001, the United States was the target of two major terrorist attacks. One of these was the coordinated hijacking of four commercial aircraft, which were used as flying bombs against high-profile targets – the World Trade Center in New York City and the Pentagon in Washington, DC. Over 3,000 people died in these devastating events, the worst terrorist incident in American history. In contrast, the anthrax letter bomb attacks of October–November 2001 killed five of the twenty-two persons infected with *Bacillus anthracis*. Despite this low fatality rate, many Americans were traumatized by this unique terrorist incident, including the thousands who had to be medically treated for possible anthrax exposure and the tens of thousands more who believed that future attacks were imminent.[17]

The Amerithrax crisis had several stages. On 18 September envelopes containing *Bacillus anthracis* were sent to five prominent TV and print media addresses in New York City and Miami, Florida. In all cases the language of the letters was meant to convey the impression that they had been sent by radical Islamic organizations. The next wave of attacks occurred three weeks later, when anthrax-contaminated letters were received by two prominent Democrats – Senator Thomas Daschle, Senate majority leader, and Senator Patrick Leahy, chairman of the Senate Judiciary Committee. This time, the *Bacillus anthracis* payload was more lethal, containing a high concentration of spores that were carefully milled for maximum aerosol dissemination.[18] This reinforced suspicion among CDC experts and the FBI that the perpetrator might be a well-trained scientists who had access to the militarized Ames strain of anthrax and "sophisticated equipment ... [for] the use of special additives or coatings."[19] Yet despite this emerging profile, eight years passed before FBI agents had sufficient evidence to launch legal proceedings. All this changed in July 2008 when the Bureau charged Dr Bruce Ivins, a prominent Detrick scientist, with criminal conspiracy; but since Ivens committed suicide before a trial could take place, the question of his guilt remains unresolved.[20]

The consequences of the long Amerithrax investigation had a profound psychological impact on defence scientists in the United States and Canada. This was not surprising given the FBI's shifting list of possible suspects at Detrick (USAMRIID) and Dugway.[21] Not surprisingly, because of the level of collaboration between CANUKUS and the TTCP, anthrax specialists at DRDC Suffield and Porton Down were added to the list of suspects. In the case of Suffield, for example, investigators discovered evidence that small samples of the Ames anthrax strain had been received from Bruce Ivins, who also served as an adviser during the upgrade of the DRDC BSL-3 laboratory prior to 2001.[22] While no public discussions about the possible involvement of Suffield scientists took place, there were other aspects of the Canada-US relationship that received considerable attention. Of particular importance were the security dimensions of the northern border, reinforced by allegations that some of the 9/11 terrorists had come from Canada. Many of these concerns were expressed by Secretary of State Colin Powell when he warned that "some nations need to be more vigilant against terrorism at their borders if they want their relationship with the U.S. to remain the same"[23]

In reality, Canadian authorities had already adopted many counterterrorist measures. These included an immediate response to the closing of American air space on 11 September when some 224 commercial US aircraft were diverted to various Canadian airports, providing sanctuary for more than 33,000 passengers and air crew. Prime Minister Jean Chrétien also established the Ad Hoc Committee on Public Security and Anti-Terrorism to provide an official response to the crisis. But most important of all was the government's Anti-Terrorist Plan, which had five major objectives: prevent terrorists from entering Canada; protect Canadians from terrorist acts; develop tools to identify, prosecute, and convict terrorists; keep the Canada-US border open and secure; and cooperate with the international community to counter terrorist threats. In order to finance these undertakings, Ottawa announced that it would invest $7.7 billion for counterterrorist purposes over five years, with $280 million being immediately allocated to the Royal Canadian Mounted Police and the Canadian Security Intelligence Agency.[24]

Most of the Canadian government's security initiatives during the fall of 2001 had two goals: to protect Canadians from CBRN terrorism and to reassure Americans that cross-border bioterrorist attacks could not take place.[25] A key dimension of this dialogue was Canada's efforts to ensure that the disastrous closing of the US-Canada border on 9/11 was not repeated.[26] To achieve this goal, intense negotiations took place between Foreign Affairs Minister John Manley and Tom Ridge, the defacto coordinator of US homeland security, which eventually produced the Smart Border Agreement of December 2001.[27] At the same time, improvements were made in the various bilateral military cooperative arrangements, both through the MOU/CANUKUS system of biodefence research and under the umbrella of the North American Aerospace Agreement (NORAD). Another important development was the October 2002 creation of the US Northern Command (USNORTHCOM), whose mandate included assistance to civil authorities in crises "related to terrorism threats, critical infrastructure protection, cross-border information sharing, public health and marine security."[28]

During this critical period, Health Canada's Centre for Emergency Preparedness and Response (CEPR) was also given expanded manpower and financial resources.[29] Under the leadership of Ron St John CEPR emphasized three programs: upgrading the national emergency stockpile of antibiotics and vaccines (NESS); fast-tracking the shipment of anthrax vaccine from its Michigan-based producer; and securing a guaranteed supply of ciproflaxin (CIPRO), an effective antibiotic against anthrax.[30] In addition, a number of external health security initiatives were instituted, such as the Global Surveillance System (CPHIN II), whereby disease outbreaks were tracked throughout the world in real time, with the relevant data being disseminated in six languages. But most important of all was the November 2001 formation of the G-7 Global Health Security Initiative "to strengthen the public response to the threat of international biological, chemical and radio-nuclear terrorism."[31]

Another major strategy was to recruit the country's leading life sciences to Health Canada's biodefence programs. An important aspect of this campaign was the January 2002 symposium "Biological Terrorism: Canadian Research Agenda," jointly sponsored by CEPR and the quasi-independent Canadian Institutes of Health. Held in Toronto, the meetings attracted a virtual who's who of North America's academic microbiologists and virologists, along with key biodefence specialists from the Centers for Disease Control (Atlanta), the US Army Medical Research Institute of Infectious Diseases (Fort Detrick), and the National Microbiology Laboratory (Winnipeg). The symposium addressed three key questions: What unique research contributions can Canadian scientists make in dealing with bioterrorism? How can Canada best interact with international research efforts in this field? What specific research initiatives should Canadian researchers pursue?[32] After much debate, a report card of Canada's biodefence potential was prepared. On the positive side, the delegates applauded the level of expertise among scientific workers in the fields of bacterial, viral, and vaccine research in Canadian universities, reinforced by their counterparts in private biotechnology companies such as Aventis Pasteur and Cangene. Other assets included such cooperative programs as the Canadian Bacterial Disease Network (CBDN) and the Canadian Network for Vaccines and Immunotherapeutics (CANVAC), which brought together researchers from academia, industry, and government.

On the negative side, the delegates were reminded that Canadian health security programs faced formidable problems in terms of financial resources, jurisdictional barriers, and structural divisions among the three major groups of stakeholders. For example, most research in the life sciences was academic, with only limited connections with either government or industry, and the country's vaccine capabilities were underdeveloped, driven by corporate rather than security goals. Significantly, the delegates agreed that the most urgent goal was to expand the research links between Canada's life scientists and their research colleagues in the United States "in order to forge an integrated biosecurity-related agenda."[33]

Meanwhile the Chrétien government was seeking legislative approval for its expanding security system. The most important component was the Anti-Terrorism Act (Bill C-36), which strengthened Ottawa's counterterrorist capabilities by providing severe penalties for terrorist offences; facilitating police use of electronic surveillance devices; and justifying the arrest and detention of suspected terrorists, on the basis of secret information.[34] Justification for these draconian measures was provided by Minister of Justice Anne McLellan in her 20 November 2001 appearance before the House of Commons Justice Committee: "Perhaps the greatest gap in the current laws is created by the necessity of preventing terrorist acts ... Bill C-36 will allow us to convict those who facilitate, participate in and direct terrorist activity and these must include preventative measures which are applicable whether or not the ultimate terrorist acts are carried out."[35] In contrast, Canadian civil liberties spokespeople denounced Bill C-36 as a dangerous violation of democratic principles, worse even that the US Patriot Act, which was "not known for its leniency."[36]

The second part of the government's security agenda was contained in the Public Safety Act (Bill C-42), an omnibus measure that sought to amend eighteen federal laws associated with national security. However, this sweeping and ill-conceived measure soon encountered strong opposition in the House of Commons, with critics challenging the sections on airport security, the terms relating to the transportation of hazardous material, and the proposal for domestic legislation consistent with the Biological Weapons Convention. In the latter case, the issue was eventually resolved on 6 May 2004, when the Biological and Toxins Weapons Implementation Act (C-7, Part 23), came into force, with a mandate "to prohibit biological weapons, as well as biological agents of types and quantities that have no justification for peaceful purposes ... [and] provide more legal basis for the regulation of dual-use biological agents."[37]

Global and Canadian Health Security

The establishment of the Global Health Security Initiative in December 2001 demonstrated how personal relations between Canadian and American officials were important factors in the development of important bilateral security policies. The process began in October 2001 with the Washington-based meetings between Alan Rock, Minister of Health Canada, and Tommy Thompson, US Secretary of Health and Human Services, designed to ensure that adequate supplies of anthrax and smallpox vaccines would be available if a bioterrorist attack should occur in either country. Rock and Thompson also took advantage of this opportunity to discuss the agenda of the forthcoming meeting of G7 ministers of health in Ottawa, both agreeing that a broad system of biodefence should be established.[38] As a result of their combined efforts, G7 ministers of health were convinced of the advantages of creating the Global Health Security Initiative (GHSI) as a way of improving the response of both individual countries and the organization as a whole in dealing with the threat of biological, chemical, and nuclear attacks.

Administratively, the GHSI was a loosely structured collaborative agreement that operated on the principle that each country would develop its own health security strategies, consistent with the organization's basic principles and procedures. Various patterns of cooperation soon emerged, with Japan and Italy focusing on chemical weapons, France specializing in radiological devices, and Mexico contributing its epidemiological expertise.[39] Initially, Canada, Germany, the United Kingdom, and the United States were primarily responsible for biodefence research, in part because most of the world's fifteen high-containment laboratories (BSL-4) were located in these countries.[40] Significantly, Director General Frank Plummer of the Winnipeg National Microbiology Laboratory assumed a major role in developing and coordinating the Global Health Security Laboratory Network for "improving collaboration among high level laboratories, expanding linkages in order to strengthen the ability to rapidly and accurately diagnose diseases whether naturally or intentionally occurring, and strengthening overall global public health capacity."[41]

Since 2001, eleven GHSI ministerial meetings have been held annually in different capital cities, reinforcing the high status the organization enjoys within its eight member states. Equally important was the work of the Global Health Security Action Group (GHSAG), an ad hoc executive body that directs the organization's defensive BW research and testing projects. Initially, the GHSAG was primarily concerned with possible bioterrorist assaults involving smallpox, anthrax, or plague, as well as developing a consultative system that involved regular laboratory consultations and cooperative field trials. One of the most successful of these was the 2003 Ottawa-based Exercise Global Mercury, based on a simulated bioincident in which "self-inoculated terrorists ... [became] vectors to spread smallpox in target countries."[42] During the past nine years, however, GHSAG has become extensively involved with the challenges posed by emerging infectious diseases such as SARS, which crippled Toronto's health care system during the spring of 2003. Avian influenza (H5N1) has been another major concern, given its high fatality rate (60 per cent), a grim situation that has greatly concerned scientists of the GHSAG Working Group on Pandemic Influenza Preparedness.[43] In this undertaking valuable assistance was received from the World Health Organization, particularly after 2005 when the WHO branded pandemic influenza as one of the greatest threats to global health.[44]

North American WMD Exercises: The TOPOFF Series

An essential dimension of Canada's biodefence strategy was its involvement with similar US programs, particularly the congressionally mandated TOPOFF series of exercises. Unlike the ad hoc TOPOFF-1 experience, planning for the new set of exercises in 2003 was firmly under the control of the newly created Department of Homeland Security. It also received strong endorsement from prominent members of the Bush cabinet, notably Attorney General John Ashcroft and Secretary of State Colin Powell, who described the event as the "embodiment of international cooperation that surfaced in the aftermath of September 11th."[45] For their part, Homeland Security officials issued a press release outlining the major goals of the exercise: namely, to strengthen the ability of all government departments to deal with WMD terrorism and to coordinate domestic counterterrorism strategies with an international response system.[46] This broader approach was evident when representatives of Canada's health, defence, and law enforcement and intelligence agencies were invited to participate in TOPOFF-2, working closely with their US counterparts at the departmental and political levels. The first stage of the exercise took place on 10 May, with a simulated pneumonic plague attack on three major public facilities in Chicago: the central railway station, O'Hare International Airport, and the United Center, where a hockey game between the Chicago Blackhawks and the Vancouver Canucks was under way. Since there was no prior warning of this plague attack, all those infected in those locations became disease carriers, thereby establishing a chain of disease transmission throughout the United States and Canada. Two days later (12 May), the simulated incident featured the release of a radiological device in Seattle, causing numerous casualties. And to keep things interesting, British Columbia was faced with two terrorist threats: a chemical weapons assault on Vancouver, and the possibility that a "dirty bomb" would be launched against Victoria by a foreign trawler steaming towards the BC coast.[47]

Did high-ranking government officials and first responders meet the challenge of working together effectively during this type of public health disaster? In the case of the Chicago plague attacks, the American organizers were rather disappointed with the collective response. But in some ways this was the nature of the terrorist weapon since plague has a relatively long gestation period, which meant that several days passed before local health officials realized they faced an epidemic. By this stage in the exercise, the hospital system was inundated with seriously ill patients, and there were reports of severe shortages in supplies of vaccines and antibiotics.[48] On the other side of the border, Canadian scientists and security officials could not be effectively evaluated because of the sudden advent of the SARS epidemic, which not only drained the resources of Health Canada, but also forced the Ontario government to totally withdraw from the exercise. But despite this unfortunate situation, CEPR planners regarded their participation in this simulated North American health crisis as a valuable experience. Not only did it force federal ministers to appreciate the devastating consequences of a bioterrorist attack, it also enhanced the status of the National Counter-Terrorism Plan (NCTC) and facilitated close cooperation between Canadian and US biodefence officials.[49]

These connections were reinforced through joint participation in TOPOFF 3 (April 2005), TOPOFF 4 (October 2007), and the new Tier 1, National Level Exercises of 2009–11.[50] The most important of these events was TOPOFF 3, which was the last major US exercise to focus on the bioterrorist threat. It featured a secret BW attack on New Jersey, along with a separate but complementary Canadian operation (Triple Play).[51] Despite this duality, Canadian public health officials such Ron St John, Director General of the Centre for Emergency Preparedness and Response, who directed the Canadian program, the TOPOFF experiences represented "an important learning curve for Canadian counter-terrorist specialists."[52]

Canada and International Biodefence Initiatives

Because of the international dimensions of the terrorist threat Canada worked closely with a number of other countries in developing shared counterterrorist strategies.[53] Of particular importance were arrangements with the European Union, which in October 2001 established a

Health Security Committee as part of its program of preparedness and response against biological and chemical attacks (BICHAT).[54] These initiatives were reinforced by the activities of a fourteen-member task force consisting of CBW experts from high-profile laboratories such as Germany's Robert Koch Institute and France's Pasteur Institute, who were determined that the EU countries should not fall behind the United States in biodefence capabilities.[55] Not surprisingly, Canada's closest EU contacts were with the United Kingdom, because of their mutual involvement with the CANUKUS system and the Tripartite Technical Cooperation Program (TTCP). In addition, British civil defence officials already had impressive counterterrorist credentials, gained from their long experience in dealing with the Irish Republican Army (IRA), which fortunately did not use WMD devices. In 2001 most of the essentials of the UK biodefence program were outlined in the Ministry of Defence's report, *The Strategic Defence Review: A New Chapter*, which stressed "working out the right structures, the range of equipment and the people needed to deal with the evolving technologies and methods available to terrorist groups."[56]

Canada was also involved with the CBRN programs of the North Atlantic Treaty Organization. Cooperation between Ottawa and Brussels had been established during the 1950s, when the military dimensions of a possible biological or chemical attack were of primary concern. And this remained the focus of the Alliance until 2000 when, under pressure from Washington, it created the Centre for WMD under the chairmanship of Canadian diplomat Ted Whiteside. NATO's transition towards a more coordinated biodefence strategy was reinforced at the Prague Summit of May 2002 when it was decided that an integrated reaction team should be established to deal with bioterrorist incidents, that the biodefence stockpiles of all member nations should be upgraded, and that training programs would be provided by the NATO Bio/Chemical Defence School.[57] These initiatives were complemented by a series of programs adopted by NATO's Civil Emergency Planning Directorate (CEPD), with its emphasis on the all-hazards civil defence approach rather than military models in dealing with CBW terrorism.[58]

In carrying out this work NATO biodefence experts worked closely with infectious disease specialists at the World Health Organization, despite its traditional stance of political neutrality.[59] In June 2002, for example, the World Health Assembly passed a resolution calling upon the Director General "to examine the possible development of new tools, within the mandate of WHO ... concerning the global public health response to prevent, contain or mitigate the effects of deliberate use of biological, chemical or radiological agents to cause harm."[60] Another major development was the 2005 revision of the International Health Regulations (IHR), which established the legal obligation for all WHO member states to maintain core surveillance, reporting, and response capabilities "against all types of biological threats, be they natural or deliberate."[61]

Despite Canada's range of biodefence commitments, arms control specialists at DFAIT still regarded the Biological Weapons Convention as the most important instrument in preventing the proliferation of biological agents and delivery systems. By the time of the Fifth Review Conference of 2001 there was considerable confidence in Ottawa that the verification protocol would provide the necessary machinery for controlling the BW ambitions of rogue states or terrorists groups. The proposed system had a number of impressive enforcement strategies: specific lists of prohibited agents / toxins; systematic investigation and reporting procedures; expanded confidence-building measures; and encouragement for countries to pass national implementation legislation consistent with the principles of the BWC.[62] These assets were not, however, sufficient in convincing the world's most powerful states that the Protocol should be adopted. In particular, the administration of George W. Bush regarded the inspection system as deeply flawed, because it both undermined US national security and threatened the prosperity of the country's pharmaceutical and biotechnology sectors. Nor was the US government swayed by the arguments of its closest allies to give the new system an opportunity to deal with BW proliferation problems. Indeed, during the December 2001 Review Conference John Bolton, Undersecretary of State for Arms Control and International Security, outraged the 145 delegates by demanding that the Protocol be immediately terminated. In a flash, ten years of creative international negotiations came to an end.[63]

The response of Canada's diplomats to the American knockout punch was mixed. At the formal level, it was business as usual; but behind the scenes there was bitterness over Washington's secretive and negative policies. On the other hand, DFAIT officials were not convinced by a proposal from pro-Protocol spokesmen that the verification process should continue without the United States, citing the precedent that the 1999 Land Mines Convention (the Ottawa Process) had achieved its goals without Washington's support.[64] How was it possible, the Canadians asked, to operate the Biological Weapons Convention without the United States, given that country's dominant position in the pharmaceutical / biotechnology fields and in the biodefence sector? As a compromise, however, Ottawa supported informal discussions between interested groups about how to protect the BWC system from total collapse.[65]

During the past nine years Canada has encouraged the incremental approach towards improving the operation of the Biological Weapons Convention through a series of annual meetings.[66] The first of these took place in August 2003, when Experts of States Parties recommended that all member countries adopt national legislation consistent with the terms of the BWC, while also establishing government oversight procedures for pathogenic microorganisms and toxins.[67] The following year, the Experts Group explored how international investigations of alleged use of biological weapons, or suspicious outbreaks of disease, could be improved both through greater cooperation among member countries and with the assistance of the WHO.[68] The final set of meetings, held in June 2005, were dedicated to the study of international scientific codes of conduct, based on the previous work carried out by the British Royal Society, the International Committee of the Red Cross, and the US National Academy of Sciences. All of these groups addressed the crucial question of how to balance scientific freedom and national security concerns within the context of rapid changes in the fields of biotechnology and molecular biology.

Canada and the Global Partnership Program

In July 2002, the G8 summit at Kanasaksis, Alberta, created the Global Partnership against the Spread of Materials of Mass Destruction from the former Soviet Union. This new organization had three major goals: to ensure the safe storage and transfer of CBRN materials; to safeguard former weapons facilities from sabotage or theft; and to prevent the illicit transfer of weapons or agents[69] Since 2002 Canada has spent over $1 billion carrying out its obligations, which were administered by the Global Partnership Program (GPP/IPX), located in the Department of Foreign Affairs and International Trade. Overall, this system has been quite successful in providing other professional opportunities for former BW scientists who had previously worked at the approximately sixty-five military and "civilian" Soviet-era research institutes.[70] In return, Canada and its allies posed several questions to these former BW weaponeers: How was their scientific expertise utilize under the Soviet system? How many of the eighty viruses, bacteria, fungi, and toxins that the CDC listed as dangerous to human health or agriculture were studied in Russian laboratories? And to what extent were Russian life scientists aware of the consequences of genomic sequencing and genetic engineering?[71]

Biodefence and the Challenges of Dual-Use Biotechnology

The year 2003 brought significant changes to the biodefence policies and priorities of the United States and its allies. The second invasion of Iraq, ostensibly to destroy Saddam Hussein's weapons of mass destruction, had a number of important ramifications. On the one hand, the failure to discover the reputed caches of biological, chemical, and nuclear weapons discredited the administration of George W. Bush and undermined the reputation of Secretary of State Colin Powell. At the same time, fears of reprisals against the US homeland from supporters of Iraq dramatically increased the level of US security and an acceleration of biodefence programs. Much of this activity was coordinated by the newly created (March 2003) Department of Homeland Security (DHS), with its broad mandate to assume the leading role in civil emergency planning. By 2007, the DHS in partnership with the Department of Health and Human Services (DHHS), had developed an extensive biodefence system, costing upwards of $50 billion.[72] Among the many new programs was the Biomedical Advanced Research and Development Authority (BARDA), with its emphasis on an all-hazards approach, which equated terrorist attacks with major natural disasters such as Hurricane Katrina and the outbreak of pandemic influenza.[73]

Scientific responsibility and ethics were another dimension of the American biodefence dialogue because of growing concern about high-risk recombinant DNA experiments. The severity of this problem was revealed in several major controversies. One of these was the Australian Interluken 4 experiments with mousepox when, instead of protecting the mice involved in the experiment, the drugs turned off their immune system and caused instant death. Another incident involved scientists at the State University of New York (Stony Brook), who used the genetic map of the polio virus to construct a synthetic version of this terrible disease.[74] Public apprehension over these types of scientific experiments created demands that either the scientific community improve its self-monitoring programs or that the US government establish restrictive research guidelines. Not surprisingly, this latter proposal was bitterly opposed by microbiologists, virologists, and other medical specialists, who resented any form of blanket scientific censorship, particularly if carried out by ill-informed politicians and heavy-handed law enforcement officials.[75] These fears were intensified by reports that the FBI, because of its sweeping powers under the Select Agent guidelines, was targeting certain high-risk scientific researchers. The most controversial case involved Dr Thomas Butler, a microbiologist at Texas Tech University, who was charged under the Patriot Act with illegally storing and exporting samples of *Yersinia pestis* (plague). Despite appeals from the presidents of the National Academy of Sciences and the Institutes of Medicine, who warned that this case would deter many outstanding US scientists from becoming involved in biodefence research, Butler was sent to jail.[76] Other critics pointed out that intrusive biosecurity measures directed against foreign-born researchers would also detract from the country's ability to develop scientifically advanced medical counter-measures.

In 2003 the United States National Academy of Sciences published an influential report on the challenge of dual-use research entitled *Biotechnology in an Age of Terrorism* (Fink Report). Fundamentally, the Fink Committee sought to address problems associated with the misuse of biotechnology in order to prevent "potential catastrophic consequences." While many areas of "dangerous" research were considered, special emphasis was placed on experiments that "1. demonstrate how to make vaccines ineffective; 2. create recombinant organisms with increased resistance to vaccines and antibiotics / antivirals; 3. increase the virulence of known pathogens or make non-pathogens virulent; 4. increase the ability of pathogens to transmit from host to host; 5. expand the pathogen's host range; 6. allow a pathogen to escape diagnosis or detection; 7. facilitate weaponization of any biological agent or toxin."[77] In 2004, based on the recommendations of the Fink Report, the US Department of Health and Human Services created the National Science Advisory Board on Biosecurity (NSABB), with a mandate to identify research "that may require special attention and security ... [and] professional codes of conduct for scientists and laboratory workers."[78] It was also assumed that the NSABB, because of the high scientific status of its twenty-five permanent members, would work effectively with the country's 400 Institutional Biosafety Committees in dealing with possible dual-use problems.

Overall, the response of the US life science community to the formation of the Board was quite positive, although it was noted that undue attention was being placed on academic scientists while their counterparts in the biotechnology-pharmaceutical and defence sectors were judged by different standards. But these criticisms have generally been allayed during the past eight years largely because the National Science Advisory Board for Biosecurity has provided a valuable forum where questions about dual use could be discussed in an open and informed fashion.[79] Above all, the Board has stressed the need to balance the rights of scientific inquiry with the dictates of national security and public safety, while avoiding being "trapped in the examples and mindset of the past ... of what constitutes a potential risk."[80] According to David Franz, former commander of USAMIRID, this was particularly important for scientists like himself, who had spent much of their earlier careers involved with military biodefence planning, when "we were thinking about a cloud of Soviet made bugs coming across the Fulda gap against our forces."[81] Franz and his NSABB colleagues did, however, concede that the threat of bioterrorism necessitated constant vigilance, particularly since developments in the biosciences made it possible for them to establish "a footprint of substantial proportion today that they couldn't have accomplished 20 years ago before aerosol particle technology."[82] In their opinion, this meant that enhanced BW delivery systems represented "a clear and imminent danger" and that traditional BW agents could be made more virulent and resistant to antibiotics through genetic engineering.

Many of these concerns were discussed in a follow-up study by the National Academy of Sciences, *Globalization, Biosecurity and the Future of the Life Sciences* (2006). Unlike the Fink Report, this new study stressed the value of developing global biosecurity strategies, rather than taking a narrow "US first" approach.[83] Another key recommendation was for the establishment of an interdepartmental scientific advisory group that would assist US security agencies in developing "concepts, plans, activities, and decisions ... about advancing technologies and their potential impact on the development and use of future biological weapons."[84] The new NAS study believed that the Select Agent system, with its rigid classification of dangerous pathogens, should not be continued because it ignored rapid changes in the life sciences and enabling technologies, thereby making the US vulnerable to the next generation of biological agents.[85]

One of the most controversial new fields of research was synthetic organisms, which was the focus of a special 2008 NSABB report. Here, three important issues were explored: How can possible risks associated with the generation of novel organisms be addressed?[86] What strategies can be employed to safeguard against the misuse of synthetic biology and associated technologies?[87] How can global cooperation in addressing related biosecurity concerns be encouraged? While the Board strongly endorsed accelerated research in the field of synthetic genomics by creating synthesizing and then combining fragments of DNA, it also warned about the dangers of creating dangerous pathogens "*de novo,* because of serious threats to American public health, the environment, and national security.[88] The report also pointed out that while researchers had successfully created infectious chimeric viruses using combinations of genomic material from various select agents, "these novel organisms did not fit into traditional classification schemes."[89] But the most controversial aspects of synthetic genomic research was the 2005 sequencing of the genome of the 1918 influenza A virus (H1N1), which many microbiologists predicted would increase the possibilities of a BW terrorist attack, since the Spanish flu virus was "perhaps the most effective bioweapons agent now known."[90]

While the National Science Advisory Board on Biosecurity was an exclusively American scientific watchdog, it served as model for other countries.[91] In Canada, for instance, virologist Grant McFadden of the University of Western Ontario was commissioned to conduct a survey of about "possible steps that could be taken in Canada to minimize potential risk from biotechnology misuse in the future."[92] In carrying out this 2005 research project, McFadden consulted with a wide range of experts from academia, industry, and government to find out whether problems existed, and, if so, what role Ottawa should assume in providing solutions. Overall, there was a consensus on a number of issues. First, while Canada should improve its monitoring of dual-use biotechnology, it was felt that a legislative approach would be ineffectual, given rapid changes in the life sciences. Second, there was general agreement that American initiatives, such as the NSABB, would not work in Canada because of its perceived bias against academic scientists. As one respondent put it, "We need to exchange science ... but not copy their laws."[93] But opinion was divided on what kind of "Made in Canada" monitoring system should be created. For some scientists, the preferred option was to expand the responsibilities of Health Canada / Public Health Agency of Canada, while others favoured the establishment of an independent Office of Dual Use Technology that "could serve both national and international agendas ... as a criterion for participating as an international player."[94] All of these issues were subsequently discussed at the March 2006 National Forum on Dual Use Biotechnology, jointly organized by Health Canada and DRDC Ottawa. This Ottawa-based symposium featured an impressive group of biotechnology experts from academia, industry, and government, who were asked to consider three major questions: How should the federal government foster awareness of dual-use problems? What specific research and ethical issues should be addressed? Was it possible to develop guidelines that would "ensure the safety and security of Canadians while not impacting freedom to operate"?[95] Unfortunately, despite this promising start, the Forum had a very short life span: within two months the entire process was scuttled by the hierarchy of the health and defence departments for reasons that were not explained.[96]

While Canada, unlike the United States, did not develop a national biotechnology oversight system, officials did take some ad hoc measures in this direction. In 2009, for example, the Human Pathogens and Toxin Act (HPTA) came under parliamentary review, with its goal of upgrading Canada's biosafety / biosecurity standards in order to deal with the biotechnology challenges of the twenty-first century.[97] According to health officials, all pathogens, whether domestic or imported, should be under the regulatory control of the Public Health Agency of Canada (PHAC) to ensure that "no person may carry on activities with these dangerous substances without a licence and without complying with the laboratory biosafety guidelines."[98] Under this system, five different categories of pathogens were created, with most bacterial BW agents being placed in risk group three, while the fourth category contained deadly viruses such as Ebola, Marburg fever, and Lassa fever. [99] The next stage in the implementation process required each of the 4,500 laboratories targeted by the legislation to provide an inventory of its collection of pathogens and toxins. Once this was completed, PHAC officials would begin monitoring the issuance of biosafety licences, the appointment of biological safety officers, and the creation of security screening protocols.[100]

Initially, Canada's life scientists viewed the Human Pathogens and Toxins Act with considerable scepticism.[101] This was evident during the February 2009 hearings of the House of Commons Standing Committee on Health when representatives of the country's leading universities, research institutes, and scientific organizations complained about PHAC's lack of prior consultation and expressed outrage that relatively safe risk group two pathogens were being included within the proposed system.[102] Nor did these scientists approve of the HPTA's schedule of penalties for non-compliance, which was denounced as a misguided attempt to impose US types of restrictions on Canada's life science community.[103] Another set of criticisms came from representatives of the Ontario Agency for Health Protection, who predicted that the legislation would cause unnecessary duplication, expanded costs, and inefficiencies since Ontario laboratories "already have security checks in place for a number of individuals who are working in laboratories … who have access to certain types of pathogens."[104] While the bill was eventually passed in July 2009, it remained an inadequate attempt to deal with the problems of dual-use biotechnology[105]

Canada and Terrorist Threats since 2001

Since 2001, Canada has experienced only three terrorist incidents, compared with over 150 in the United States. It has also avoided the types of devastating attacks that occurred in Madrid in March 2004 or in London in July 2005.[106] But even more serious terrorist threats soon emerged. In May 2007, for example, British and American security officials narrowly averted a coordinated terrorist plot to use liquid explosives against a number of commercial aircraft travelling between London's Heathrow airport and North American destinations.[107] Concerns about airport security were reinforced on 24 December 2009 when an al Qaeda operative tried to ignite an explosive on Northwest Airlines Flight 253, just before the plane landed at Detroit's international airport.[108] That same month, Ontario's Attorney General announced that six of the remaining "Toronto 18" terrorist group members would be tried in 2010, bringing closure to a legal process that had begun three years earlier.[109]

Canada's counterterrorist system was institutionalized in April 2004 with the publication of the federal government's policy statement, *Securing an Open Society: Canada's National Security Policy*.[110] Under these new guidelines, four security organizations were given authority to deal with the terrorist threat: the Department of Public Safety, the Integrated Threat Assessment Centre, the National Security Advisory Council, and the Cabinet Committee on Security, Public Health, and Emergencies. In addition, considerable attention was focused on security at the Canada-US border, largely through the new Canada Border Services Agency and more rigorous enforcement of the 2001 Immigration and Refugee Protection Act.[111] Another way of dealing with suspected radicals was through the use of security certificates, whereby non-citizens living in Canada could be detained and deported on the grounds that they presented a threat to national security.[112] An even more controversial practice was Canada's clandestine involvement with the US system of extraordinary rendition of Canadian citizens of Muslim background.[113] According to Justice Dennis O'Connor, who supervised the 2004 Arar Commission of Inquiry, Canadian security agencies routinely exchanged incriminating and incorrect information with their counterparts in the United States. As a result, a number of innocent people were subsequently detained by US officials without charges, and then secretly transported to countries such as Syria, Jordan, and Egypt, where they endured sustained torture.[114] As Justice O'Connor explained,

> The RCMP provided American authorities with information, including the entire database from the aforementioned terrorist investigation, in ways that did not comply with RCMP policies requiring screening for relevance, reliability and person information ... The RCMP provided American authorities with information about Mr. Arar without attaching written caveats, as required by RCMP policy, thereby increasing the risk that the information would be used for purposes which the RCMP would not approve, such as sending Mr. Arar to Syria ... In November 2002, CSIS received information about Mr. Arar from the Syrian Military Intelligence (SMI) and did not do an adequate reliability assessment as to whether the information was likely the product of torture ... In March and April 2003, DFAIT failed to take steps to address the statement by Syrian officials that CSIS did not want Mr. Arar returned to Canada.[115]

There are several explanations for this abuse of power by Canadian security officials.[116] First was the powerful incentive to reassure US authorities that Canada was ferreting out Islamic terrorists such as Ahmed Ressam, the 1999 millennium bomber, who wanted to use Canada as a launching pad to attack the United States.[117] A second factor was the growing consensus, on both sides of the border, that well-educated and westernized Muslim terrorists represented a special threat, since they were more qualified to use asymmetrical forms of warfare. This viewpoint was reinforced by the findings of the United States 9/11 Commission, which warned "that al Qaeda has tried to acquire or make weapons of mass destruction for at least ten years ... [and] there is little doubt that the United States would be the prime target."[118] A third element consisted of deficiencies in the counterterrorist and intelligence capabilities of the RCMP, who had re-entered this field after 9/11, despite severe criticism of its performance during the 1970s.[119] And finally, there was the general belief among security officials that the 2001 terrorist bombings represented only the first wave of assaults, with more devastating attacks on North America and Europe already being planned by al Qaeda and its allies. These fears appeared vindicated by the August 2006 Heathrow airplane bomb plot, which, according to British Home Secretary John Reid, "was potentially bigger than the September 11 attacks ... Had this plot been carried out, the loss of life to innocent civilians would have been on an unprecedented scale."[120]

By this stage, Canada had discovered its own homegrown terrorists – the so-called "Toronto 18."[121] On 2 June 2006 it was announced that a combined police operation, involving the RCMP, the Ontario Provincial Police, and Toronto regional police forces, had arrested a number of alleged saboteurs and militants of Muslim middle-class backgrounds. According to news reports, these Toronto-based terrorists were planning to use fertilizer truck bombs against high-profile targets in the province, such as the Toronto Stock Exchange, the regional headquarters of the Canadian Security Intelligence Service, and a regional military establishment. But the most bizarre part of their agenda was an armed takeover of the Parliament buildings in Ottawa, followed by the beheading of Prime Minister Stephen Harper.[122] Public reaction to these disclosures varied. For some journalists the threats were exaggerated and the "plotters" pathetically incompetent.[123] But for the most part the Canadian media and the general public accepted Ottawa's claim that Canada was now a major al Qaeda target.[124] These arrests also attracted the attention of American commentators, with one self-appointed expert on Canadian affairs declaring that this counterterrorist experience would "give Canadians an opportunity to shed their complacency ... and [with] the close cooperation that now exists between the United States and Canada, Canada is more secure than it was before September 11, 2001 – and so are we."[125]

A more complete picture of the Toronto-based terrorists emerged during the series of lengthy trials. It was revealed, for instance, that efforts had been made to organize jihad training exercises in northern Ontario, to smuggle firearms into the province, and to construct bombs through the purchase of three tons of ammonium nitrate "from a truck driven by an undercover police officer."[126] Equally revealing was the terrorists' motivation for being involved in this organization: in some cases, they felt a general sense of alienation from Canadian society, reinforced by specific opposition to NATO's involvement in Afghanistan, while the ringleaders demonstrated a pathological desire to carry out a devastating attack – "an act of Al Qaeda in Canada."[127] The legal proceedings against the accused ended on 23 June 2010, when a Brampton jury found Asad Ansari and Steven Chad guilty of participating in a terrorist organization. By this stage, eleven of the eighteen adults originally charged had been convicted of crimes.[128]

Significantly, news of these judicial results had a marked impact on how Canadians viewed these cases since, "at the time of the arrests, few believed home grown terrorism was a reality in this country or that Muslim youth were being radicalized." This reality check was reinforced in June 2010 when Richard Feddon, head of CSIS, announced that his agency was "tracking more than 200 people with possible links to as many as 50 terrorist groups."[129]

But what if the Toronto 18 had decided to use biological, chemical, or radiological weapons? Could Canada's security system have prevented a bioterrorist incident or dealt with its consequences? These challenging questions received considerable attention from Canada's politicians and law enforcement officials throughout the first decade of the twenty-first century. In December 2004, for example, the Standing Senate Committee on National Security and Defence prepared a critique of the federal government's counterterrorist performance. It was a mixed review. On the positive side, the committee praised the work of the Ottawa-based Threat Assessment Centre in coordinating the security programs of different departments and agencies. Another plus was the Safe Third Country Agreement with the United States, which not only expedited the claims of bona fide refugees, but also allowed immigration officials on both sides of the border to pool their resources in excluding suspected radicals from North America. But the committee also discovered serious deficiencies in Ottawa's performance.[130] These included the systematic underfunding of the Canadian Security Intelligence Service (CSIS), ineffectual security surveillance on the St Lawrence Seaway / Great Lakes system, and continuing problems of airport security.[131]

Unfortunately, improvements were slow in coming. Indeed, it was only in August 2007 that the Emergency Management Act was finally passed, authorizing the Minister of Public Safety "to exercise leadership relating to emergency management."[132] Among Public Safety's top priorities was the establishment of its Government Operations Centre, which had responsibility for coordinating the security needs of twenty specific departments and maintaining close contact with law enforcement and health security authorities at the provincial level.[133] Yet despite these upgrades, in 2008 the Senate Committee on National Security gave Public Safety a failing grade for its inability to reassure Canadians "that essential government operations will function during emergencies."[134] The Senate report also found flaws with the Public

Health Agency's biodefence operation, particularly its ineffective distribution of essential medical supplies through the National Emergency Stockpile System (NESS), which was described as being "more appropriate to the Korean War era than to the needs of first responders today."[135] Nor were critics impressed with delays in the implementation of the National Health Incident Management System (NHIMS), since this was supposed to improve PHAC's response in dealing "with known and emerging infectious disease outbreaks ... either naturally or as a result of a malicious release of a biological agent ... [including] smallpox and pandemic influenza."[136]

Pandemics and Canadian National Security, 2001–2010

In Canada the connection between public health and biodefence is longstanding.[137] Indeed, the overlap between these two health security systems was clearly evident during the influenza pandemics of 1957 and 1968, and during the World Health Organization's long campaign to eradicate smallpox. After 2001, these dual disease threats assumed even greater importance, as was evident in the response towards the anthrax-contaminated letters, the work of the Global Health Security Initiative, the impact of the 2003 SARS epidemic, and the challenge of the 2009 swine flu pandemic.

The SARS Crisis, March–July 2003

The March 2003 outbreak of severe acute respiratory syndrome (SARS) provided a catalyst for changes in Canada's health security system.[138] Toronto was one of the few places in the world where this new and deadly virus gained a powerful foothold, infecting 438 people and killing 44.[139] This emerging infectious disease demonstrated an alarming ability to transmit between people, particularly in Toronto's hospitals, where 40 per cent of those infected were health care workers.[140] According to Dr James Young, who helped direct Ontario's containment program, the initial response was extremely difficult since "we did not know that the cause was coronavirus ... We did not know the duration of the incubation period. We did not know whether it was spread by droplet or by air. We had no reliable diagnostic test, no vaccine, and no treatment."[141] The SARS crisis also demonstrated the clash of priorities between Canadian public health officials and the World Health Organization. This was evident on 23 April when the WHO issued a controversial global travel advisory against visiting Toronto, based on questionable scientific evidence and an apparent indifference towards the severe economic consequences for the city.[142] Worse was to follow.[143] In late May, there was a second SARS outbreak at the North York General Hospital, affecting both patients and medical first responders.[144]

Once the SARS outbreak had abated, a series of studies attempted to determine the lessons to be learned from this major health crisis.[145] The most devastating critique came from Ontario Commissioner Justice Archie Campbell, who concluded that "SARS was contained only by the heroic efforts of dedicated front line health care and public health workers ... with little assistance from the central provincial public health system that should have been there to help them."[146] Based on these observations, Campbell concluded that Ontario's public health system required drastic change and improvement. This was also the opinion of public health officials who had combated the epidemic, and whose opinions strongly influenced the deliberations of the Ontario Expert Panel on SARS and Infectious Disease Control (Walker Commission).[147] As a result, in June 2007 the Ontario Agency for Health Protection and Promotion (OAHPP) was established with a mandate to link "public health practitioners, front-line health workers and researchers to the best scientific intelligence and knowledge from around the world."[148] Similar developments occurred at the federal level, particularly after the recommendations of the National Advisory Committee on SARS and Public Health (Naylor Commission) that the federal government should "ensure not only that we are better prepared for the next epidemic, but that public health in Canada is broadly renewed."[149] Within a year three important administrative changes were implemented: the establishment of the Public Health Agency of Canada, the appointment of a Chief Medical Officer for Canada, and the promotion of Frank Plummer, scientific director of the National Microbiology Laboratory, to the position of Associate Deputy Minister in PHAC.

Did the SARS outbreak provide any guidance in dealing with a possible bioterrorist incident? According to James Young, Ontario's Commissioner of Public Safety, there was a definite correlation between the two forms of disease transmission since SARS had shown itself as "a potential form of bioterrorism." And, indeed, in terms of its impact on the province's hospital system, SARS shared many of the characteristics associated with the bioterrorist scenarios of TOPOFF 2.[150] These linkages were further explored in a 2004 study by the prestigious US Institute of

Medicine, which recommended "that efforts to address microbial threats should encompass and be enriched by existing strategies for defense against bioterrorism ... [since] authorities do not know until well into an outbreak if it is a naturally occurring or man-made threat."[151] Another perspective was provided by Stanley Falkow, a prominent American geneticist: "We face a dilemma that there will be a future time, and it is coming closer and closer, when in wrong hands biotechnology making use of genome information could create a novel pathogen with unique properties ... the availability of the complete genome sequence of human isolates of SARS could be used by a very sophisticated bioterrorist as a pathway to synthesize a new version of the SARS virus. Which do we fear most, nature or bioterrorism?"[152]

Pandemic Influenza: H5N1 (Bird Flu) and H1N1 (Swine Flu)

During the past fourteen years influenza has been regarded as one of the greatest threats to global health – in both its avian and human strains. In 1996, the highly pathogenic form of H5N1 influenza A virus (bird flu) was first detected in geese in Guandong province, China, spreading to poultry in Hong Kong the following year.[153]

What made the latter incident so alarming was not only evidence of human infection, but the fact that six out of eighteen cases proved fatal. By March 2009, there were reports of 411 confirmed human cases in 17 countries, with the total death toll reaching 254, an alarming ratio of 62.4 per cent fatalities.[154] Fortunately, almost all of these cases were associated with sustained high exposure to infected birds, with little evidence of human-to-human transmission.[155] But there was no guarantee that this situation would continue. As a September 2006 report of Health Canada explained, "There are concerns that this virus could mutate – or if someone infected with human influenza also became infected with H5N1 avian influenza, the viruses could 'mix' creating a new strain ... causing an influenza pandemic."[156]

International cooperation in dealing with an influenza pandemic was considerably enhanced in September 2005 when the UN General Assembly launched the Partnership on Avian and Pandemic Influenza. In short order a number of key guidelines were established: increasing awareness of the threat of avian influenza at the national level, developing more effective surveillance systems, and improving transparency in disease reporting.[157] A related trend was the development of health security arrangements at the regional level. At the March 2006 Cancun Summit, for example, leaders of the United States, Canada, and Mexico recommended the establishment of a trilateral pandemic influenza emergency management system based on "a comprehensive, science-based and coordinated approach within North America."[158] It was also stipulated that while the system would be based on mutually acceptable health security principles, domestic influenza response plans would take priority, given the great variance among the three countries.[159] For Canada, the basic guidelines were embodied in the Canadian Public Health Agency's 2004/06 Report on Pandemic Influenza. For Mexico, it was the August 2006 Plan nacional de preparacion y respuesta ante una Pandemia de Influenza.[160] And in the United States, President Bush's National Strategy was issued on 1 November 2005, with the warning that "If we wait for a pandemic to appear, it will be too late to prepare."[161]

On 23 August 2007 Canadian Prime Minister Stephen Harper, President José Calderón of Mexico, and President George W. Bush of the United States issued a series of carefully orchestrated joint statements.[162] One of the most important of these was the announcement that a trilateral health security system had been drafted, under the auspices of the Security and Prosperity Partnership, entitled *The North American Plan for Avian and Pandemic Influenza* (NAPAPI). Among its many features, NAPAPI pledged that if a major health crisis should occur, all three countries would coordinate their efforts "to contain a novel strain of human influenza at its source, slow its spread ... [but] allow the appropriate movement of people and cargo across mutual land borders and ports of entry in a way to achieve the public health objective with minimal social and economic impact."[163] The NAPAPI was quickly approved at the Montebello summit. But to ensure that the plan would not gather dust on the shelf, provision was made for the establishment of a North American Coordinating Body for Avian and Pandemic Influenza, composed of "senior officials from most of the key agencies that would ... play a significant role in promoting coordination among the three countries at senior official levels."[164] During the next eighteen months incremental improvements were made to the NAPAPI operating system.

On 18 May 2009 Dr Margaret Chan, Director General of the World Health Organization, warned delegates at the World Health Assembly in Geneva about an imminent threat from a novel influenza A virus:

> For five long years, outbreaks of highly pathogenic H5N1 avian influenza in poultry, and sporadic fatal cases in humans, have conditioned the world to

> expect an influenza pandemic, and a highly lethal one … As you now know, a new influenza virus with great pandemic potential, the new influenza A (H1N1) strain, has emerged from another source on another side of the world. Unlike the avian virus, the new H1N1 virus spreads easily from person to person, spreads rapidly within a country once it establishes itself … We expect this pattern to continue.[165]

Dr Chan also warned that a number of factors increased the dangers of a deadly influenza outbreak. These included the possibility that it would become more pathogenic through mutation, the prospect that the virus could evade existing antivirals such as Tamiflu, and the challenge of obtaining sufficient vaccine to protect the world's 6.8 billion inhabitants, particularly those living in developing countries.[166]

By June 2009 WHO officials were convinced that swine flu represented a serious global crisis, since the novel virus had spread to 74 countries, with more than 28,700 cases, and 144 confirmed deaths. There was also speculation among some of the world's leading virologists about the possibility of a fusion between the swine and bird flu viruses since "H5N1 is in more than 60 countries … [and] kills half the people it infects."[167] On 10 June Geneva raised the threat level to phase 6, a maximum-response category that required all member nations to be prepared for an emergency situation. Dr Chan was not dissuaded by arguments from WHO critics, who claimed that the pandemic alert system "must consider how deadly the virus is, not just how fast it is spreading."[168]

The American and Canadian responses to the 2009 influenza pandemic reveal a number of interesting similarities.[169] While each country assumed that an early delivery of the vaccine would take place, this did not materialize, given their contractual arrangements with national pharmaceutical companies. Despite the relatively low level of morbidity and mortality during the pandemic, both countries discovered that their hospitals and other health care systems were overwhelmed by infected patients, a situation that was further aggravated by problems of communication between experts in the public health and medical science fields. Despite the aggressive immunization campaign launched by the Public Health Agency of Canada and the US Department of Health and Human Services, many North Americans regarded the newly prepared influenza vaccine as unsafe.[170] In addition, there were concerns about the effectiveness of antibiotics such as ciprofloxacin and doxycycline, on the grounds that while they worked well against anthrax and other Category A biological agents, they had "poor activity against …

Staphyylococcus aureus and Streptococcus pneumoniae, 2 common post-influenza pathogens."[171]

Despite these shortcomings, Canada, the United States, and Mexico avoided serious loss of life during the pandemic. In part, this can be attributed to the positive impact of the North American Plan for Avian and Pandemic Influenza, which was instrumental in keeping borders open and facilitating medical cooperation among the three countries.[172] This was particularly important during the early stages of the outbreak when Mexican public health officials received valuable laboratory assistance from experts at the Winnipeg National Microbiology Laboratory, who were able to sequence and identify the H1N1 virus within twelve hours of receiving the Mexico City isolates.[173] Certainly the three political leaders viewed the NAPAPI as a powerful tool, as was evident in their October 2009 press release at the Guadalajara summit: "North America's coordinated response to the initial outbreak of the H1N1 flu virus has proved to be a global example of cooperation … of a joint response and transparent response."[174]

Did the swine flu pandemic of 2009 compare in any meaningful way with a possible bioterrorist attack? While there was some sporadic discussion of this correlation, overall the linkages between the two disease systems appeared very tenuous, perhaps because the H1N1 outbreak was so mild. Indeed, there was a consensus among medical experts that this outbreak was "the least lethal modern pandemic … [killing] about one of every 2,000 people who get it … [while] the Spanish flu of 1918 killed about 50 of every 2,000, and the 1957 and 1968 pandemics killed about 4 out of every 2,000." On the other hand, some critics pointed out that, given the many failures in the Canadian and American emergency response systems, "if influenza's Big One had struck in 2009, we would have been in a world of hurt."[175]

The Bioterrorist Threat: A 2010 Perspective

On 26 January 2010 the United States Commission on the Prevention of Weapons of Mass Destruction Proliferation and Terrorism warned President Barack Obama that a weapon of mass destruction would "be used in a terrorist attack somewhere in the world by the end of 2013 … [and] that weapon is more likely to be biological than nuclear.[176] This subject was also the focus of Senate bill (S.1649), "The Weapons of Mass Destruction Prevention and Preparedness Act of 2009: "In contrast to nuclear weapons, the technological hurdle to develop and disseminate bioweapons ... and terrorists' known interest in bioterrorism combine to produce an even greater menace ... in part because we are unlikely to realize that an attack has occurred before it begins to kill many of its victims."[177]

These pronouncements could not be dismissed as bioterrorist fear mongering, since these warning had been confirmed by other influential organizations. In 2008, a study prepared by the Office of the Director of National Intelligence, entitled *Global Trends 2015: A Transformed World*, predicted that terrorists would launch a biological weapons attack against the United States during the next five years, since "the diffusion of technologies and scientific knowledge will place some of the world's most dangerous capabilities within their reach."[178] Similar warnings were included in several biodefence studies published in January 2009 by the Department of Health and Human Services, notably *Public Health Preparedness: Strengthening CDC's Emergency Response.*[179] Another dimension of this debate over dangerous pathogens was the June 2010 decision of the World Health Assembly that remaining stocks of *Variola major* would not be destroyed at this time because of US arguments that its biodefence system still required access to the actual virus.[180]

In Canada, the major counterterrorist challenge was not in dealing with abstract threats, but rather in protecting the February 2010 Vancouver Winter Olympic Games from CBRN terrorism. Although the Royal Canadian Mounted Police were given the lead role in directing the work of the Integrated Security Unit, most of the country's security and police agencies were represented.[181] A particular concern was that terrorists might attack the Olympic site with biological weapons, given their portability, ease of use, and high killing potential.[182] For this reason, the Winnipeg National Microbiology Laboratory (NML) agreed to provided a highly specialized mobile emergency response laboratory, which carried out systematic aerosol samples in the restricted Olympic site, using advanced BW detection devices. In case a bioterrorist incident occurred, the NML team had already worked out an elaborate cooperative system with the BC Centres of Disease Control, while at the same time using the resources of its 24/7 operations centre at the National Microbiology Laboratory. At the national level, NML Scientific Director Frank Plummer regularly consulted with officials of the Centre for Emergency Response and Preparedness in Ottawa and the Cabinet Committee on Security, Public Health and Emergencies.

This successful exercise in protecting a vulnerable site from terrorists also involved other security stakeholders. High on the list was the biodefence backup role provided by experts at DRDC-Suffield, although most of their attention was focused on the threat of chemical weapons.[183] Behind the scenes a steady stream of advice was offered by BW experts at the Atlanta-based Centers for Disease Control and the United States Army Medical Research Institute of Infectious Disease (Fort Detrick).

In 2008, however, this easy relationship between the Suffield and Detrick defence science communities was disrupted by dramatic changes in the long-standing Amerithrax case. Not only did FBI spokesmen claim that Bruce Ivins, a senior Detrick scientist, was responsible for the 2001 anthrax attacks, but they also raised questions about his relationship with DRDC Suffield personnel during this period. In seeking more information on these subjects, in March 2012 the author requested clarification from the Canadian Department of Defence about Ivins's role in the transfer of highly refined anthrax from Detrick to Suffield, and whether on-site investigations of the two laboratories had been carried out by either Canadian or American security agencies.[184] Although the DND official response did not address either of these questions, it did provide some useful comments about Canada's contemporary biodefence system:

> During the period of 1997 to 2005, the USAMRIID provided DRDC Suffield with samples of the Ames strain of *Bacillus anthracis*. This exchange was part of a normal collaboration in the development of medical countermeasures ... At no times in that period, or since, did DRDC Suffield hold or possess any strain of anthrax in the dry form such as was used in the US anthrax letters case ... Level 3 organisms are handled at DRDC Suffield in BioSafety Level 3 laboratories (BSL-3) ... Great care is taken to ensure all required measures are respected. Inventories are verified and certified annually. Also, the Biological and Chemical Defence Review Committee (BCDRC) verifies this inventory on an annual basis.[185]

Conclusion

The present and future threat of biological weapons, either from states parties or terrorists, has been a subject of considerable scholarly debate since 9/11. One contentious issue has centred on whether traditional biological weapons, as illustrated by anthrax letter bombs, could still be used effectively against civilian targets, or whether improvements in biodefence capabilities throughout the past decade would prevent a repeat of this kind of BW crisis. Another perplexing question involves the creation of lethal new BW weapons, through genetic engineering, that would not only evade vaccines and antibiotics / antivirals but also have unique aerosol dissemination properties. The most celebrated discussion of this subject was the so-called "seven deadly sins" of irresponsible and dangerous research that were featured in the US National Academy of Sciences study *Biotechnology in an Age of Terrorism* (2003). In turn, this seminal report provided a a catalyst for a wide range of government reports, books, and articles, including the NAS's sequel, *Globilization, Biosecurity, and the Future of the Life Sciences* (2006).[186]

Significantly, biodefence experts did not agree on whether terrorists could obtain and use advanced forms of bioweapons. In recent times one of the most thoughtful proponents of the worse-case scenario has been British BW expert Malcolm Dando, who argues that if terrorist groups develop "a much more systematic application of the new biology to hostile purposes ... [they] may gain the capabilities to cause more human casualties."[187] Barry Kelman, in his book *Bioviolence* (2007), develops an even stronger case, claiming that "bioweapons should be compared to nuclear weapons ... in terms of potential casualties ... [but] are far easier to make than nuclear weapons ... [and] are well within many people's capabilities ... [since] pathogens are naturally available, and refined seed stocks of potentially weaponizable agents are found widely in laboratories around the world."[188]

Other scholars are more sceptical of these "dark fantasies." As one author noted, "despite the appearance of mass destruction scenarios in books, broadcasts and screenplays for 30 years, terrorists have not tried to implement most of those scenarios."[189] American political scientist Milton Leitenberg goes even further by criticizing the failure to differentiate between the intentions of terrorist groups and their capabilities, pointing out that "advanced genetic engineering capabilities are not likely to become available to real world terrorist groups in the near future."[190] Greg Koblentz's recent book *Living Weapons* (2009) agrees, stating that existing terrorist organizations lack the scientific and financial resources to acquire sophisticated bioweapons and that "natural pathogens are lethal and terrifying enough for their purposes."[191]

Unfortunately, there has been little public discussion of these important themes in Canada. In part, this can be explained by the difficulty of obtaining relevant BW information, given Ottawa's rigid "right to know" policies. Another problem is the country's lack of high-quality and independent think tanks, such as the US National Academy of Sciences or the British Royal Society, to provide in-depth studies that Canadian lawmakers would take seriously. Instead, we are left with self-serving and superficial reports from Ottawa's consultant community or periodic scare-mongering media exposés.

This author certainly faced many challenges in trying to understand Canada's contemporary biodefence policies. But fortunately other research options became available, notably the opportunity of interviewing many of Canada's leading defence scientists, who demonstrated a refreshing willingness to discuss key questions. In particular, they were asked whether, in their personal opinion, Canada would be the target of a bioweapons attack. For Frank Plummer, Scientific Director of the National Microbiology Laboratory, the answer was affirmative. He envisaged a scenario whereby terrorists infected with smallpox or Ebola would launch a BW suicide attack because of "the level of fear that bioweapons generate."[192] In contrast, Ron St John, Director General of the Centre for Emergency Preparedness and Response, felt that it was unlikely that Canada would experience a catastrophic BW attack since "terrorists groups lack the necessary scientific skills; it is easier to use readily available explosives ... [and] intelligence and policing capabilities have greatly improved, and medical countermeasures are more effective."[193] Hopefully, neither thesis will be tested.

Conclusion

In 1992 the Department of Foreign Affairs and International Trade, as part of a BWC international exercise in transparency, submitted a summary of Canada's involvement with biological warfare since the Second World War. Unfortunately, this document was seriously flawed in terms of historical accuracy. Instead of providing a reliable account of the country's BW legacy, it attempted to minimize the role that Canadian defence scientists had assumed in the Allied offensive programs during the Second World War and the Cold War. More recent portrayals of Canada's biodefence activities by the federal government, either in its official statements or commissioned studies, have also been unsatisfactory. To make matters worse, few of the country's scholars have shown much interest in this important subject.

Pathogens for War has attempted to rectify this situation by examining the impact of bioweapons, bioterrorism, and pandemics on Canadian society within the North American context. The primary focus of the book has been the evolving threat of biological weapons since 1939, both from state parties and from terrorist organizations. Closely related was the discussion about the connection between offensive and defensive research, or, more specifically, how Canada and its allies developed effective biodefence measures in dealing with their perceived vulnerability during the past seventy-two years. By the 1990s bioterrorism was also regarded as a serious threat to existing Canadian and American biodefence policies. Another major theme of this book has been the symbiotic relationship between the threat of bioweapons and disease pandemics throughout the twentieth century and the development of more holistic health security strategies since 9/11.

Canada's Biodefence Experience during the Twentieth Century: Major Trends

The Second World War transformed the threat of biological weapons from a theoretical menace to an evolving weapon of mass terror. This trend was evident in a number of ways. First was the creative role that scientists in Canada, the United Kingdom, and the United States assumed in alerting their political leaders about enemy BW activities and in directing their respective national BW programs. A second trend was the establishment of dedicated laboratory facilities in these three Western countries for the dangerous work of weaponizing some of the world's most fearsome pathogens for use against enemy cities. Third, all three countries experienced a gradual militarization of biological warfare programs, in which decision making about the strategic application of biological weapons shifted from academic scientific administrators to the army chiefs of staff. This transformation was most pronounced in the United States, where, by the end of hostilities in August 1945, the War Department had created a virtual BW Manhattan Project, the US becoming the world's only bioweapons superpower. A fourth factor was the civil defence dimensionsof biological warfare as exemplified by the joint Canada-US Rinderpest Project at Grosse Ile and by the coordinated North American response in 1944–5 to the threat of Japanese war balloons carrying human and animal BW agents.

Why did Canada and its allies continue their BW programs after 1945 when these weapons were not used strategically during the war? The most important factor was the consensus among the military hierarchy in Washington, London, and Ottawa that this weapon system was on the verge of realizing its potential. Of particular importance were the "lessons" of the joint US-UK anthrax cluster bomb project, with its stated goal of destroying enemy cities; other BW munitions such as "bot tox," tularaemia, and brucellosis also demonstrated considerable weapons potential. During the early stages of the Cold War improved use of cluster bombs and the development of the large-area concept of aerosol dissemination further enhanced the offensive dimensions of biological weapons. On the other side of the equation, Canadian and American civil defence planners developed a number of strategies in anticipation of a Soviet bioweapons attack, either as part of a concerted WMD assault or by saboteurs in covert BW operations.

While these broad themes demonstrate the various ways in which Canada responded to the challenge of biological warfare prior to 1969, it is also important to consider the specific contributions of the biodefence scientists themselves. During the Second World War, for example, prominent medical experts provided invaluable assistance to Canada's war effort. As Wilder Penfield later recalled in a 1973 letter to Harold Ettinger, "I remembered all the times when we met in Ottawa ... Banting, [Betrum] Collip, [Charles] Best, Solandt and various prima-donnas – and I might even say other prima-donnas! Those were great days and with Mackenzie at the helm the research man-power of Canada was drawn out ... There was a wonderful friendship and a great dedication to a cause ... that changed the whole course of the life of each of us for a time but that is the way with a war."[1]

This same pattern prevailed in the wartime experiences of Guilford Reed (Queen's), James Craigie (Toronto), and Everitt Murray (McGill), who served on a part-time basis with the Department of Defence's C-1 Committee, the organization that coordinated Canada's offensive and defensive bioweapons programs. The only exception was Murray, whose 1943 secondment from McGill, to serve as chairman of the Committee, meant a dramatic change in his professional and personal life, characterized by a steady commute between Montreal, Ottawa, Suffield, Grosse Ile, and Washington, DC.[2] After 1945 this critical liaison role was assumed by Dr Omond Solandt, the newly appointed Director General of the Defence Research Board, who quickly recognized that Canada had unique assets in the bioweapons field. This was evident in 1947 when he launched three major initiatives: creating the DRB Biological Warfare Committee, arranging for Canadian defence scientists to participate in the Tripartite CBW Agreement, and providing Suffield's vast testing facilities for American and British offensive weapons trials.

By the mid-1950s, the wartime cohort of academic scientists had been replaced by full-time military BW scientists, such as Archie Pennie, Robert Heggie, and G.R. Vavasour. The career patterns of these defence scientists included an initial assignment at either Suffield (DRES) or Ottawa (DREO), followed by an administrative appointment at DRB headquarters in Ottawa. This second-generation group also carried out the tasking role of supervising the national BW program while working closely with American and British bioweapons specialists through the Tripartite / Quadripartite system of CBW exchange, reinforced by specific bilateral arrangements. Even more challenging were the adjustments required after November 1969, when US President Nixon terminated the Pentagon's offensive biological warfare program, thereby paving the way for the establishment of the Biological and Toxin

Weapons Convention (1975). Indeed, after 1975 Canadian biodefence experts became increasingly involved with international efforts to enforce the principles of the BWC, a difficult undertaking because of Soviet violations at home and abroad.

At the same time, the operation of The Technical Cooperation Program (TTCP) was transformed through the reduction of the number of major subgroups from seventeen to eight, along with the downgrading of the status of bioweapons research. Instead, only Chemical Defence was included in the traditional Subgroup E, while biodefence became merely one of the technology liaison groups (TLG-1).[3] In addition, the termination of the Defence Research Board in 1974 had a negative impact on the status of Canadian defence scientists within the TTCP, creating the possibility that the country would be excluded from the advanced biodefence research being carried out at Fort Detrick and Porton Down. The situation was resolved, however, by Canada's 1982 admission to the CANUKUS Memorandum of Understanding.[4]

The 1980s witnessed the emergence of Canada's third generation of biodefence specialists, featuring scientists such as Kent Harding, Bill Kournikakis, and Cam Boulet of DRES, and Ken Johnson of DREO. Their ambitious agenda included: CBW research for Canada's national defence purposes; involvement with MOU projects to deal with bioweapons threats from the Soviet Union; and monitoring the CBW threats from "rogue" states such as Iraq, Iran, South Africa, and Libya. In addition, after the 1991 Gulf War, these experts participated in the verification work of the United Nations Special Committee (UNSCOM), with a mandate to destroy Iraq's BW capabilities, an experience that greatly enhanced their individual and collective knowledge about bioweapons vertification systems. This proved useful during the early 1990s when DRES scientists, working closely with DEA arms control officials, strongly supported attempts to establish a BWC verification protocol.

Many of these issues remain essential components of Canada's contemporary biodefence system. But today's BW experts also face some unique challenges, notably in preventing terrorists groups from acquiring bioweapons because of the availability of dual-use biotechnology as well as ensuring that high-risk experiments do not have unintended harmful consequences. Because of the diversity of these threats, a number of government organizations are now involved with Canada's biodefence response system, "integrating two policy realms previously separate from one another – security and public health."[5] While Defence Research and Development Suffield remains an important player, with its high-quality scientific staff and facilities, valuable assistance is also provided by Public Health Agency of Canada, both through the Centre of Emergency Preparedness and Response and the Winnipeg-based National Microbiology Laboratory. Indeed, the careers of Ron St John (CEPR) and Frank Plummer (NML) personify many of the major trends in Canada's biodefence developments since 9/11, when bioterrorism replaced concerns over state-based biowarfare attacks.

Despite their small numbers, Canada's biodefence experts have carried out high-quality research in meeting domestic and external commitments. Prior to Nixon's 1969 declaration, they focused on helping the United States develop a potent bioweapons program, while the defensive phase became institutionalized after the 1975 Biological Weapons and Toxin Convention. Unfortunately, these achievements received little praise or recognition from federal government officials, largely because of concerns that any linkage with germ warfare would tarnish Canada's international image.[6] Yet, in contrast to Ottawa's policy of denial, scientists such as Banting, Murray, Reed, Pennie, and Vavasour did not regard their involvement with BW weapons research as violating medical or ethical principles. From their perspective, since Canada was threatened by ruthless adversaries who were willing and able to use biological weapons, it was necessary to develop a range of protective policies. And this included having a BW retaliatory capability to deter enemy attack.

North American Biosecurity: Change and Continuity

In July 2004 the Canadian Cabinet Committee on Security, Public Heath and Emergencies outlined its counterterrorist priorities in the publication *Securing an Open Society*.[7] While this document contained useful information about Canada's future biodefence strategies, there was virtually no discussion about what counterterrorist policies that had been implemented since 9/11. Nor was there any attempt to compare Canada's biosecurity policies with those adopted by the Bush administration, despite the fact that Washington had already spent upwards of $30 billion protecting Americans from external and internal BW threats. In contrast, the total cost for all Canadian counterterrorism programs was about $2 billion, which established a Canadian civil defence record but was far below the expenditures of even Homeland Security by itself.[8]

There are many problems in analysing recent trends in North American biodefence policies, particularly in obtaining relevant documentation.

However, there are important reasons why this research agenda should not be ignored. First, it provides an opportunity to apply the central themes of the historical sections of this study to the contemporary situation, thereby providing a degree of scholarly continuity. Second, while archival material has many advantages, other sources can also be used. These include government-generated reports, media reports, in-depth studies by research foundations, relevant books and articles, and personal interviews. Indeed, this book has greatly profited from the information and interpretations provided by many of the leading Canadian, American, and British bioscientists. Third, since the Canadian and American biodefence programs are so closely connected, there are many advantages in adopting a comparative model of analysis. Moreover, despite the huge disparity in information between the two national biodefence experiences, there are sufficient numbers of Canadian case studies to justify this comparative approach. Because of space constraints, three major thematic areas were selected for detailed examination, since they have relevance for both the US and Canadian biodefence experiences since 2004.

Pathogens, Vaccines, BL-4 Laboratories, and Scientific Ethics

After the terrorist incidents of 2001, protecting civilians from bioweapons assumed many forms. One of the most important was the need to understand the lethality level of different BW agents and their patterns of dissemination in order to achieve timely medical treatment. Significantly, groundwork for understanding why weaponized anthrax represented such a serious threat for civilian targets had been established during the summer of 2001, when researchers at DRDC Suffield discovered that highly refined anthrax spores could be easily aerosolized by air conditioning systems, causing widespread contamination and infection in large buildings and subway systems.[9] The anthrax letter attacks of October–November further validated these findings, forcing a reassessment of the traditional military model of how many *Bacillus anthracis* spores were required to produce fatalities. On the positive side, the Amerithrax crisis demonstrated that ciprofloxacin and other antibiotics were effective in treating cases of pulmonary and cutaneous anthrax.[10]

A related development was the attempt after 2001 to create high-performance vaccines and antibiotics / antivirals, the most ambitious initiative occurring in the United States under Project BioShield (2003). Taking advantage of its vast funding and political leverage, Project

BioShield was able to enlist the research capabilities of American pharmaceutical and biotechnology companies in the development of vaccines for anthrax, smallpox, tularaemia, and other category A pathogens. Many Canadian-based drug companies also profited from this unique funding opportunity through corporate partnership and joint projects, becoming valued partners in this elaborate US biodefence system. On the other hand, some of the most innovative vaccine work took place at the National Microbiology Laboratory (Winnipeg), where researchers have carefully analysed Ebola and Marburg fever, two deadly viral hemorrhagic fevers with recorded fatality rates of 90 per cent during earlier outbreaks in Zaire and the Congo. Focusing on the molecular structure and pathogenesis of the Ebola virus, the Winnipeg team successfully developed a comprehensive diagnostic test and a promising experimental vaccine.[11]

Yet, despite considerable progress, there are still serious deficiencies in the American and Canadian vaccine strategy. Of particular concern has been the failure to develop user-friendly anthrax vaccines, which would help overcome civilian and military resistance towards compulsory immunization.[12] Even more controversial was the Bush administration's unsuccessful attempts in 2003 to vaccinate 500,000 civilian first responders against smallpox because of concerns over adverse reactions and because of low threat perceptions.[13] This legacy of public suspicion about unsafe vaccines also surfaced during the 2009 swine flu pandemic when thousands of people, on both sides of the border, rejected the immunization option.[14]

Another contentious aspect of North American biodefence has been the proliferation of high-containment laboratories (BSL-4), particularly in the United States, where the number of labs has increased from five to fourteen, with more on the drawing boards.[15] From the perspective of US security officials, there are a number of reasons for this expansion: the ability to conduct sophisticated research on the most dangerous bacterial and viral pathogens; the risks associated with synthetic biology work; and the growing importance of microbial forensics, which involves "the use of advanced genetic, chemical and physical techniques to characterize a pathogen or toxin used in a biological attack."[16] Another factor has been increased scientific focus on the effects of zoonotic (animal) diseases; experts claim that about 40 per cent of human illness are caused by these pathogens.[17] Consequently, there has been growing demand for integrated high-containment facilities that can examine both human and animal disease agents within the same facility, like the one at the Canadian

Science Centre for Human and Animal Health. Indeed, the costly and controversial National Bio and Agro-Defense Laboratory in Manhattan, Kansas, draws extensively from the experiences of the Canadian Science Centre for Human and Animal Health (Winnipeg).[18]

Many life scientists, however, have expressed concern that these BSL-4 laboratories cause more problems than they solve. One dimension of this debate involves the concept of biosecurity; or, more specifically, whether the old adage "guards, gates, guns, and two man rule" effectively prevents the illegal removal of biological agents, or deter terrorists from sabotaging laboratory facilities.[19] In the United States there was also growing concern about the possibility "that scientists will become terrorists [rather] than that terrorists will become scientists."[20] These fears were reinforced by the 2008 FBI report that Bruce Ivins, a prominent research scientist at USAMIRID, was responsible for the 2001 anthrax letters because of his unique access to this deadly pathogen.[21]

Another dimension of this investigation was the fact that Ivins, in his role as Detrick's leading anthrax scientist, maintained close communication with scientists at DRDC Suffield and shared samples of his concentrated Ames strain of *Bacillus anthracis.* This connection persuaded the author to pose three question to Suffield administrators: How did the exchange system with USAMRIID operate? What biosafety and biosecurity procedures were in place at the Suffield BSL-3 laboratory? Did representatives of the US Federal Bureau of Investigation visit the station as part of their Amerithrax investigation?

On the biosafety side of the equation, there were also a number of well-publicized incidents at various BSL-4 facilities in North America, including the 2004 incident at Boston University, where scientific researchers were exposed to tularaemia, a traditional BW agent.[22] The Winnipeg National Microbiology Laboratory was not free of such controversies. In March 2005, for example, a courier truck carrying samples of anthrax and other pathogens was involved in a collision near the laboratory's downtown location.[23] Even more serious was the May 2009 arrest of NML researcher Konan Yao at the Canada-US border, when it was discovered that he had illegally removed twenty-two vials of the Ebola vaccine from the laboratory in order to advance his own scientific career. Although the local media accepted assurances by Scientific Director Plummer that the theft involved only the harmless vaccine, not the actual Ebola virus, there were still serious questions about the laboratory's pathogen audit systems and its security screening system.[24]

Issues of scientific responsibilities and codes of conduct were another aspect of the North American dialogue about biosecurity / biosafety.[25] Of particular concern was the possibility that some advanced DNA research projects might have unintended and unfortunate consequences. For example, the 2001 Australian Interluken 4 mousepox experiments were regarded as the "model" disastrous event. Other candidates for this dubious honour included the publication of the decoded genomic sequence of the 1918 Spanish flu, the reconstruction of the polio virus, and the growing number of synthetic biology experiments.[26] Yet suggestions that government should carefully monitor high-risk experiments and limit the dissemination of scientific information produced vigorous oppositions from scientists around the world, particularly in the United States, where some charged that certain right-wing politicians wanted to reactivate the scientific "McCarthyism" of the 1950s.[27] Many of these concerns were allayed by the 2005 establishment of the US National Science Advisory Board on Biosecurity (NSABB), whose careful review of advanced biotechnology research during the past six years has impressed academic and industrial scientists alike.[28] Regrettably, Canada has not adopted this sensible approach, and its oversight system for dual-use biotechnology remains incomplete, despite the passage of the Human Pathogens and Toxins Act (2009).[29] But how long can this luxury of inaction be sustained?

In December 2011, researchers on both sides of the border were shocked when the NSABB recommended that two scientific articles that explored the possibilities of genetically altering the deadly avian influenza virus (H5N1) should not be published in their present form.[30] As the *New York Times* observed, "the research should never have been undertaken because the potential harm is so catastrophic and the potential benefits from studying the virus so speculative."[31] In rebuttal, many US virologists have stated that these restrictions on mutant flu research would deter the country's leading influenza experts from working in this field, thereby undermining attempts to develop effective medical countermeasures.[32] As Yoshihiro Kawaoka, one the principals in this controversy, pointed out, "because H5N1 mutations that confer transmissibility in mammals may emerge in nature, I believe that it would be irresponsible not to study the underlying mechanisms."[33] For its part, the Public Health Agency of Canada's response was limited: it mandated that all research on mutant flu be confined to the high-containment facilities (BSL-4/C-4) of the National Microbiology Laboratory.[34]

Expanding the Biodefence Agenda

The development of improved detection / surveillance technology and integrated health security response systems was another important aspect of North American biodefence planning.[35] Some of the most ambitious projects were associated with the US civilian BioWatch program, along with the Pentagon's attempts to develop a multilayered national detection system designed to protect US cities against germ warfare.[36] Sceptics have questioned whether these detection systems, despite their technological sophistication, are sufficiently reliable, given the difficulties of differentiating between virulent and innocuous strains of bioagents.[37] Even more controversial was the 2006 establishment of the National Biodefence Analysis and Countermeasures Center (NBACC) at Fort Detrick with the goal of understanding "classical, emerging, and genetically engineered pathogens ... across the spectrum of potential attack scenarios."[38] This project has been denounced for its high costs and pervasive secrecy and because it violates the spirit, if not the terms, of the Biological Weapons Convention.[39] On the other hand, American biodefence officials claim that the Detrick undertaking is an essential counterterrorist initiative that provides essential scientific knowledge not only about the present generation of bioweapons but also about genetically altered pathogens.

While Canada is not directly involved with the NBACC, scientists at DRDC Suffield and the Winnipeg National Microbiology Laboratory carefully monitor its many projects.[40] Indeed, the Winnipeg Microbiological Emergency Research Team, working closely with specialists at DRDC Suffield, used some of this advanced American BW detection equipment during their three-month vigil of the 2010 Vancouver Winter Olympics.[41] In addition, the Ottawa-based Canadian Centre for Security Science, with its mandate to service the security needs of many departments, has an elaborate biodefence technological exchange with the US Department of Homeland Security.[42]

Given past trends, it is not surprising that effective cooperation between the major stakeholders in the North American counterterrorist system remains a challenge. During the 2001 Amerithrax crisis, for instance, there was a bitter and well-publicized turf war between public health specialists from the Centers for Disease Control, who wanted to contain anthrax contamination and treat its victims, and FBI experts, who were primarily focused on microbial forensic investigations for law enforcement and prosecution purposes.[43] Closely related were problems in convincing intelligence agencies to share their top-secret data about suspected terrorist activity with research scientists, public health experts, and even law enforcement officials. While this narrow right-to-know approach was roundly criticized by both the US 9/11 Commission and the National Academy study, *Globalization, Biosecurity, and the Future of the Life Sciences*, effective interaction across institutional boundaries remains difficult.[44]

Of course, Canada has its own turf wars, as is evident in a number of well-publicized incidents involving the Canadian Security and Intelligence Service and the Royal Canadian Mounted Police.[45] In addition, public health officials periodically complain that the security agencies (CSIS and the RCMP), as well as the Department of Defence, often exclude them from the intelligence-sharing loop, particularly when it involves information obtained from American and British sources. Fortunately, some of these jurisdictional problems were rectified by the 2009 memorandum of understanding between the National Microbiology Laboratory and DRDC Suffield, which gave Frank Plummer and his Winnipeg-based life scientists primary responsibility in dealing with bioterrorist threats.[46] For their part, Suffield's scientists continue to expand training programs offered by the Counter-Terrorist Technology Centre (2003) as part of the innovative programs implemented by Director General Camille (Cam) Boulet.[47] These include the specialized CB Forensic Reference Laboratory and its chemical / biological exposure chamber, where first responders experience an actual CBW environment, while specially designed mannequins are used for more dangerous experiments.[48]

During the past decade, there has been increased awareness in Washington and Ottawa that health security issues should be given greater emphasis in the formulation of foreign policy priorities. This approach has been particularly popular with the US State Department, which in 2002 established a separate health security division on the grounds that bioterrorism prevention and response should "be truly multi-sectoral with critical and leading roles played by the public health, agricultural and environmental sectors of our government."[49] During the last eleven years State Department officials, working closely with their counterparts in DHS and DHHW, have established close links with the World Health Organization and with the special United Nations enforcement agency created by Security Council Resolution 1540 " to prohibit WDM proliferation to non-state actors."[50] An even more elaborate arrangement, involving diplomats and health officials from Canada and Mexico, was the August 2007 North American Plan for Avian and

Pandemic Influenza (NAPAPI), intended "to contain a novel strain of human influenza at its source, slow its speed ... [but] allow the appropriate movement of people and cargo across mutual land borders and ports of entry in a way to achieve the public health objectives with minimal social and economic impact."[51] Significantly, NAPAPI's first major challenge – dealing with the 2009 influenza pandemic – was widely praised by the leaders of all three countries at the North American summit of August 2010 for having facilitated an effective medical and public health response to this serious crisis while preventing serious economic disruption at the borders.[52]

Bioterrorism: The Public Health and Psychological Dimensions

How effective are North American public health facilities in dealing with sudden surges of infected patients, either from natural disease outbreaks or a bioterrorist attack? During the 2000 TOPOFF-1 exercise, for example, the simulated release of pneumonic plague overwhelmed Denver's public health system, causing widespread anxiety and panic. An even more realistic pattern of first responder burnout was observed in March 2003 when SARS suddenly hit certain Toronto hospitals.[53] Both of these cases illustrate a number of common problems: faulty hospital emergency systems; inadequate protective equipment and medication; inferior surveillance and decontamination capability; and ineffective communication between health officials from different levels of government.[54] There was also growing criticism that the US government's emphasis on applied biodefence research diverted researchers away from potential breakthroughs in basic research, thereby undermining the country's long-term ability to develop new vaccines and other medical countermeasures.[55]

Psychological problems, including post-traumatic stress disorder, are another dimension of major public health crises caused either by terrorists attacks or natural disease outbreaks.[56] Although this is a relatively underdeveloped field of scientific inquiry, a number of useful studies have dealt with terrorist incidents during the past two decades.[57] One of these was the March 2002 NATO comparative study of the 1995 Tokyo subway incident and the 2001 US anthrax letter bomb attacks, which concluded that "the purpose of CBN terrorism is not primarily to take lives or destroy property ... [but] to weaken the cohesion that binds communities together ... to sow distrust, fear and insecurity."[58] These observations were reinforced by more detailed studies of the Amerithrax crisis, including the use of interviews which demonstrated that over one-quarter of those polled claimed that they were *very* worried about being victims of a terrorist attack.[59] Another approach has been to examine how people who have been exposed to dangerous pathogens responded to their experiences.[60] Perhaps the most studied group were US postal employees located in anthrax-contaminated mail sorting plants, who reported a deep sense of alienation about being treated "more as case studies than people in need of treatment."[61] This sense of vulnerability and powerlessness was also a common complaint of Toronto hospital workers infected during the SARS crisis. In 2009 they launched a class action suit against the Ontario Ministry of Health and several Toronto area hospitals on the grounds that they were not given adequate protection against this lethal virus.[62]

Are threats of an imminent and deadly biological weapons attack exaggerated? Is there a profitable counterterrorism industry composed of politicians, private consultants, and other threat / risk entrepreneurs who "just happen to have stuff to sell"?[63] Not surprisingly, these controversial questions have gained considerable attention in the United States, where federal authorities have spent over $50 billion since 2001 on just the civilian side of biodefence. At the same time, neither the FBI nor any other US security agency has produced evidence of any serious terrorist activity, leading some sceptics to claim that the threat from terrorists "seems as fanciful as some of their schemes."[64] Canada has also developed its own "insecurity entrepreneurs" – many Ottawa-based consulting companies, like their Washington counterparts, have successfully used former national security "insiders" to obtain generous government contracts.[65] In 2003, Gwynne Dwyer, a popular journalist and historian, addressed many of these themes in an article entitled "Canada's War on Terror": "Terrorism is not an enormous threat to life as we know it ... [but] a marginal nuisance that some governments find useful to inflate into an enormous bogeyman. We should get a grip on reality and stop worrying so much."[66]

While it would be reassuring to embrace Dwyer's arguments, there are powerful arguments that suggest biological warfare remains a serious threat. First, while bioweapons have never been used operationally in any international conflict, this does not mean that nation-states or terrorist groups have not considered their strategic deployment. Certainly the 1944–5 Anglo-American project to develop 6 million anthrax cluster bombs was based on the premise that this weapon could be used either to end the Second World War or to provide a WMD deterrent during the early Cold War years. Even more important was the fact that by 1969 the United States, with the assistance of its Canadian and British allies, was able to develop an impressive offensive capability with both lethal and incapacitating BW munitions. But the most convincing arguments about the viability of biowarfare was the massive Soviet biological weapons program, with its genetically altered super-pathogenic agents, its sophisticated delivery system, and its incorporation into the country's strategic planning.

Another major concern was the possibility that terrorist organizations will use lethal biological weapons against urban targets in North America and Europe. This viewpoint was clearly expressed in the 2010 report of the US Commission on Weapons of Mass Destruction, which argued that bioweapons and "Nukes" should be regarded as equivalent to weapons of mass terror. Closely related is the impact of biotechnology on future bioweapons development, creating the spectre of drug-resistant bacteria and viruses, the possibility of combining BW agents into a single weapon, and the immense potential of synthetic biology. Many of these concerns were aptly summarized by US Secretary of State Hillary Clinton in her 7 December 2011 address to the seventh BWC Review Conference in Geneva:

> There are some in the international community who have their doubts about the odds of a mass biological attack or major outbreak ... but that is not the conclusion of the United States ... [since] terrorist groups have made it known they would want to acquire and use these weapons ... [and] the emerging gene synthesis industry is making genetic material widely available ... [that] could also potentially be used to assemble the components of a deadly organism.[67]

List of Biological Agents and Toxins

AGENT	FORM	DISEASE
Bacillus anthracis (N)	Bacteria Cutaneous Gastronomical Pulmonary	Anthrax (3 types)
Bacillus globigii (U)	Bacteria	Simulant for anthrax
Brucella suis (US)	Bacteria	Brucellosis (Malta fever)
Burkholderia mallei (LA)	Bacteria	Glanders*
Chlamydia psittaci	Bacteria / *Rickettsia*	Psittacosis (Parrot fever)
Clostridium botulinum	Toxin (Botulinum)	Botulism (3 types A-B-C)*
Coxiella burnetii (OU)	Bacteria / *Rickettsia*	Q-fever
Ebola	Virus	Hemorrhagic fever*
Foot and mouth	Virus	Deadly animal disease*
Francisella tularensis (UL)	Bacteria	Tularaemia (Rabbit fever)*
Lassa fever	Virus	Lassa fever
Marburg	Virus	Hemorrhagic fever*
Melioidosis (HI)	Bacteria	Whitmore's Bacillus*
Ricin (castor bean)	Toxin	Ricin
Rift Valley fever	Virus	Rift Valley fever
Rinderpest (R)	Virus	Rinderpest (animal disease)
Saxitoxin (TZ)	Toxin	Shell-fish poisoning
Staphylococcal enterotoxin	Toxin	Staphylococcus (SEB)

Variola major	Virus	Smallpox*
Variola minor	Virus	Smallpox*
Venezuelan equine	Virus	Encephalitis
Yellow fever (OJ)	Virus	Yellow fever
Yersinia pestis (LE)	Bacteria Bubonic	(Pneumonic) Plague*

*Listed as part of the eleven most dangerous pathogens under the 2011 US Select Agent List.

Biographical Profile of Key Biowarfare Scientists

Frederick Banting was Canada's most famous scientist during the interwar years because of his Nobel Prize-winning work in the discovery of insulin. In 1938 Banting attempted to convince Canadian and British military officials about the threat of a German BW attack. While he was initially unsuccessful in this task, he and his colleagues at the University of Toronto conducted pioneering research in the field of biological warfare, establishing the foundation for Canada's wartime bioweapons program. He died in an air crash in February 1941 while travelling to England to discuss greater Anglo-Canadian BW cooperation.

H.M. Barrett joined the University of Toronto's chemistry department in 1932. During the war years he was research director of chemical research at the Suffield Experimental Station. He later served two terms as Chief Superintendent (1947–9, 1951–2) and was Director General of the DRB in the late 1950s.

Camille Boulet began his specialized work in chemical and biological warfare (CBW) research at Defence Research and Development Canada (DRDC) Suffield during the 1980s. After serving in several important positions in Canada's biodefence system, Boulet returned to Suffield in 2009 as director general, with a mandate to meet the CBW requirements of the country's military and health security communities.

E.L. Davies occupied a senior position at Porton Down before becoming superintendent of the Suffield Experimental Station in 1941. After the war he became a prominent member of the Defence Research Board.

Paul Fildes, a highly respected microbiologist, was director of British biological warfare research between 1940 and 1945, and a major advocate of an Allied retaliatory BW capability. After the war, he was a member of the influential UK Biological Research Advisory Board.

Armand Frappier was the long-time head of the biological department of the University of Montreal and the Montreal Institute of Microbiology, Canada's second-largest producer of biologics and vaccines. In 1959 he was chairman of the DRB Advisory Committee on Biological Warfare.

Riley Housewright was scientific director of Fort Detrick during the 1960s and a vocal opponent of President Nixon's decision to terminate the US offensive bioweapons program.

Alex Longair joined the Defence Research Board in the late 1940s after working on nuclear issues at the British embassy in Washington. As a prominent member of the DRB's Joint Special Weapons Committee, his specialty areas included atomic, biological, and chemical warfare.

Otto Maass became a member of the McGill chemistry department during the early 1920s and soon established himself as one of the country's most productive scientists. During the war he directed Canada's chemical and biological warfare programs under the auspices of the Canadian Army. In 1947 he was appointed to carry out similar duties for the Defence Research Board.

C.J. Mackenzie was Dean of Engineering at the University of Saskatchewan during the interwar years. He became acting president of the National Research Council in 1939 and full-time head in 1944. After the war, he supervised Canada's atomic energy program.

A.G.L. McNaughton was a McGill engineering graduate who achieved considerable distinction during the First World War. In the interwar years, he served as Chief of the Defence Staff and president of the National Research Council. Between 1939 and 1943 he was Canada's senior military officer in Britain. In 1944 McNaughton became Minister of Defence and strongly endorsed plans for the postwar operation of Experimental Station Suffield.

George Merck was the president of the influential Merck pharmaceutical company until 1943, when President Roosevelt asked if he would participate in the US biodefence program. He served as chairman of the WRC Committee in 1943–4 and then as special BW consultant to the Secretary of War. His official report of January 1946 provided a major survey of American BW activities during the war years, including cooperation with Canada and the United Kingdom.

Charles Mitchell was chief of animal pathology at the Department of Agriculture and the C-1 Committee's expert on rinderpest during the Second World War. Between 1947 and 1956 he participated in the joint Canadian-American BW animal research program at Grosse Ile.

E.G.D. (Everitt) Murray was chairman of McGill's Department of Bacteriology and Immunology between 1930 and 1956. As an internationally recognized bacteriologist for his work on dysentery, gas gangrene, and listeria, Murray maintained close contact with the leading British and America scientists in this field. These links proved invaluable during the Second World War when Murray was chairman of the C-1 Committee, which directed Canada's biological warfare program. Although he did return to his McGill laboratory after the war, he was a member of the BW advisory committee of the Defence Research Board until the early 1950s.

Archie Pennie was educated as a chemical engineer in Scotland and served in the Royal Air Force during the war. He joined the Defence Research Board in 1947 and subsequently became Superintendent of its operation at Churchill, Manitoba (1954–6) and Defence Research Establishment Suffield (1957–63). He subsequently served at DRB headquarters until the organization was terminated in 1975.

Frank Plummer became a recognized international authority on HIV/AIDS during the 1980s because of his fieldwork with patients in Kenya and his follow-up laboratory work at the University of Manitoba. In 2000 he was appointed Scientific Director of the Winnipeg-based National Microbiology Laboratory, which is Canada's only high-containment facility. Plummer's involvement with Canada's biodefence planning was expanded after 2001 with the establishment of the Global Health Security Initiative and the 2003 formation of the Public Health Agency of Canada.

Guilford Bevil Reed joined Queen's University in 1915 after completing a PhD in bacteriology at Harvard University. Until his retirement in 1954, Reed was one of Canada's foremost microbiologists because of his research on tuberculosis, gas gangrene, and the functions of toxoids / vaccines. During the Second World War Reed carried out a number of important experiments on anthrax, brucellosis, and botulinum toxin in his Kingston laboratory and at the weapons-testing facilities at Suffield, Alberta. Between 1945 and 1955, Reed was a dominant member of the Defence Research Board's BW advisory committee and carried out important research on insect vectors and shellfish toxins.

Richard Shope was one of the leading American virologists during the 1930s because of his role in the discovery of the influenza virus. Between 1942 and 1944 he was the scientific director of the joint Canadian-American rinderpest project at Grosse Ile that successfully developed an effective vaccine against this animal disease.

Omond Solandt pursued his studies in physiology at the University of Toronto and Cambridge. During the war he became one of Britain's leading experts in operational research and was a member of the 1945 UK atomic bomb mission to Japan. In 1947 he became the first director general of the newly created Defence Research Board, a position he held until 1956. Solandt subsequently became the first president of the Science Council of Canada (1966–72).

Ron St John was the first Director of Health Canada's Centre for Emergency Preparedness and Response that was established in 2000, as part of Canada's response to the threat of bioterrorism. St John assumed an important role in coordinating Canada's health security programs after 9/11 and helped coordinate the activities of the Global Health Security Initiative, established by the G-8 (plus Mexico) in 2001.

G.R. Vavasour was a leading authority on biological warfare for the Defence Research Board during the 1960s and 1970s.

Notes

Introduction

1 The Games began on 12 February 2010 and ended two weeks later. Michael Zekulin, "Olympic Security: Assessing the Risk of Terrorism at the 2010 Vancouver Winter Games," *Journal of Military and Strategic Studies* 12, no. 1 (Fall 2009): 2–25.
2 At the 1972 Munich games Palestinian militants killed eleven Israeli athletes and coaches.
3 The Atlanta incident killed one person and injured 111; it was followed by over 100 bomb hoaxes. Zekulin, "Olympic Security."
4 Philip Boyle and Kevin Haggerty, *Privacy Games: The Vancouver Olympics, Privacy and Surveillance: A Report of the Privy Commissioners of Canada under the Contributions Program* (Edmonton, March 2009), 32.
5 Donald Avery, interview with Dr Frank Plummer, Scientific Director, National Microbiology Laboratory, Winnipeg, 13 October 2010.
6 The VP Bio Sentry system was based on the cooperation between DRDC Suffield and General Dynamics Canada (Calgary) in the CBW detection field dating back to the 1990s and upgraded after 2001 with the Stand-off Integrated Bioaerosol Active Hyperspectral Detection apparatus, which used light detection and ranging technology (LIDAR). Much of this research was carried out in cooperation with the US Biowatch program. Equally important was the DRDC Valcartier Compact Atmospheric-Sounding Interferometer Development Model, and Research Model (CATSI-EDM) that had already demonstrated its ability to detect poison gas at an 80 per cent success rate at a distance of 5 kilometres. Both of these systems were "effectively deployed in support of OP PODIUM at the Vancouver Olympics." *The*

Dragon Din (Directorate of Chemical Biological Radiological Nuclear Defence), Spring and Summer, 2010.

7 *Ottawa Citizen* 8 April 2009.

8 "U.S. Notes Al Qaeda Olympic Threat," *Globe and Mail* 14 January 2010.

9 This non-partisan Committee, under the leadership of former senator Bob Graham (chairman), and senator Jim Talent (vice chairman), was a continuation of the work authorized by the 9/11 Commission Act of 2007. *Prevention of WMD Proliferation: Report Card*, US Commission on the Prevention of Mass Destruction and Terrorism (Washington, DC, January 2010), www.pharmathene.com/WMD_report_card.pdf.

10 This report was a continuation of a December 2008 report, *World at Risk*, which, in turn, was closely associated with the September 2009 introduction of Senate bill (S.1649), "The Weapons of Mass Destruction Prevention and Preparedness Act of 2009." US, *Congressional Record*, 8 September 2009, S9135-39.

11 World Health Organization, "Concern Over Flu Pandemic Justified," Dr Margaret Chan, Director-General of the World Health Organization, Address to Sixty-second World Health Assembly, Geneva, Switzerland, 18 May 2009.

12 *New Scientist* 21 November 2011. One of the research teams was located at Erasmus Medical Centre in Rotterdam, while the other was located at the University of Wisconsin-Madison.

13 "Bioterrorist Fears Could Block Crucial Flu Research," *New Scientist*, 21 November 2011; "New Bird Flu Research Causes Stir," *Toronto Star* 21 November 2011. The prestigious journals *Nature* and *Science* were asked to respect the ruling by the NSAAB.

14 "An Engineered Doomesday," *New York Times* 15 January 2012.

15 Information about Canada's role within the Tripartite system, with its annual meetings and specialized working groups, were of great value in preparing this manuscript. In addition, records of these meetings provided a fascinating window of analysis for understanding key developments in the US and UK offensive and defensive biological programs at the operational level.

16 Some useful studies dealing with Canada's involvement with WMD and the Cold War include: William Barton, *Research, Development and Training in Chemical and Biological Defence within the Department of National Defence and the Canadian Forces* (Ottawa: Queen's Printer, 1989); James Eayrs, *In Defence of Canada*, vol. 1: *Peacekeeping and Deterrence* (Toronto: University of Toronto Press, 1972); Wilfrid Eggleston, *Scientists at War* (Toronto: Oxford University Press, 1950), and *The NRC, 1916–1966* (Toronto: Clarke, Irwin, 1978); D.J. Goodspeed, *A History of The Defence Research Board of Canada* (Ottawa: Queen's

Printer, 1958); Robin Ranger, *The Canadian Contribution to the Control of Chemical and Biological Warfare* CIIA Paper (Wellesley Paper 5/1976); John Sawatsky, *Men in the Shadows: The RCMP Security Service* (Toronto: McClelland and Stewart, 1980); Denis Smith, *Diplomacy of Fear: Canada and the Cold War, 1941–1948* (Toronto: University of Toronto Press, 1988); Reg Whitaker and Gary Marcuse, *Cold War Canada: The Making of a National Insecurity State, 1945–1957* (Toronto: University of Toronto Press, 1994).

17 Donald Avery, "Biological Warfare," in *The Encyclopedia of War,* 1st ed., ed. Gordon Martel (Malden, MA: Wiley-Blackwell, 2012).

18 Gregory Koblenz, *Living Weapons: Biological Warfare and International Security* (Ithaca, NY: Cornell University Press, 2009), 5.

19 Barry Kellman, *Bioviolence: Preventing Biological Terror and Crime* (New York: Cambridge University Press, 2007), 208.

20 Elizabeth Fenn, *Pox Americana: The Great Smallpox Epidemic of 1775–8* (New York: Hill and Wang, 2001), 85–7.

21 Mark Wheelis, "Biological Sabotage in World War I," in *Biological and Toxin Weapons: Research, Development and Use from the Middle Ages to 1945,* ed. Erhard Geissler and John Ellis van Courtland Moon (Oxford: Oxford University Press, 1999), 35–62.

22 *League of Nations Treaty Series,* vol. 94 (1929), "Protocol for the Prohibition of the Use in War of Asphyxiating, Poisonous or Other Gases, and of Bacteriological Methods of Warfare."

23 Gregory Koblenz has divided biological weapons development into four major stages: 1) pre-germ theory; 2) applied microbiology; 3) industrial microbiology and aerobiology; and 4) molecular biology and biotechnology. Koblenz, *Living Weapons,* 10–12.

24 Peter Fischer and Carol Lipson, *Thinking about Science: Max Delbruck and the Origins of Molecular Biology* (New York: Norton, 1988); James Watson, *DNA: The Secret of Life* (New York: Alfred Knopf, 2004).

25 G. Rattray Taylor, *The Biological Time Bomb* (New York: New American Library, 1968), 13.

26 The DNA saga was followed by two other major developments: the 1973 discovery of a recombinant DNA technique whereby genetic material was transferred from one organism to another, and the culmination in 2000 of projects to sequence the entire human genome. Gary Zweiger, *Transducing the Genome: Information, Anarchy, and Revolution in the Biomedical Sciences* (New York: McGraw-Hill, 2001), 201–15.

27 Matt Ridley, "Foreword," in *Inspiring Science: Jim Watson and the Age of DNA,* ed. John Inglis et al. (Cold Spring Harbor, NY: Cold Spring Harbor Laboratory Press, 2003), xviii–xix; Donald Fredrickson, *The Recombinant DNA*

Controversy, A Memoir: Science, Politics, and the Public Interest 1974–1981 (Washington, DC: National Academies of Sciences, 2001), 11.

28 Richard Swiderski, *Anthrax: A History* (London: McFarlane, 2004), 4–5.

29 Charles Rosenberg, *Explaining Epidemics and Other Studies in the History of Medicine* (New York: Cambridge University Press, 1992); Christopher Willis, *Plagues* (London: Flamingo, 1997); Paul de Kruif, *Microbe Hunters* (London: J. Cape, 1930); Hilary Koprowski and Michael Oldstone, eds., *Microbe Hunters – Then and Now* (Bloomington: Medi-Ed Press, 1996).

30 Geoffrey Bilson, *A Darkened House: Cholera in 19th Century Canada* (Toronto: 1990); John Barry, *The Great Influenza: The Epic Story of the Deadliest Plague in History* (New York: Penguin, 2005); Elizabeth Reid, *HIV/AIDS: The Global Inter-connections* (West Hartford: Kiamarian Press, 1995).

31 Carol Moberg and Zanvil Cohn, *Launching the Antibiotic Era: Personal Accounts of the Discovery and Use of the First Antibiotics* (New York: Rockefeller University Press, 1990); Jonathan Tucker, *Scourge: The Once and Future Threat of Smallpox* (New York: Atlantic Monthly Press, 2001); Georgina Feldberg, *Disease and Class: Tuberculosis and the Shaping of Modern North American Society* (New Brunswick, NJ: Rutgers University Press, 1995); David Oshinsky, *Polio: An American Story* (New York: Oxford University Press, 2005).

32 Jared Diamond, *Guns, Germs and Steel: The Fates of Human Societies* (New York: Norton, 1999), 357–8.

33 Some of the more useful studies are Barry Kellman, *Bioviolence: Preventing Biological Terror and Crime* (2007); David Fidler and Lawrence Gostin, *Biosecurity in the Global Age: Biological Weapons, Public Health, and the Rule of Law* (Stanford: Stanford University Press, 2008); and Mark Wheelis, Lajos Rozsa, and Malcolm Dando, eds., *Deadly Cultures: Biological Weapons since 1945* (Cambridge, MA: Harvard University Press, 2006). This latter work features seventeen articles by leading experts in this field writing on various dimensions of bioweapons.

34 Theodore Rosebury, *Peace or Pestilence? Biological Warfare and How to Avoid It* (New York: McGraw Hill, 1949); Robert Harris and Jeremy Paxman, *A Higher Form of Killing: The Secret Story of Gas and Germ Warfare* (London: Arrow, 2002); Stockholm International Peace Research Institute (SIPRI), *The Problem of Chemical and Biological Warfare*, vol. 3 (New York: Humanities Press, 1973); Edward Spiers, *Chemical and Biological Weapons: A Study of Proliferation* (Houndsmills: Macmillan, 1994); Leonard Cole, *The Eleventh Plague: The Politics of Biological and Chemical Warfare* (New York: W.H. Freeman, 1997); Brian Balmer, *Britain and Biological Warfare: Expert Advice and Science Policy, 1930–1965* (London: Palgrave, 2001); Edward Regis, *The Biology of Doom: The History of America's Secret Germ Warfare Project* (New York: Holt, 1999);

Jonathan Tucker, ed., *Toxic Terror: Assessing Terrorist Use of Chemical and Biological Weapons* (Cambridge, MA: MIT Press, 2000); Tom Mangold and Jeff Goldberg, *Plague Wars: A True Story of Biological Warfare* (New York: St Martin's Press, 1999).

35 Ken Alibek, *Biohazard: The Chilling True Story of the Largest Covert Biological Weapons Program – Told from Inside by the Man Who Ran It* (New York: Random House, 1999); Frederick Sidell, Ernest Takafusi, and David Franz, eds., *Medical Aspects of Chemical and Biological Warfare* (Washington, DC: Office of the Surgeon General, Department of the Army, 1997); Raymond Zilinkas, ed., *Biological Warfare: Modern Offense and Defense* (London: Lynne Reiner, 2000); Susan Wright, *Preventing a Biological Arms Race* (Cambridge, MA: MIT Press, 1990); Malcolm Dando, *Biological Warfare in the 21st Century: Biotechnology and the Proliferation of Biological Weapons* (London: Brassey's, 1994); Judith Miller, Stephen Engleberg, and William Broad, *Germs: Biological Weapons and America's Secret War* (New York: Simon and Schuster, 2001).

36 Joshua Lederberg, ed., *Biological Weapons: Limiting the Threat* (Cambridge, MA: MIT Press, 1999); Jim Davis and Barry Schneider, eds., *The Gathering Biological Warfare Storm* (Westport, CT: Praeger, 2004); See also *Biological Weapons: From the Invention of State-Sponsored Programs to Contemporary Bioterrorism* (New York: Columbia University Press, 2005); *Biotechnology in an Age of Terrorism* (Washington: The National Academies Press, 2004);

37 Among the scholarly works are Anne Clunan, Peter Lovoy, and Susan Martin, eds., *Terrorism, War, or Disease: Unravelling the Use of Biological Weapons* (Stanford: Stanford University Press, 2008); Gregory Koblenz, *Living Weapons: Biological Warfare and International Security* (Ithaca, NY: Cornell University Press, 2009); D.A. Henderson, *Smallpox, The Death of a Disease: The Inside Story of Eradicating a Worldwide Killer* (New York: Prometheus Books, 2009); and Milton Leitenberg, *Assessing the Biological Weapon and Bioterrorism Threat* (Strategic Studies Institute, U.S. Army War College, 2005).

38 There have also been a number of documentaries about the threat of biological warfare, particularly in the United States and the United Kingdom. In June 2010, however, aspects of the Canadian BW experience were revealed in "*Project N,*" broadcast by the television section of Radio-Canada, which examined the historical legacy of the Anglo-Canadian project to produce weaponized anthrax spores at Grosse Ile, Quebec.

39 *Outbreak* was based on Richard Preston's popular novel *Hot Zone;* since 1995 the film has grossed over $190 million internationally.

40 *Contagion* was directed by Steven Soderbergh, written by Scott Z. Burns, and featured Matt Damon, Laurence Fishburne, Jude Law, and Kate Winslet. Many of the themes of the movie were drawn from incidents such as the

anthrax letter attacks (2001), Hurricane Karinia (2005), and the swine flu pandemic (2009). *New Scientist* 15 September 2011; *Toronto Star* 9 September 2011.

41 Michael Woroboys, *Spreading Germs: Disease Theories and Medical Practices in Britain 1865–1900* (Cambridge: Cambridge University Press, 2000); W.D. Foster, *A History of Bacteriology and Immunology* (London: William Derck, 1970); A.M. Silverstein, *A History of Immunology* (San Diego, CA: Academic Press, 1989).

42 Woroboys, *Spreading Germs,* 6, 14, 235.

43 Ibid., 3, 234–8.

44 Olga Amsterdamska, "Standardizing Epidemics: Infection, Inheritance, and Environment in Prewar Experimental Epidemiology," in *Heredity and Infection: The History of Disease Transmission,* ed. Jean-Paul Gaudillière and Ilana Löwy (London: Routledge, 2001), 139.

45 Robert Murray, "Developing Streams Affecting Biomedical Sciences, esp. "Microbiology" (unpublished paper, February 2007, in possession of the author. Correspondence with the author, 10 February 2008.

46 Woroboys, *Spreading Germs,* 3. In 1914 it was not possible to detect or isolate viruses because of their small size and the lack of electron microscopes.

47 Ibid., 235.

48 Andrew Mendelsohn, "Medicine and the Making of Bodily Inequality in Twentieth Century Europe," in Gaudillière and Löwy, eds., *Heredity and Infection,* 21–2.

49 Amsterdamska, "Standardizing Epidemics," 142–4; Lise Wilkinson and Anne Hardy, *Prevention and Cure: The London School of Hygiene and Tropical Medicine: A 20th Century Quest for Global Public Health* (London: Kegan Paul, 2001), 125–9.

50 Peter Nara, "Humoral Immunity," in *Immunotherapy for Infectious Disease,* ed. Jeffrey Jacobson (Totowa, NJ: Humana Press, 2002), 3–22.

51 The mercury-based Salvarsan was perhaps the most notable chemical treatment of a major disease prior to 1930, given its extensive use against syphilis. Moberg and Cohn, *Launching the Antibiotic Era,* 45–93.

52 Joan Austoker and Linda Bryder, eds., *Historical Perspectives on the Role of the MRC: Essays in the History of the Medical Research Council of the United Kingdom and Its Predecessor the Medical Research Committee, 1913–1953* (New York: Oxford University Press, 1989).

53 See Everett Mendelsohn et al., eds., *Science, Technology and the Military,* 2 vols. (Boston: Kluwer Academic Publishers, 1988); Stuart Leslie, *The Cold War and American Science* (New York: Columbia University Press, 1993); Guy Hartcup, *The Challenge of War: Britain's Scientific and Engineering Contributions*

to World War Two (New York: Palgrave, 1970); Guy Hartcup, *The Effects of Science on the Second World War* (New York: Palgrave, 2000).

54 John S.C. Blair, *In Arduis Fidelis: Centenary History of the Royal Army Medical Corps*, 2nd ed. (Edinburgh: Lynx Pub, 2001), 18–93. In 2004, the Canadian government created the Royal Canadian Army Medical Corps. Bill Rawling, *Death Their Enemy: Canadian Medical Practitioners and War* (Ottawa: AGMU Marquis, 2001), 63–101.

55 See Vincent Cirillo, *Bullets and Bacilli: The Spanish-American War and Military Medicine* (New Brunswick, NJ: Rutgers University Press, 2004); Richard Gabriel, *A History of Military Medicine* (Boulder, CO: Greenwood, 1992).

56 G.W.L. Nicholson, *Seventy Years of Service: A History of the Royal Canadian Army Medical Corps* (Ottawa: Borealis, 1977), 68–113. Rawling, *Death Their Enemy*, 159–205.

57 Michael Bliss, *Banting: A Biography* (Toronto: McClelland and Stewart, 1984), 41.

58 For a full account of the wartime experiences of E.G.D. Murray, see Donald Avery and Mark Eaton, *The Meaning of Life: The Scientific and Social Experiences of Everitt and Robert Murray, 1930–1964* (Toronto: The Champlain Society, 2009), xli–xlix.

59 Estimates of the number of deaths during the 1918 influenza pandemic range between 40 and 100 million. In the United States, about 450,000 died, compared with about 16 million in India. Recently, samples of the 1918 virus strain (H5N1) have been subjected to genome sequencing, allowing researchers to understand why it was so lethal. See Jeffrey Taubenberger and David Morens, "1918 Influenza: The Mother of All Pandemics," *Emerging Infectious Diseases* 12, no. 1 (January 2006), http://www.cdc.gov/ncidod/EID/vol12no01.

60 During the 1918–19 outbreak Guilford Reed, who had joined Queen's University in 1915, attempted to identify the causal agent and prepared a crude vaccine that he injected into 200 convalescent soldiers in the university's military hospital. In 1920 he was appointed head of the newly appointed Department of Bacteriology, a position he held until his death in February 1955. "Guilford B. Reed, Bacteriologist," Queen's University Archives, Loc 163.12, 4/13, Department of Health Sciences Papers, Queen's University.

61 Joseph Hanaway, *McGill Medicine*, vol. 2: *1885–1936* (Montreal and Kingston: McGill-Queen's University Press, 2006); Alison Li, *J.B. Collip and the Development of Medical Research in Canada* (Montreal and Kingston: McGill-Queen's University Press, 2003); and Wilder Penfield, *No Man Alone: A Neurosurgeon's Life* (Boston: Little Brown, 1977). See also Avery and Eaton, *The Meaning of Life*.

62 LAC, National Research Council Papers (NRC) 88/89, vol. 5-4-M-4-7, ACMR organizational report, 19 April 1938.

63 National Research Council Archives (Ottawa), Frederick Banting Papers, Box 3, Dr C.B. Stewart, Assistant Secretary ACMR to Dr L.S. Klinck, president of the University of British Columbia, 10 November 1938.

64 Cited in Donald Avery, *The Science of War: Canadian Scientists and Allied Military Technology during the Second World War* (Toronto: University of Toronto Press, 1998), 37.

65 Cited in Susan Wright, ed., *Preventing a Biological Arms Race* (Cambridge, MA: MIT Press, 1990), 368.

66 The relationship between biodefence and natural pandemics is explored in Donald Avery, "The North American Plan for Avian and Pandemic Influenza: A Case Study of Regional Health Security in the 21st Century," *Global Health Governance* 3, no. 2 (Spring 2010).

1 Canada's Role in Allied Biological Warfare Planning in the Second World War

1 R.D. Cuff and J.L. Granatstein, *Ties That Bind: Canadian-American Relations in Wartime from the Great War to the Cold War* (Toronto: Samuel Stevens Hakkert, 1977), 69, 101.

2 Colonel Stanley Dziuban, *Military Relations between the United States and Canada* (Washington: Office of the Chief of Military History, US Army, 1959), 18–47.

3 Mel Thistle, ed., *Mackenzie-McNaughton Wartime Letters* (Toronto: University of Toronto Press, 1975), 151.

4 John Bryden's study *Deadly Allies: Canada's Secret War, 1937–1947* (Toronto: McClelland and Stewart, 1989) provides a popular, if somewhat inaccurate, account of Canada's involvement in biological warfare during the Second World War.

5 "Protocol for the Prohibition of the Use in War of Asphyxiating, Poisonous or Other Gases, and of Bacteriological Methods of Warfare," signed at Geneva, 17 June 1925; entered into force 8 February 1928. Susan Wright, *Preventing a Biological Arms Race* (Cambridge, MA: MIT Press, 1990), 369. Significantly, while both the United States and Japan signed the Protocol, neither country officially ratified the document until the 1970s.

6 Cited in Donald Avery, *Science of War: Canadian Scientists and Allied Military Technology during the Second World War* (Toronto: University of Toronto Press, 1998), 18.

7 The spectre of germ warfare had already been featured in the 1921 racist novel by Hilda Glynn-Ward, *Writing on the Wall* (Toronto: University of

Toronto Press, 1974, Reprint), which portrayed a sinister takeover of British Columbia by its Asian population using a deadly typhoid pandemic which killed off the white population.

8 Major Leon Fox, "Bacterial Warfare: The Use of Biological Agents in Warfare, "*The Military Surgeon* (March 1933): 576–77.

9 *Toronto Daily Star* 27 March 1935.

10 During the 1930s, Banting directed one of Canada's largest medical research laboratories, employing upwards of forty researchers. In 1935 his prestige was further enhanced when he was elected to the prestigious Royal Society of London. Michael Bliss, *Banting: A Biography* (Toronto: McClelland and Stewart, 1984), 214–22, 236–46.

11 Banting placed the agents into three categories: (1) water-borne – typhoid, cholera, bacillary dysentery, and amoebic dysentery; (2) air-borne – spinal meningitis, virulent haemolytic streptococcus, influenza; (3) insect-borne – bubonic plague, yellow fever, malaria, sleeping sickness. Library and Archives Canada (LAC), National Research Council Papers (NRC), 87/88, 104, vol. 69, McNaughton memorandum, 17 September 1937.

12 Ibid., Banting to McNaughton, 16 September 1937.

13 It does not appear that Banting was aware of the biological warfare research which was being conducted by special units of the Japanese army in northern Manchuria under the direction of Colonel Shiro Ishii. See Sheldon Harris, *Factories of Death: Japanese Biological Warfare 1932–45 and the American Cover-up* (New York: Routledge, 1995).

14 NRC, vol. 69, McNaughton to Lieutenant Colonel L.R. LaFleche, Deputy Minister of National Defence, 1 October 1937.

15 *The Times* (London) 14 October 1937.

16 NRC, vol. 69, N.E. Gibbons to Dr Newton (NRC), 23 September 1937.

17 Banting had previously received assurances from University of Toronto President Cody that his top-secret biological research could be conducted on university property. Ibid., Banting to McNaughton, 16 September 1938; University of Toronto Rare Book Room, War Diary of Frederick Banting (Diary) Box 20, Hunter to Banting, 14 February 1938.

18 Canadian BQ planners also consulted the 1938 British report "Proposals for an Emergency Bacteriological Service to Operate in War." NRC/87-88/104, volume 69, file 36-5-0-0, Banting to McNaughton, 16 September 1938; ibid., L.R. LaFleche to McNaughton, 21 September 1938; ibid., McNaughton to Banting, 24 September 1938.

19 In 1937 Sir Maurice Hankey, Secretary of the CID, had been instructed to create the Biological Warfare Sub-Committee, which, in turn, was instrumental in establishing the Microbiological Research Establishment (MRE)

at Porton. Stephen Roskill, *Hankey: Man of Secrets*, vol. 3: *1931–1963* (London: Collins, 1974), 321–4. Brian Balmer, *Britain and Biological Warfare: Expert Advice and Science Policy, 1930–1965* (London: Palgrave, 2001), 3–42; Gradon Carter, *Chemical and Biological Defence at Porton Down 1916–2000* (London: Palgrave, 2000).

20 Balmer, *Britain and Biological Warfare*, 22–5.

21 NRC/87-88/104, vol. 69, 36-5-0-0, Report of Conversation with Sir Edward Mellanby in the office of Dr Wodehouse, Deputy Minister, Department of Pensions and Health, 29 September 1938; ibid., McNaughton to Major General LaFleche, 3 October 1938.

22 Bliss, *Banting*, 254–97. Dr I.R. Rabinowitch, a toxicologist at McGill University, was also part of this mission, given his expertise in chemical warfare. Avery, *Science of War*, 154–6.

23 Balmer, *Britain and Biological Warfare*, 33.

24 Banting Diary, 3, 8 January 1940.

25 Ibid., 11, 12 January 1940.

26 Microbiologist Topley provided the most extensive critique of Banting's report, claiming that apart from the threat of sabotage, it was unlikely that bacteriological warfare would "produce the devastating results suggested by Sir Frederick Banting." Cited in Balmer, *Britain and Biological Warfare*, 33.

27 Banting Diary, 11, 12 February 1940.

28 In his campaign for Canadian involvement in BW work, Banting informed Sir Edward Mellanby and Air Vice Marshall H.E. Whittingham of the British Air Ministry about the successful experiments which his colleague, virologist James Craigie, had been carrying out in developing a vaccine for typhus, a traditional scourge of battlefields. Banting Diary, 31 December 1940; NRC, vol. 69, H.E. Whittingham Air Vice Marshall to C.J. Mackenzie, 13 November 1940.

29 Bliss, *Banting*, 283–4.

30 LAC, E.G.D. Murray Papers, vol. 29, Mitchell to Murray, 9 June 1942.

31 Murray to Banting, 25 June 1940, cited in Donald Avery and Mark Eaton, *The Meaning of Life: The Scientific and Social Experiences of Everitt and Robert Murray, 1930–1964* (Toronto: Champlain Society, 2008), 101.

32 Banting Diary, 17 May 1940.

33 LAC, NRC, 87/88/104, vol. 69, Banting to Mackenzie, 24 June 1940; ibid., Mackenzie to Banting, 25 June 1940; Banting Diary, 27 June 1940.

34 NRC, 87-88/104, vol. 169, Colonel H. Desrosiers (Deputy Minister DND) to Secretary NRC, 12 July 1940.

35 The largest donations came from Canada's economic elite, including the Eaton, Bronfman, and Massey families. By Order in Council PC 4260, the

War Technical and Scientific Development Committee was established in July 1940 to administer the fund, consisting of more than $1.3 million. Avery, *Science of War*, 47–9.

36 Banting Diary, 3, 31 December 1940.

37 Brian Balmer claims that most of the work in creating the British bioweapons program was carried out by Lord Hankey, behind the backs of the prime minister and cabinet. Personal note to author, 28 August 2011.

38 In 1940 Paul Fildes, head of the MRC Bacterial Chemistry unit at Middlesex Hospital, London, became the driving force behind Britain's BW program. Balmer, *Britain and Biological Warfare*, 80–3.

39 Jeffrey Legro, *Cooperation under Fire: Anglo-German Restraint during World War II* (Ithaca, NY: Cornell University Press, 1995), 144–77; Richard Price, *The Chemical Weapons Taboo* (Ithaca, NY: Cornell University Press, 1997), 82–156; Edmund Russell, *War and Nature: Fighting Humans and Insects with Chemicals from World War I to Silent Spring* (New York: Cambridge University Press, 2001), 80–183.

40 For a more detailed discussion of Canada's CW activities in the Second World War see Avery, *Science of War*, 122–50.

41 HQS, 4354-11-1, vol. 2, "Report of Canadian Mission," 16 October 1940.

42 Under PC1/6687 of 2 July 1941 the terms of the joint Anglo-Canadian CW operation were clarified. LAC, J.L. Ralston Papers, vol. 396, file 54, Hankinson to N. Robertson, 1 March 1941.

43 Ibid., Ralston to Hon. C.G. Power, 31 March 1941.

44 On 9 April 1941 by Order in Council PC 2508, arrangements were made to secure the ESS site from the province on a ninety-nine-year lease. HQS, 4354-2, Hankinson to Robertson, 22 March 1941.

45 LAC, War Diary of C.J. Mackenzie, 27 March 1941; HQS, 4354-2, Power to Ralston, 26 March 1941.

46 HQS 4354-2. Order in Council 2 July 1941.

47 Ibid., Colonel H.S. DesRosiers, Acting Deputy Minister DND to Sir Edward Beatty, 24 May 1941.

48 Ibid., Colonel Morrison, memorandum, 8 May 1941.

49 NRC, 4-C-9-19, Mackenzie to Hon. J.A. Mackinnon, Chairman, Committee of the Privy Council on Scientific and Industrial Research, 2 June 1941; Directorate of History, Department of National Defence (DHist), 745.043, D3, "Minutes of the Second Meeting of the Inter-Service Board," 2 July 1941.

50 At the age of forty-five Davies became the first superintendent of ESS, after having spent the pre-war years as a chemist with the chemical warfare research unit at Porton Down, England (near Salisbury). After 1945, he became a prominent member of the Canadian Defence Research Board.

51 HQS, 4354-1-8, Flood to Deputy Minister Pensions and National Health, Ottawa, 29 December 1941; LAC, Privy Council Papers (PCO), vol. 10, 4C-9-22, Watson to Davies, 8 December 1941. Frederic Brown, *Chemical Warfare: A Study in Restraints* (Princeton: Greenwood Press, 1968), 207–30.
52 HQS, 4354-6-5-1 Lieutenant General Stuart, CGS, to Minister of Defence 13 February 1942.
53 HQS, 4354-1-8, "Minutes of the Meeting held 20 January 1942"; ibid., "Minutes of Meeting on Chemical Warfare on 9 Feb. 1942."
54 Ibid., Inter-Service Board Recommendations on Canadian Chemical Warfare Policy for Consideration of the Chiefs of Staff, 21–2 March 1942.
55 Brown, *Chemical Warfare*, 207–30; Leo Brophy and George Fisher, *The Chemical Warfare Service: Organizing for War* (Washington: Department of the Army, Office of Military History, 1959), 49–51.
56 Maurice Pope, *Soldiers and Politicians: The Memoirs of Lt. Gen. Maurice A. Pope* (Toronto: University of Toronto Press, 1962).
57 National Academy of Sciences Collection (hereafter NASC), Washington, DC, File CWS Chief, 1942–4, Lt. Col. M.E. Baker, Progress Report No. 54, 15 August 1941.
58 Ibid. The code letter WBC had no special meaning, except to provide a necessary security cover. Ibid., Suggestions for WBC Program, 23 July 1942.
59 NASC, File CWS, "Suggestions for WBC Program." Barton Bernstein, "Origins of the U.S. Biological Warfare Program," in Wright, ed., *Preventing a Biological Arms Race,* 9–25. National Archives and Record Administration of the United States, Washington, DC (NARA), RG 165, Entry 488, Files of George W. Merck, Special Consultant, 1942–6 (Merck Papers), Box 187, E.B. Fred File, Fred to George Merck, 16 August 1943.
60 NARA, Records of the Joint Chiefs of Staff, 1948–50 (385.2, file 12-17-43, "Report to the Secretary of War by Mr. George Merck, Special Consultant for Biological Warfare," November 1945.
61 Cited in Ed Regis, *Biology of Doom: The History of America's Secret Germ Warfare Project* (New York: Henry Holt, 1999), 25.
62 "Paul Gordon Fildes,1882–1971" by G.P. Gladstone et al., *Biographical Memoirs* (Royal Society of London, 1973), 317–41. United Kingdom National Archives (UKNA), Cabinet Records (CAB) 120/782, 58921, Hankey to Lord Ismay, GCOS, 6 December 1941.
63 NAUK, War Office Records (WO), 188/695, Paul Fildes, "Memorandum on Toxic Dust: Statement for Plans Division, Naval Staff, 13 May 1942."
64 NASC, WBC Committee, US-UK Cooperation, Report of Visit of Lt Col. James Defandorf to United Kingdom, 18 April 1942.

65 NRC, vol. 69, Valdee to Defries 23 December 1940; ibid., Banting to Valdee, 30 December 1940; Valdee to Banting, 9 January 1941.
66 NASC, Washington, Canada-USA file, Fred to Murray, 17 December 1941.
67 Fred also recruited two of his colleagues in the Department of Bacteriology: Ira Baldwin and William Sarles. Both became important BW administrators during the Second World War and the Cold War. Ira Baldwin, *My Half Century at the University of Wisconsin: Adapted from an Oral History Interview* (Madison, WI: privately published, 1995), 122–38.
68 NASC, E.B. Fred File, Fred to Frank Jewett, 19 February 1942.
69 NRC, vol. 69, A. Landsborough-Thomson, British Medical Research Council, to Collip, 16 December 1941.
70 Ibid., Mackenzie to Murray, 19 December 1941.
71 The committee also discussed the possibility that cities such as Vancouver, with its questionable sanitary standards and large rat population, would be vulnerable to an outbreak of bubonic plague. NAC, R.L. Ralston Papers, vol. 30, "Recommendations to the Department of Defence of Canada by the Committee of Project M-1000," included in letter Collip to Colonel A.A. Mageem, Assistant, Minister of National Defence, 29 December 1941.
72 The usual sources of rinderpest vaccine were the Veterinary Research Institute, Muktesar, U.P., India, or the Veterinary Research Laboratory, Onderstepoort, Transvaal, South Africa.
73 NRC, vol. 69, Minutes of the Meeting on Project M-1000, 19 December 1941.
74 Reed described some of his own experiments with botulinum toxin with special reference to the persistence of its toxicity in natural waters and the fact that "chlorination does not appear to destroy the toxin."
75 NRC, vol. 69, Minutes of the First Meeting of Sub-Committee on project M-1000 ON, 19 December 1941.
76 LAC, Murray Papers, vol. 29, Murray to Fred, 30 January 1942.
77 The delegates also discussed the possibility of using herbicides against the Axis powers, since "destruction of the rice crop would cripple the Japanese just as destruction of the potato crop would the Germans." NRC, vol. 69, Minutes of the Meeting, 19 December 1941.
78 In February 1942 the NRC received a secret British report of German biological warfare tests which showed "that toxins can be carried by bomb shells without being destroyed," while anthrax spores were "of particular efficiency." NRC, vol. 69, Enclosed in letter H.C. Bazett to Mackenzie, 4 February 1942.
79 LAC, Colonel R.L.Ralston Papers, vol. 30, Ralston to the Hon. James Gardiner, 10 March 1942; ibid., General Murchie to Ralston, 6 March 1942.

80 In May 1942 Charles Mitchell and A.E. Cameron carried out a site visit to Grosse Ile. Although they believed extensive renovations were required, they felt that the former immigrant isolation hospital could be used for rinderpest research. In addition, the island (2 miles long, 1 mile wide) could be effectively divided, thereby isolating healthy and test animals. Murray Papers, vol. 29, Report, 21 May 1942.
81 Ibid., Murray to Fred, 30 January 1942.
82 Ibid., Fred to Murray, 31 July 1942; ibid., Kelser to Murray, 3 August 1942; ibid., Murray to Kelser, 5 August 1942.
83 NRC, vol. 69, Second Report of the Committee M-1000 on Biological Warfare, 12 June 1942.
84 The Canadian commissioners were Murray (chair), C.A. Mitchell, J. Craigie, and G.B. Reed; the US members were General R.A. Kelser (chair), E.B. Reed, R.E. Dyer (NIH), and H.W. Schoening (Dept. Agriculture).
85 It was not, however, until 30 December 1942, by Order in Council PC69/11742, that the Canadian government gave its official authorization. LAC, HQS, 4354-20-1, Review of Canadian Chemical Warfare Policy.
86 Shope's innovative research on influenza during the 1930s made him a world leader in this field. NASC, Fred memorandum, 20 July 1942; Minutes of the Joint America-Canadian Commission, 31 August 1942. The Rockefeller University, "Richard E. Shope (1901–66), 1957 Albert Lasker Clinical Medical Research Award," www. Rockefeller.edu.
87 All US equipment for GIR entered duty-free and remained American property. Ibid., Report of Colonel Defandorf and Lt. Col. Peake on Visit to Grosse Ile, 29 October 1942.
88 Queen's University Archives, Building Committee, 29 September 1938–16 June 1945, Colin Drever, Architect to Dr W.E. McNeill, Vice Principal and Treasurer, 11 April 1942. In 1931, Reed personally designed the bacteriology laboratories in the New Medical Science Building (1907), which remained the focal point of his research activity until 1953, when the Kingston DRB Laboratory opened at nearby Barriefield.
89 Reed's MA and doctoral students in bacteriology could obtain their degrees by working on biodefence topics. This arrangement had previously been implemented by Otto Maass at McGill University. See Avery, *Science of War*, 134–9.
90 Reed's laboratory also worked closely with a specialized facility at nearby Belleville in raising millions of fly pupae – Musca domestica and Drosophia – required for the fly bait projects (LPM).
91 Bernstein, "Origins of the U.S. Biological Warfare Program," 9–25.

92 Rexmond C. Cochrane, *History of the Chemical Warfare Service in World War II (1 July 1940–15 August 1945*, vol. 2: *Biological Warfare Research in the United States* (Historical Section, Plans, Training and Intelligence Division, Office of the Chief, Chemical Corps, November, 1947) (hereafter *Biological Warfare Research*).
93 Murray Papers, vol. 29, Reed to Murray, 16 February 1942; ibid., Murray to Reed, 20 February 1942.
94 Mackenzie Diary, 26 October 1942.
95 Balmer, *Britain and Biological Warfare*, 74–102.
96 UKNA, CAB, 106/1 Report of a Visit by Dr Fildes and Dr Henderson to U.S.A. and Canada in November-December 1942. According to Lord Stamp, Fildes was "one of the great scientific autocrats," who thoroughly dominated the Porton BW group. Occasional Papers on Biological Warfare No. 8: A Chapter from the Unpublished Memoirs of the Third Lord Stamp, edited by G.B. Carter (Porton Down, February 1988).
97 UKNA, CAB, 106/1, Report of a Visit by Dr Fildes and Dr Henderson to U.S.A. and Canada in November-December 1942.
98 HQS, 4354-1-23, Murray to Maass, 19 October 1942.
99 Murray Papers, vol. 29, Maass to Murray, 21 August 1942.
100 Mackenzie Diary, 24 November 1942.
101 HQS, 4354-8-4-1, Murray to Maass, 2 December 1942; ibid., 4354-33-13-3, Reed to Maass, 1 December 1942.
102 Ibid., Murray to Maass, 8 December 1942.
103 HQS, 4354-33-13-3, Murray to Maass, 30 August 1943.
104 Bernstein, "Origins of the U.S. Biological Warfare Program," 14.
105 NAUK, WO, 188/699 Fildes Reply to Murray's Report, 30 March 1943.
106 WO, 188/699 Fildes to Colonel J.H. Defandorf (US), 22 April 1943.
107 Grosse Ile scientists used the African bovine strain of rinderpest.
108 J.A. Baker and A.S. Greig, "Rinderpest. XII. The Successful Use of Young Chicks to Measure the Concentration of Rinderpest Virus Propagated in Eggs," *American Journal of Veterinary Medicine* 7 (April 1946): 196–8. (Now *Journal of the American Veterinary Medical Association.*)
109 HQS, 4354-33-13-5, Murray to Maass, 26 December 1942.
110 Ibid., Shope to Murray, 18 July 1943.
111 Ibid., Murray to Shope, 20 July 1943.
112 Ibid., Shope to Maass, 26 October 1943.
113 In September 1943 a breakdown of personnel at the two research centres showed that fifteen scientists, technicians, and support staff were connected with the GIN project and twenty-five personnel with the GIR. Ibid.

114 HQS, 4354-33-13-3, Murray to General Kelser, 27 January 1943; Reed to Fildes, 3 March 1943.
115 HQS, 4354-27-1-1, General Young to Ralston, 16 February 1943.
116 HQS, 4354-1-23, Murray to Maass, 23 July 1943; ibid., 4354–33–13–3, Major Bishop to Paterson, 23 April 1943; ibid., Murray to Reed, 18 June 1943.
117 Regis, *Biology of Doom*, 42. HQS, 4354–33–13–5, Murray to Maass, 26 December 1942.
118 HQS, 4354-33-13-3, Murray to Maass, 8 December 1942.
119 DRDC-Suffield, 44-5002 (HQS, 4354-33-14-7) D.C.W. & S Project MS 23, "Final Summary Report of Mass Production of N at W.D.C.S. October 1943 to August 1944," by Major R.C. Duthie and Staff.
120 HQS, 4354-1-23, Murray to Maass, 7 August 1943.
121 HQS, 4354-33-13-3, Murray to Major R.C. Duthie, 12 July 1944.
122 Ibid., Murray to Duthie, 31 July 1944. "Final Summary Report of Mass Production of N."
123 Ibid., "Final Summary Report of Mass Production of N," 18–22.
124 Ibid.
125 NARA, Merck Papers, Box 186, Minutes of the meeting of the CI Committee together with representatives of the UK and Suffield, held in Ottawa, 8 January 1945.
126 "Final Summary Report of Mass Production of N."
127 In January 1946 the US-Canadian Commission agreed to turn most of its rinderpest stockpile over to the United Nations Relief and Rehabilitation Agency "for use in China." HQS, 4354-1-23, Reed to Maass, 14 January 1946; Robert Paterson, US Secretary of War, to Douglas Abbott, Minister of Finance, 8 May 1946. NAW, State / War / Navy Coordinating Committee, Lot 57, Box 74, file 15 M, Fred to Dr Richard Tolman, 26 May 1946.
128 Directorate of History, file 91/364, 'Progress Report for the Experimental Station, Suffield, Alberta: Period April 1st to September 30, 1947."
129 Ibid. The professional staff came from the following disciplines: chemistry, physics, meteorology, mathematics, pharmacology, pathology, entomology, veterinary science, engineering (mechanical, chemical). At least 75 per cent of the total staff were in uniform.
130 NAUK, WO, 188/699, "Fildes report on meeting held on 28.4.43 to discuss Canadian co-operation in the development of Bacteriological Warfare weapons."
131 Ibid., Lord Stamp to Fildes 12 May 1943. Murray had already secured permission from the federal Department of Health to carry out BW trials at the site, with the understanding that there would be "a safety buffer of at least five square miles." HQS, 4354–27–1-1, Murray to Maass, 12 December 1942.

132 Ibid., George Merck to J. Davisdon Pratt, Controller of Chemical Defence Development, Ministry of Supply, 25 June 1943. WO, 188/699, Fildes to George Merck, 8 July 1943.

133 DRDC Suffield – Bursting Chamber Files (2-0), Report on Trials 1943 (n.d.); Report of Trial 24 February 1944; Report of Trial 15210, 14 November 1944.

134 DRDC Suffield Document (43–3178). Paterson had also supervised a related trial "Swine Fever Virus-Peat Mixtures Dispersed Statically from the Piston-Based Ejection Bomb." DRDC-Suffield TS-77 (SR-97).

135 WO, 188/699, D. Henderson to Lord Stamp, 20 August 1943.

136 Ibid. By 1943 Porton scientists were aware that there could be quantitative differences in sensitivity of various animals to "N" spores, which meant that many experiments used monkeys rather than sheep.

137 DRDC, 2003-2724 TS-49, Porton Report, 10 October 1943 (BDP 29); WO, 188/699, Fildes to ESS, 27 September 1943.

138 HQS 4354-29-17-5 (C-5017) Director DCW&S to The Secretary, Department of National Defence (Air), 5 April 1943.

139 WO 188/699, Draft of Suggested Programme for Canada (BIO/2461), 27 September 1943.

140 Ibid., E.B. Fred to Fildes, 28 September 1943.

141 Ibid., Fildes to Maass, 29 September 1943.

142 Ibid., Fildes to Maass, 21 October 1943.

143 HQS, 4354-33-23 (Reel 5019), G.B. Reed to Major J.C. Paterson, 12 November 1943. Ibid., Davies to Reed, 18 November 1943.

144 HQS, 4354-33-17-1 (Reel 5010), Murray to Colonel M.B. Chittick, 23 December 1943.

145 HQS, 4354-33-17-1 (Reel 5019), Murray to Colonel M.B. Chittick, 23 December 1943.

146 HQS, 4554-20-18, Maass to Major General J.V. Young, Master General of Ordnance, 3 January 1944.

147 HQS, 4354-33-13-2, Minutes of Meeting on Cooperative Insect Vector Program, 15 November 1944.

148 Assessment by CWS scientist Dr W.J. Nungester, cited in Cochrane, *Biological Warfare Research*, note 79.

149 WO, 188/699, Fildes to Brigadier General E.L. Sibert, ETO, London, 28 January 1944.

150 WO, 188/701, "Practical Development of the N Bomb Project, by P. Fildes.," 9 December 1943 (BDP-36).

151 WO, 188/699, Fildes to William Sarles, Secretary WRC Committee, 30 December 1943.

152 WO, 188/699, Stamp to Fildes, 11 February 1944.
153 Regis, *Biology of Doom*, 42–62.
154 HQS, 4354-33-1, Review of Present Position of B.W. in Canada by the C-1 Committee (E.G.D. Murray, Chairman), 15 February 1944, cited in Avery and Eaton, *Meaning of Life*, 125–6.
155 WO, 188/654, Fildes Response to Canadian Memorandum dated 15.2.44, signed E.G.D.Murray.
156 WO, 188/699, Stamp to Murray, 9 February 1944.
157 WO, 188/699, J. Davidson Pratt to Otto Maass, 15 April 1944. HQS, 4354-33-1 (5018), "Review of Present Position of the C-1 Committee," EGD Murray, 15 February 1944.
158 HQS, 4354-1-23, Murray to Paterson, 31 July 1944.
159 The task of weaponizing type A botulinum toxin involved an elaborate process of nutrition, harvesting, storage, and preparation for dissemination, as either a liquid or a dry powder. HQS, 4354-33-13-2, Murray to Davies, 14 August 1944.
160 Ibid., Davies to Fildes, 25 November 1944 (telegram).
161 Bernstein, "America's Biological Warfare Program," 308–10.
162 Field tests at Suffield had used the virulent strain 19 of Brucella abortus in a type F 4-lb bomb. NARA, Merck Papers, Box 186, "Minutes … CI meeting Ottawa … January 8, 1945."
163 The shot-gun shell, or SS, was a Camp Detrick innovation which worked on the principle "that the smaller the bomb, the larger the burst-cloud relative to the amount of the filling." Cochrane, *Biological Warfare Research.*
164 Ibid.
165 HQS, 4354-33-16-2, Fothergill to Maass, 15 February 1945; ibid., Maass to Fothergill, 20 February 1945.
166 Ibid., "Report of a Meeting held at Camp Detrick on February 11 1944."
167 WO, 188/654, Fildes to Philip Allen, Offices of the War Cabinet, 12 April 1944.
168 WO, 188/699, "Notes on visit by Brigadier General Prentiss to Biology Section, Porton, on 28.2.44".
169 WO, 188/654, Fildes to Philp Allen Secretary of the War Cabinet, 10 February 1944.
170 WO, 188/699, Fildes to Stamp 12 February 1944; ibid., Fildes memorandum (BIO 4317), 20 March 1944.
171 WO, 188/699, "Notes on visit by Brigadier General Prentiss to Biology Section, Porton, on 28.2.44."
172 NARA, RG 112, vol. 3, file 13, Digest of Information Regarding Axis Activities in the Fields of Bacteriological Warfare, 8 August 1943.

173 Ibid., Report J.E. Zanetti, Special Projects, CWS, to Colonel Defandorf, London, 30 May 1944.
174 Ibid., Defandorf to Chief, CWS Technical Division, 26 April 1943.
175 NASC, Washington DC, Canadian-USA File, E.G.D. Murray to Dean Fred, 20 July 1942.
176 Ibid., John Gordon to Colonel Lunderberg 2 June, 7 June, 24 June 1944. RG 112, vol. 2, file 3, Data on Examination of Serum of Prisoners of War for Antibodies to Various Antigens – 28 June 1944 to 30 May 1945. In addition, there were tests for plague (34), tularaemia (40), brucellosis (4), cholera (30), bacillus dysentery (36), and ricin (34).
177 NARA, RG112, vol.1, file 3, Merck to Colonel Bayne-Jones, 17 September 1943.
178 Ibid., Box 19, Joint Committee on New Weapons and Equipment: Implications of Recent Intelligence Regarding Alleged German Secret Weapon, 14 January 1944.
179 Ibid.
180 Ibid., Buxton, Acting Head of OSS to JCS, 17 December 1943.
181 NARA, RG 112, vol. 3, file 13, Minutes of Meeting Barcelona Sub-Committee (Joint Committee on New Weapons and Equipment), 12 January 1944.
182 Ibid., Recommendations of the "Barcelona" Sub-Committee, 14 January 1944.
183 Ibid. Minutes of the Second Meeting, Barcelona Sub-Committee; Report of the Barcelona Committee, April 1944.
184 NARA, Records of the United States Joint Chiefs of Staff, Record Group 385. 2, File 12-17-43, Report by The Joint Committee on New Weapons and Equipment, 20 February 1944.
185 For a more detailed discussion of the D-day bioweapons controversy see Avery, *Science of War*, 165–9.
186 LAC, HQS, 4354-33-13-6, Young to Ralston, 17 December 1944.
187 Cochrane, *Biological Warfare Research*, 120–230.
188 HQS, 4354-33-13-6, Maass to C.J. Mackenzie, 15 February 1944; Murray to Maass, 20 March 1944.
189 Ibid., Reed to Colonel Morris, 23 March 1944.
190 Ibid., Murray to Maass, 6 April 1944.
191 WO, 188/654, Philip Allen to Paul Fildes, 20 January 1944.
192 CAB, 136/7. Fildes to The Rt. Hon. Ernest Brown, Whitehall, 28 January 1944.
193 WO, 188/699, William Sarles to Fildes, 14 February 1944.
194 WO, 188/481, Special Projects Division, Intelligence Branch, Periodic Intelligence Report No. 2, 24 March to 15 May 1944.

195 CAB, 130/11, Minutes of meeting 18 April 1944. CAB 136/1, Lord Rothschild to Philip Allen (War Cabinet).
196 On 8 March Prime Minister Churchill indicated that he fully supported the military policy that if Germany initiated BW warfare, "the only deterrent would be our power to retaliate." CAB, 136/1, Churchill to Duff Cooper, Chancellor of the Duchy of Lancaster, 8 March 1944.
197 Avery and Eaton, *The Meaning of Life*, lxxvii–lxxii.
198 On 28 May Dr Brock Chisholm, Director General of Canadian Army Medical Services, requested an interview with high-ranking US military officials in London to formally challenge the controversial report prepared by Dr Boles, an apparently unqualified consultant. NARA, RG 112, Box 2, File 9, Colonel Lunenberg memorandum to the Surgeon General US Army, 30 May 1944.
199 LAC, Ralston Papers, vol. 30, Stuart to Murchie, 19 May 1944.
200 CAB, 136/8-9-11, Note on meeting held at SHAFE on 27 May 1944 regarding immunization against "X"; ibid., General Brunskill Report to the War Cabinet, 2 June 1944; ibid., The War Office to B.A.S. Washington, 3 June 1944. British officials also claimed that if immunized Canadian troops were captured by the Germans, it might foster suspicions that the Allies were preparing to launch germ warfare attack, thereby fostering a CBW crisis.
201 LAC, Ralston Papers, vol. 30, Stuart to Murchie, 8 June 1944.
202 HQS, 4354-33-19, Minutes of the sub-committee of the CWISB, 20 January 1945.
203 Adolph Hitler opposed using biological weapons until the Allied bombing of Dresden in February 1945. Erhard Geissler, "Biological Warfare Activities in Germany, 1923–1945," in *Biological and Toxin Weapons: Research, Development and Use from the Middle Ages to 1945*, ed. Erhard Geissler and John Ellis van Courtland Moon (Oxford: Oxford University Press, 1999), 91–124.
204 NARA, George Merck Papers, 187, Marquand File, Marquand to Harvey Bundy, 31 August 1943; ibid., William Sarles to Merck, 20 April 1944.
205 Merck Papers, Box 187, Colonel Paget to Sarles, 16 November 1944.
206 Merck Papers, Box 187, Sarles File, Sarles to Merck, 30 September 1944.
207 Ibid., Sarles to Merck, 7 August 1944.
208 Maass had been recommended for membership in the American BW Committee by General Porter of the CWS. Ibid., Box 186, Sarles to Merck, 20 November 1944.
209 Ibid., Note for Lord Stamp, 17 November 1944; ibid., Sarles to Merck, 20 November 1944.
210 Ibid., Merck to Stimson, 2 December 1944.

211 HQS, 4354-33-1 (5018), Colonel W Goford, Colonel DSD (W) Memorandum on Biological Warfare: War Cabinet Documents, 30 October 1944.
212 Ibid. Goforth also criticized the tendency to treat all BW activity as top secret since it prevented both military and civilian medical experts from preparing effective defensive measures.
213 Fildes was impressed by Gorforth's memorandum, calling it "the brightest thing I have seen from Service sources, and I am sending copies to our Committee [ISSBW]." WO, 188/654, Fildes to Stamp, 8 September 1944.
214 HQS, 4354-33-1 (5018), Lieutenant General Murchie to R.L. Ralston, Minister of Defence, 23 August 1944.
215 HQS, 4354-33-1 (5018), Murray to Major Skey, 2 September 1944.
216 HQS, 4354-33-15, Lt. Col. H.N. Worthley, CWS Special Projects to Murray, 27 November 1944; Murray to Worthley, 27 February 1945.
217 HQS, 4354-33-1 (5018), Skey to Murray, 8 October 1944. One major development was the creation of the BW Intelligence Committee, which included representatives from the UK, the US, and Canada.
218 HQS, 4354-33-1 (5018), Murray to Major A.J. Skey, 19 October 1944.
219 Robert Harris and Jeremy Paxman, *A Higher Form of Killing: The Secret Story of Gas and German Warfare* (London: Arrow, 2002), 131–4.
220 Cochrane, *Biological Warfare Research*, 221–56.
221 WO, 188/695, Paul Fildes, "Appreciation of the Value of the 4 lb HE/Chem, Bomb Type F, Charged N," 14 August 1944.
222 DHist, 740, Minutes of the Chemical Warfare Inter-Service Board, 13 January 1944.
223 NARA, Papers of the Office of Scientific Research and Development, vol. 227, General Porter to James Conant, 16 October 1944.
224 NARA, Merck Papers, vol. 186, Lt. Col. Woolpert, memorandum to William Sarles, 25 October 1944.
225 In August 1944, when Prime Minister Churchill had requested an assessment of a large-scale BW retaliatory attack on Germany, Fildes calculated that 4 million 4-lb bombs would be required. WO, 188/695, Fildes Report (BIO/5154), 14 August 1944.
226 For the BW attack on German cities, Fildes estimated that there should be about 30 anthrax cluster bombs per square mile so that "an unprotected man will be subjected to a 50 % risk of death." WO, 188/695, Fildes Report BIO/5154, 14 August 1944; ibid., BIO/7326, 6 November 1944.
227 According to Fildes's estimates, the 12 Vigo fermenters each had a capacity of 20,000 gallons, so that about 40 runs would be required to produce enough anthrax for 1 million bombs. WO, 188/695, Files Report BIO/5136, 14 August 1944.

228 By the end of November 1944, the US/UK team achieved a 98 per cent success rate in dropping trials.

229 Fildes proposed that the UK should postpone the movement of crated and assembled clusters from the US until after an enemy BW attack had occurred, on the grounds that the Vigo plant could produce half a million bombs within a month. WO 188/ 695, Paul Fildes, "Report to the Inter-Service BW Sub-Committee, 17 July 1944."

230 Ibid., Fildes report, "Stability of Charging of N Bombs," BIO/4982, 17 July 1944.

231 HQS, 4354-33-1, "Notes on an Informal Meeting in Washington August 7 1944."

232 DRDC Suffield, Canada Book, BW Trials Carried Out at ESS during the Second World War. Examined at DRDC Toronto May 2004.

233 WO 188/699, Fildes to Stamp, 21 August 1944; Fildes to Stamp, 4 November 1944.

234 WO, 188/165, Memorandum R.J.W. Lefevre, Air Ministry, to Porton, 30 March 1945.

235 Moon, "US Biological Warfare, Planning, and Preparedness," in Geissler and Moon, eds., *Biological and Toxin Weapons*, 239–40. In addition, there were four anti-animal agents and six plant inhibitors.

236 The DEF Committee was placed under the administrative control of the US Biological Warfare Committee (USBWC), chaired by George Merck. Moon, "U.S. Biological Warfare Planning and Preparedness," in Geissler and Moon, *Biological and Toxin Weapons*, 239–40. In addition, there were four anti-animal agents and six plant inhibitors.

237 NASC, DEF Committee, file Project N, Meetings 1944–5, Colonel H.N. Worthey, Chief Special Projects Division, CWS, to Dr O.H. Pepper, Chairman of DEF Committee, 26 December 1944. Ibid., Washington Report of Meeting, 11 January 1945.

238 By 1945, Vigo had over 1,500 employees, along with other personnel from the different contractors, such as the Detroit company Electro, which made the 4-lb shells and cluster devices. Regis, *Biology of Doom*, 71–3.

239 DEF Committee, file Project N, Meetings 1944–45: Washington Report of Meeting, 11 January 1945.

240 Ibid., Colonel HN Worthley, Chief, SPD, to Dr Pepper, 13 July 1945.

241 For a more detailed account of this crisis see Avery, *Science of War*, 172–3.

242 NARA, Office of Scientific Research and Development Collection (ORSD), Box 7, War Department Report, Military Intelligence Division, 1 September 1945.

243 Everitt Murray presented the Canadian perspective on the Japanese balloon crisis at an emergency meeting of US Western Command Headquarters, 9–10 March 1945. NARA, RG 112, Box 10, File 44.

244 HQS, 4354-32-1, Minutes of the Thirty Fifth Meeting of the Canadian Chemical Warfare Inter-Service Board, 1 March 1945; OSRD, file 227, Vannevar Bush to George Merck, 2 July 1945.

245 On 25 July 1945 Murray submitted his resignation as chairman of the CI Committee, recommending Guilford Reed as his successor. HQS, 4354-33-17, Murray to Maass, 25 July 1945.

246 Another area of trilateral cooperation occurred in the field of intelligence gathering, where Canadian, British, and American code breakers developed an elaborate system of interchange that was continued during the Cold War. Richelson Jettrey and Desmond Ball, *Ties that Bind: Intelligence Cooperation between the UK/USA* (London: Allen and Unwin, 1985).

247 See Avery and Eaton, *The Meaning of Life,* 145–52. On 28 May 1943 Reed was named an Officer of the Order of the British Empire for his wartime work. Murray had already received this award in the First World War.

248 The rice pathogens were brown spot of rice (code named E) and rice blast (R). Cochrane, *Biological Warfare Research.*

249 Merck Papers, vol. 186, Report J. Davidson-Pratt, 1 March 1945; ibid., Minutes of the Inter-Services Committee on Chemical Warfare, 13 February 1945.

250 John McCloy, Assistant Secretary of War to Stimson, 29 May 1945, cited in John Ellis van Courtland Moon, "Project SPHINX: The Question of the Use of Gas in the Planned Invasion of Japan," *The Journal of Strategic Studies* (September 1989): 304.

251 Moon, "U.S. Biological Warfare, Planning and Preparedness," 253.

2 Bioweapons in the Cold War: Scientific Research, Civil Defence, and International Controversy

1 The MRD had replaced the wartime Biology Department Porton in 1946 and retained this title until 1957, when the organization was renamed Microbiological Research Establishment (MRE). In 1979 the present Cenre for Applied Microbiology and Research (CAMR) was created. Peter Hammond and Gradon Carter, *From Biological Warfare to Healthcare: Porton Down 1940–2000* (New York: Palgrave, 2002).

2 Both the British and French postwar bioweapons programs were abandoned after their military acquired a nuclear capability: the UK in 1956,

and France in 1964. Olivier Lepick, "The French Biological Weapons Program," in *Deadly Cultures: Biological Weapons since 1945,* ed. Mark Wheelis, Lajos Rozsa, and Malcolm Dando (Cambridge: Harvard University Press, 2006), 108–11.

3 Joseph Jockel, *No Boundaries Upstairs: Canada, the United States, and the Origins of North American Air Defense, 1945–1958* (Vancouver: University of British Columbia Press, 1987), 10–15.

4 National Archives of the United Kingdom (NAUK), WO, 188/660, US Report by the Biological Warfare Sub-Committee: "On the Capabilities of Biological Weapons, 1948." Considered by the British BW Sub-Committee, 2 September 1948.

5 Tom Mangold and Jeff Goldberg, *Plague Wars: A True Story of Biological Warfare* (New York, St Martin's, 1999), 28–38.

6 Donald Avery, "Advancing Medical Internationalism: Brock Chisholm and the Formative Years of the World Health Organization, 1946–1953," paper presented at the 38th Meeting of the International Society of Medical History Congress, Istanbul, 1–6 September 2002.

7 Herbert York, *The Advisers: Oppenheimer, Teller and the Superbomb* (Stanford, CA: Stanford University Press, 1976), 2–3.

8 See James G. Herschberg, *James B. Conant: Harvard to Hiroshima and the Making of the Nuclear Age* (New York: Alfred Knopf, 1993); Kai Bird and Martin Sherwin, *American Prometheus: The Triumph and Tragedy of J. Robert Oppenheimer* (New York: Vintage, 2006); Peter Goodchild, *Edward Teller: The Real Dr. Strangelove* (London: Weidenfeld and Nicolson, 2004); Solly Zuckerman, *Monkeys, Men, and Missiles: An Autobiography 1946–88* (New York: W.W. Norton, 1988); David Holloway, *Stalin and the Bomb: The Soviet Union and Atomic Energy, 1939–1956* (New Haven, CT: Yale University Press, 1994).

9 Brian Balmer, *Britain and Biological Warfare: Expert Advice and Science Policy, 1930–65* (Basingstoke: Palgrave, 2001), 20–90.

10 Nikolai Krementsov, *Stalinist Science* (Princeton, NJ: Princeton University Press, 1997), 271, 290.

11 There are major controversies about the actual numbers of Japanese killed in the two atomic attacks. See Gar Alperovitz, *The Decision to Use the Atomic Bomb, and the Architecture of an American Myth* (New York: Alfred Knopf, 1995).

12 After completing his physiology degree at Cambridge, Solandt served with the British Army, eventually becoming superintendent of the Army Operational Research Group. His involvement with the scientific British team in Japan was testimony of his high status. It also helped him get the job as the first Director General of the Defence Research Board in 1947.

University of Toronto Archives, Omond Solandt Papers, vol. 11, "Notes on the Atomic Bomb."

13 NARA, RG 30, George W. Merck to Secretary of War Robert P. Patterson, 24 October 1945. See John van Courtland Moon, "The U.S. Biological Warfare Program," in *Deadly Cultures*, 8–10.

14 Library of Congress, Papers of Robert Oppenheimer, "Note on Biological Warfare," September 1945, n.d.

15 NARA, Papers of the Joint Chiefs of Staff (385.2 (1948-50), file 12-17.

16 A report from the British Defence Committee of 5 October 1945 estimated that the United States government had spent $40 million in the development of biological weapons. CAB, 120/782, Defence Committee, 5 October 1945. Document obtained from the National Security Archives, Washington.

17 National Archives United Kingdom (NAUK), DEFE 2, 1251, 7809, Special Report by Paul Fildes, 12 November 1945.

18 DEFE 2/1252, 7823, Major General G.H.A. MacMillan, Assistant Chief of the Imperial General Staff (Chair), Chiefs of Staff Committee, Joint Technical Warfare Committee, Future Development of Biological Warfare, memorandum by A.C.I.G.S. (W), War Office, 6 December 1945. Document obtained from the National Security Archives, Washington, DC.

19 Ibid., Brigadier O.H. Wansbrough-Jones, memorandum, 6 December 1945.

20 Directorate of History, file 3-11-2-1, Colonel Goforth to General Letson (memorandum of telephone call), 8 August 1945. See Donald Avery, *The Science of War: Canadian Scientists and Allied Military Technology during the Second World War* (Toronto: University of Toronto Press, 1998), 196–202.

21 In 1947 the NRX (National Research X-perimental) reactor came on line in 1947, preparing the way for the CANDU (Canada-Deuterium-Uranium) system, which emerged in the 1960s. Robert Bothwell, *Nucleus: The History of the Atomic Energy of Canada Limited* (Toronto: University of Toronto Press, 1988), 94–287.

22 Avery, *Science of War*, 176–202, 251–5.

23 NARA, Merck Papers, vol. 186, minutes of the ISSBW meeting, 25 July 1945. The MRC motion was cited in this US report.

24 Ibid., extract of the conference of 17 July 1945.

25 In a July 1945 DND report it was estimated that the ESS capital structure was worth about $2,218,346, with the toxic storage area worth another $308,222. In addition, operating costs had averaged about $425,000 a year since the fall of 1941. HQS, 4354-26-4, A. Ross, Deputy Minister DND to Gordon Munro, Office of the High Commissioner of the United Kingdom, 6 July 1945.

26 HQS, 4354-2, McNaughton to Premier Ernest Manning, 2 August 1945.

27 Ibid., minutes of the meeting of the Chemical Warfare Inter-Service Board, Ottawa, 10 September 1945. The meeting also agreed that SES should have a postwar complement of 40 officers and 344 other ranks.

28 The status of Camp Detrick was seriously challenged in August 1945 when General Behorn Somervell, responsible for the Army's logistics and supply services called for the termination of all US biowarfare activities on the grounds that this research was now unnecessary. This policy was challenged by numerous high-profile BW administrators, notably George Merck, who proposed that "the Biological Warfare Services might be formed ... in the same basis and equal to the Chemical Warfare Services." NARA, George Merck Papers, vol. 186, Merck to H. Bundy, 13 September 1945. Ibid., Stimson, memorandum for the Chiefs of Staff, 13 September 1945.

29 HQS, 4354-33-16, Report of Major General A.E. Walford – Adjutant General for the Minister of Defence (General McNaughton), 6 August 1945.

30 Murray Papers, vol. 29, Murray "Notes," 22 February 1942.

31 *Project "N,"* documentary film shown on Radio-Canada (CBC) 1 June 2010. This account was verified by researchers of DRDC Suffield, as reported in a note to the author by Clement Laforce, Deputy Director, 30 August 2011.

32 David Grenville, "Omond McKillop Solandt – A Biographical Sketch," in *Perspectives in Science and Technology: The Legacy of Omond Solandt,* ed. C.E. Law, G.R. Lindsey, and D.M. Greville (Kingston: McGill-Queen's University Press, 1995), 3.

33 Omond Solandt (1909–93) was educated at the University of Toronto in the medical field of physiology prior to working on his doctorate in physiology at Cambridge University. He remained in England during the war years, where he was involved in blood transfusion work and operational research.

34 Gordon Watson, "Defence Research Board: Policies, Concepts, and Organization," in *Perspectives in Science,* 75.

35 LAC, Records of the Defence Research Board (RG 24), vol. 4243, file Schonland Correspondence, Solandt to Schonland, 2 May 1947. Solandt indicated that the DRB annual budget for 1947 was $13 million, 40 per cent of which was designated for research purposes.

36 Ibid., Solandt to Schonland, 20 August 1947. One of Solandt's early successes was his recruitment of E.L. Davies, wartime superintendent at SES, to become deputy director of the DRB, since he was "well liked in Canada, the US and the UK."

37 Ibid., file Wansborough-Jones Correspondence, Solandt to Wansborough-Jones, 22 August 1946.

38 Ibid., Solandt to Maass, 10 March 1948.

39 For a full account of the Gouzenko spy inquiry see Avery, *The Science of War*, 228–55.
40 LAC, DND, file TS 711-270-16-1, Colonel R.E.S. Williamson to Colonel W.A.B. Anderson, director of intelligence, Canadian Army, 14 June 1946.
41 Ibid., Mackenzie to Solandt, 24 August 1946; ibid., Mackenzie to Solandt, 19 September 1946.
42 DND, HQS, 4354-26-1-1, Otto Maass, Report July 1945. Another explanation could be that Soviet CBW scientists not only were well advanced in their own chemical and biological research, but also profited from the wartime activities of their vanquished enemies. From Germany, for instance, the Kremlin inherited the vast nerve gas facilities in east Prussia, along with a number of German chemical warfare specialists. Equally important is that after its August 1945 invasion of Manchuria, the Red Army seized the records of Unit 731. Sheldon Harris, *Factories of Death: Japanese Biological Warfare, 1932–1945, and the American Cover-Up* (New York: Routledge, 1994), 120–80.
43 After 1957 the work of the Tripartite CB Agreement was complemented by the much more expansive Tripartite Technical Cooperation Program (TTCP), established by the Eisenhower administration for collective defence and mutual help in both the nuclear and non-nuclear weapons field.
44 NAUK, War Office Records, WO 188/440, Report of Meeting, 16 August 1946, Gravelly Pt. VA.
45 Ibid. In 1946 there were ten qualified BW scientists working at Suffield and the Kingston Laboratory. In contrast, Fort Detrick had 150 professional scientific positions, with another 300 technicians and support personnel.
46 In November 1947 twenty-six scientists and military officials, including Omond Solandt, attended the meeting. It was chaired by the veteran defence scientist Sir Henry Tizard. Some of the topics covered were the use of atomic energy for military purposes, biological warfare, chemical warfare, ballistic missiles, radar and sonar systems, and development of military aircraft. NA/Kew, CAB 131/6, 58921, Commonwealth Advisory Committee on Defence Science, 21 November 1947.
47 Ibid.
48 NAUK, WO, 188/660, Report to the Defence Research Policy Committee by Air Marshall Sir Norman Bottomley, Chair of the BW Sub-Committee, 28 November 1947.
49 Peter Hammong and Gradon Carter, *From Biological Warfare to Healthcare, Porton Down 1940–2000* (Basingstoke: Palgrave, 2002).
50 Balmer, *Britain and Biological Warfare*, 55–184.
51 In total there were eleven official members and twenty-nine delegates, representing the elite of CBW research in Canada, the UK, and the US.

Defence Research Board (DRB), vol. 4124, file 3-900-43, Minutes of the 2nd Meeting of the Special Weapons Advisory committee, 20–2 June 1947.

52 Ibid. Dr Woolpert, Superintendent of Fort Detrick, mentioned that his establishment was building a 40-foot sphere, later known as the bowling ball, for bursting trials of different pathogens.

53 Directorate of History, file 91/364, Report Colonel G.M. Carrie (Defence Research Liaison) to Sir Henry Tizard, War Offices, UK, 11 November 1947. This report also listed sixty-nine separate buildings at SES, including five major administrative and laboratory facilities, along with specialized facilities such as the wind tunnel, decontamination centre, spray dryer toxicity rooms, bursting chamber, and RCAF hangars.

54 Ibid., Semi-Annual Progress Report, 1 October 1947 to 31 March 1948.

55 A follow-up report by Superintendent H.M Barrett in March 1948 also lamented the difficulty of attracting BW specialists. In contrast, a large contingent of Camp Detrick scientists were, for extended periods of times, "carrying out trials of mutual interest to Canada and the U.S." Ibid., Semi-Annual Progress Report, 1 October 1947 to 31 March 1948.

56 DRB, vol. 4133, file 4-953-43-1, Glen Gay to EGD Murray, 22 January 1948.

57 Directorate of History, HQS, 715-10-37-1-3, Minutes, Army Equipment Policy committee, Twelfth Meeting, 11 September 1947 (Ottawa); ibid., Report, "General Staff Requirements for Research in the Field of Special Weapons, 1947."

58 Ibid.

59 Ibid., HQS, 430-0-37-2, Canadian Army Equipment Policy Statement no. H1, Canadian Army ABC Warfare Equipment Policy, 1 February 1949.

60 Ibid.

61 DRB, vol. 4133, DRBS 3-900-43-2, Rear Admiral F.L. Houghton, Vice Chief of the Naval Staff to Chairman, DRB, 8 April 1949.

62 HQS, 188-0-0, Solandt to Dr H.M. Barrett, Chief Superintendent, SES, 8 April 1949.

63 DRB, vol. 4133, DRBS 3-900-43-2, Wallace Goforth, 330 Bay Street, Toronto, to The Secretary, Special Weapons Advisory Committee, DRB, 14 April 1949.

64 Ibid.

65 NRC, vol. 7, file 3-12-M-4.24. Proceedings of the DRB Sessions 15017, December 1948.

66 DRB, vol. 4196, file 260-900-43, pt 2, Major H.E. Sharples, Secretary Joint Special Weapons Committee, Report, 20 February 1952.

67 NRC, vol. 7, file 3-12-M3-30, pt 2, Chairman, DRB to members of DRB Committees and panels, Care and Communication of Classified Information, 10 January 1949.

68 Members of the BW Panel included Philip Greey (University of Toronto), James Orr (Queen's University), James Paterson (University of Western Ontario), Guilford Reed (Queen's), Charles Mitchell (Department of Agriculture), Major Sawyer (DRB), Otto Maass (McGill), Murray (McGill).
69 DRB, vol. 4314, file 4-435-43-2, Minutes of the 4th Meeting of the Bacteriological Warfare Research Panel, 31 January 1948.
70 This disease was also known as Whitmore's Baccillus. There were health concerns about about testing this agent at Suffield. Ibid.
71 Charles Mitchell, Dominion Animal Pathologist (Dept of Agriculture) supported Reed's vector program, and recommended that the panel attempt to obtain information "on the insect pests found in the larger populated areas of likely enemy countries with a view to testing them as possible carriers." DRB, vol. 4314, file 4-435-43-2, Minutes of the 4th Meeting of the Bacteriological Warfare Research Panel, 31 January 1948.
72 Ibid. Despite the fact that Reed's laboratory at the New Medical Science Building was located in the centre of the Queen's University campus, there were no protests from either faculty, students, or the general public. Reed Papers, box 1, Extract from Minutes of Trustees, 20 October 1945.
73 LAC, Department of External Affairs (DEA), vol. 5919, file 50208-40, pt 1, Guilford Reed, "Canadian Appreciation of Biological Warfare," 27 June 1950.
74 Ibid.
75 LAC, DRB, 1983-84/167, vol. 7328, file 100-24/0, Lt Colonel H.W. Bishop (DCW&S), Memorandum: Biological Laboratories – Queen's University, 23 December 1946. Guilford Reed Papers, box 1, Reed to Principal Wallace, 19 December 1945.
76 Ibid., Solandt to Reed, 2 April 1947.
77 Ibid., DRB Semi-Annual Report of Chairman, 1 April–30 September 1947, Chapter Six, Kingston Laboratory.
78 Ibid., Solandt Memorandum: Defence Research Kingston Laboratory, 2 July 1953.
79 In 1947 Reed was awarded the Royal Society of Canada's Flavelle Medal for outstanding research in the life sciences; in 1952 he became president of this prestigious organization. In addition, he served as the chairman of the Fisheries Research Board of Canada between 1936 and 1953. "Forty Years a Member of Staff: Guilford B. Reed," *Queen's Review* (March 1955): 72–3.
80 Reed Papers, Omond Solandt to Principal W.A. Mackintoch, 23 January 1953.
81 Reed's funeral in Kingston attracted a wide range of scientific colleagues, including his BW associates. In his address, Principal W.A. Mackintosh praised Reed for being one of the most distinguished of Queen's scientists,

noting his presidency of the Royal Society of Canada (1952) and his honorary degree (LLD) from the University of Saskatchewan (1953) and Acadia University (1954). Reed Papers, “An Appreciation of Guilford Reed.” See also Donald Avery and Mark Eaton, *The Meaning of Life: The Scientific and Social Experiences of Everitt and Robert Murray, 1930–1964* (Toronto: The Champlain Society, 2008), 713.

82 *A History of the Defence Research Establishment Ottawa, 1941–1991*, comp. and ed. Jim Norman and Ria Crow (Ottawa: Protective Sciences Division, DREO, 1992), 105.

83 DRB, vol. 4118, file 2-936-301-3. Major Glen Gay, Secretary, SWAC, to Lt Colonel R.A. Klaehn, Canadian Technical Representative to the Chemical Corps, Edgewood, 14 April 1948.

84 Guilford Reed’s role as chairman of the federal Fisheries Board served as an excellent cover for the shellfish toxin operation. There was, however, some concern that the concentrated mussels poison, contained in an alcohol solution within 45-gallon drums, would be detected by U.S. customs. Ibid., Glen Gay, Special Weapons Research Section, to Reed, 26 April 1948; ibid., Gay to David Sim, Deputy Minister of National Revenue, 20 July 1948.

85 Ibid., Major Carl Steidtmann, U.S. Technical Representative to Chief, Chemical Corps, Washington, DC, 30 July 1948.

86 In 1977 the Church Commission of the US Senate obtained evidence that the CIA regarded shellfish toxin as an ideal weapon for its covert operations, including assassinations.

87 DRB, vol. 4118, file 2-435-172, Report of the Scientific Intelligence Division, DRB, 4 August 1949; ibid., G. de T. Glazebrook, Director, Joint Intelligence Bureau, to A.J.G. Langley, 3 August 1949, Biological Warfare in the Northern Half of the Northern Hemisphere.

88 The DRB also monitored the activities of Soviet entomologists and their Communist bloc allies to determine their level of expertise in the use of vectors and the possibilities that this research had been weaponized. Ibid., G.W. Rowley, Arctic Research, DRB, memorandum, 9 August 1949.

89 Ibid., G.M. Carrie, Defence Research member, Canadian joint Staff, London, to A.J.G. Langley, Director Scientific Intelligence, DRB, Ottawa, 2 February 1950.

90 Ibid., A.F.B. Stannard, Scientific Intelligence Division, DRB, Washington, DC, to A.J.G. Langley, 28 March 1950.

91 DRB, vol. 4224, file DRBS 1820-11, Charles Mitchell to Glen Gay, Special Weapons Research Section, DRB, 28 October 1948; Lt Col. J.C. Bond, Canadian technical representative, Edgewood, to DRB, 1 December 1949. In June 2011 the United Nations Food and Agricultural Organization

declared that rinderprest had been eradicated because of an improved vaccine and rapid diagnostic tests.

92 The advantage of using strains of rinderpest virus for growth in eggs was that its virulence could be sufficiently attenuated to allow its use as a vaccine for cattle with a high degree of immunity. Ibid., Mitchell to Solandt, 4 December 1950.

93 Ibid., Major General A.C. McAuliffe, Chief Chemical Officer, to Solandt, 9 March 1951; ibid., E.L. Davies to McAuliffe, 24 March 1951.

94 Ibid., Colonel Oram Woolpert, Director, Headquarters Camp Detrick, to E.L. Davies, April? 1951.

95 Ibid., H.M. Barrett (for chairman DRB) to S.C. Barry, Director, Production Service, DA, 21 March 1958. Among the rinderpest strains were six variants of the Kabete, along with Egypt 133-2, that had been obtained in 1950.

96 Ibid., JC Clunie, DRB, Progress Report on "R" Work at GIES, 10 April 1952.

97 DRB, vol. 4224, file 4-935-43-2, Minutes of the Meeting of the BWRP, 23 October 1954.

98 *Langley* (BC) *Advance News* 27 March 2001; *GIS Vision*, June 2002.

99 Anthony Rasporich and Max Foran, eds., *Harm's Way: Disasters in Western Canada* (Calgary: University of Calgary Press, 2004), 189; Piers Millet, "Antianimal Biological Weapons Programs," in Wheelis, Rozsa, and Dando, eds., *Deadly Cultures*, 224–35.

100 Interview with Dr R.B. Stewart, Queen's University 10 July 1998.

101 Test: 21 April 1946, Bot Tox 6 Dec 1945-US "The Casualty Producing Power of US-S when Suspended from the Bomb A/C L (14 Dec / Aqueous Slurry)

102 Defence Research Canada Suffield, (declassified), DRES, Report No 1014, "Joint Canadian-American Trials 6 December 1945.

103 Ibid., DRES, (declassified): Test: 21 April 1946, Bot Tox; 6 Dec 1945-US "The Casualty Producing Power of US-S when Suspended from the Bomb A/C L (14 Dec, 1945).

104 WO 188/695 SES, Trial Record No.247, "The Effects of Air Temperature and Relative Humidity upon Recovery of Bacteria Tularesme Dispersed from the E61 Bomb and from the E99 Bomb, 29 April 1955.

105 Porton scientists also participated in the Dugway-SES consultations, held every six months, in the development of offensive BW munitions. WO 188/695 J.D. Morton to Dr G.O. Langatroth, Superintendent SES 16 June 1955.

106 Ibid., DRES, (declassified), 2003-5172: "Canadian Report on 1954 programme for Dugway Conference on BW-CW Field Testing"; 2000-3264 (A-20) "Minutes of Joint Field Test Conference on the Winter Test Program 1956-57. (Dugway, January 1956). Statement by Dr Tervet, Deputy Commander for Scientific Activities, Dugway. (Limited Circulation)

107 Ibid., DRES, (declassified), 2003-3226 (55-2029): Notes on Meetings at Dugway to Discuss Summer Trials for 1955 (13–14 January 1955). These trials were described in the official report *U.S. Army Activity in the U.S. Biological Warfare Programs,* vol. II (24 February 1977.Table 4: Biological Field Testing Anti-Personnel Pathogenic Agents.
108 DRB Papers, vol.4220, file 700-900-267-1, Group Commander H.G. Richards, Canadian Joint Staff Mission, Washington, to Chief of the Air Staff, Ottawa, 25 October 1951.
109 NARA, Joint Chiefs of Staff Papers, 'Report ... October 1950."
110 The US military had already re-established its cold weather research station in Churchill, Manitoba. In 1951 an arrangement was also made for joint testing of chemical weapons in order to determine how war gases responded to arctic conditions, and the performance of defensive CW equipment. DRB, vol. 4220, file 700-900-267-1 G.W. Rowley, Arctic Research, to Chairman DRB 14 December 1951.
111 Ibid., Memorandum Special Weapons Test Facilities: Suffield Experimental Station, October 1951.
112 Ibid. Reference was also made to the possibilities of ground contamination and "secondary effects might be expected, especially with a persistent agent such as anthrax." (14–15)
113 Balmer, "The UK Biological Weapons Program," in Wheelis, Rozsa, and Dando, eds., *Deadly Cultures,* 56.
114 WO, 188/668, BRAD Meeting, 31 August 1951.
115 Ibid., WO, 188/660, BRAD Meeting, 29 March 1952.
116 Donald Avery, "Advancing Medical Internationalism: Brock Chisholm and the Formative Years of the World Health Organization, 1946–1953," *Proceedings of the 38th International Congress on the History of Medicine,* ed. Nil Sari et al. (Ankara, 2005), 1693–9.
117 Chisholm was particularly impressed with the research of virologist Ronald Hare, who had made two important discoveries: first, that the presence of the influenza virus could be identified by its ability to agglutinate red blood cells; and, second, that chick embryo could be used for the development of the vaccine. Ronald Hare, *The Birth of Penicillin and the Disarming of Microbes* (London: Allen and Unwin, 1970), 212.
118 R.D. Defries, *The First Forty Years: Connaught Medical Research Laboratories, University of Toronto* (Toronto: University of Toronto Press, 1968), 10–20.
119 Hare, *Birth of Penicillin,* 212–15. Unfortunately, these efforts were in vain, since the dominant influenza strain in 1944 was H1N1 strain B.
120 All Irving, *Brock Chisholm: Doctor to the World* (Markham, ON: Fitzhenry and Whiteside, 1998); Robert Berkov, *The World Health Organization: A Study in*

Decentralized International Administration (Geneva: WHO, 1957); Norman Howard Jones, *International Public Health between the Two World Wars: The Organizational Problems* (Geneva: WHO, 1978).

121 World Health Organization (WHO) Library, Geneva, Executive Board Records, 1949, Third Session (EB/3/52), Letter to Board by Dr Vinogradov, deputy minister of Public Health of the USSR, 12 February 1949. In short order Albania, Bulgaria, Czechoslovakia, Hungary, and Poland also withdrew, giving the same basic reasons, along with allegations of American domination.

122 Javed Siddiqi, *World Health and World Politics: The World Health Organization and the U.N. System* (London: Hurst, 1995), 104–25.

123 The Malaria Eradication Programme (MEP) was unique since it was the first such campaign coordinated by an international health agency. Its ultimate failure also made the WHO reluctant to support other major programs such as the eradication of smallpox, at least until the late 1960s. Siddiqi, *World Health,* 124–5.

124 Edmund Russell, *War and Nature: Fighting Humans and Insects with Chemicals from World War I to* Silent Spring (Cambridge: Cambridge University Press, 2001).

125 LAC, General Andrew McNaughton Papers, vol. 288, file 1-6, "Secretary Forrestal Issues Statement on Biolological Warfare Potentialities, 12 March 1949." *New York Herald Tribune* 14 March 1949.

126 LAC, Brock Chisholm Papers, vol. 1, file 90-92 (1949), Address to the World Union of Peace Organizations, St Cergue, Switzerland, 9 September 1949; *Ottawa Journal* 10 September 1949.

127 Tom Mangold and Jeff Goldberg, *Plague Wars: A True Story of Biological Warfare* (New York: St Martin's, 1999), 322–25; Siddiqi, *World Health,* 112.

128 Chisholm Papers, vol. 9, "To all Members of the Secretariat of the World Health Organization," 10 December 1952.

129 Siddiqi, *World Health,* 141–91.

130 Mike Brown, *Put the Lights Out: Britain's Civil Defence Service at War, 1939–1945* (Stoud: Sutton, 1999); John Dowling and Evans Harrell, eds., *Civil Defence: A Choice of Disasters* (New York: American Institute of Physics, 1986).

131 Thomas Kerr, *Civil Defense in the U.S.: Bandaid for a Holocaust?* (Boulder, CO: Westview Press, 1983); Dee Garrison, *Bracing for Armageddon: Why Civil Defense Never Worked* (New York: Oxford University Press, 2006); James Miskel, *Disaster Response and Homeland Security: What Works, What Doesn't* (Westport, CT: Praeger Security International, 2006).

132 DRB 4234, File C.J. Mackenzie Correspondence, G.C. Lawrence to Mackenzie, 19 March 1948.

133 In October 1948 Brigadier General Alex Ross, Director of Civil Defence in the Second World War, warned Worthington about the challenges of civil defence coordination, particularly the dangers of spreading defences too thinly because of public or political pressure. Indeed, since the wartime effort had been so unsatisfactory, he advised that by starting afresh "many mistakes ... can be avoided." Department of Health and Welfare (DHW), vol. 650, file 102-1-1, pt 1, Ross to Worthington, 22 October 1948.

134 Ibid., vol. 718, file 112-C-8, General Worthington to the Minister of Defence, 27 October 1949. Members of the Civil Defence Planning Committee had to ensure that recommendations about various measures were acceptable to their departments.

135 The RCMP was entrusted with the responsibility for anti-sabotage precautions "as part of their internal security measures." Ibid., General Worthington, "Emergency Planning, Civil Defence," memorandum for the Minister of National Defence, 5 November 1949.

136 Ibid., memorandum, Civil Defence Division, Department of National Health and Welfare, 1951.

137 DHW, vol. 650, file 102-1-1 pr. 2, Minutes of the Eighth Meeting of the CDCPC, 8 September 1950; ibid., Minutes of the Tenth Meeting of the CDCPC, 20 November 1950.

138 The Department of Health and Welfare also prepared a number of civil defence manuals. One of these, CD Manual No. 7, prepared in the early 1950s, provided a wide range of information about various ways of coping with an atomic bomb attack and the effects of blast, thermal, and radiation injuries. Reference was also made to the threat of biological and chemical warfare, particularly the former, with a BW assault on water supplies deemed of special concern.

139 Ibid., Charles Mitchell to Dr L.W. Billingsley, Research Co-ordination Staff (Medical), DRB, 27 September 1950.

140 Ibid.

141 Ibid., Minutes of the Eleventh Meeting of the CDCC, 19 January 1951.

142 Plans were also made for the establishment of a civil defence school, which was eventually located in a former RCAF facility in Arnprior, Ontario. Ibid., Minutes of the Twelfth Meeting of the CDCC, 2 February 1951.

143 Ibid., K.C. Charron, MD, Chief, Civil Defence Health Planning Group, DHW, to Dr Hugh Malcolmson, Director, Bureau of Industrial Hygiene, Department of Health and Public Welfare, Winnipeg, Manitoba, 25 September 1952.

144 Ibid., enclosed special report "Biological Warfare Defence Services."

145 Ibid., article cited by Milton H. Brown, MD, "Medical Aspects of Civil Defence in Biological Warfare," published in the special Civil Defence issue of the *Canadian Medical Association Journal*.

146 A list of these known agents was compiled by Canadian military and health officials, in consultation with the Special Weapons Defense Section of the United States Civil Defense Organization. The bacteria category contained nine agents, notably those causing anthrax, tularaemia, plague, and brucellosis; the virus category had seven agents, notably psittacosis, smallpox, and influenza; the *Rickettsia* category had five, fungi had two, and toxins two (bot tox, ricin).

147 National Archives and Record Administration (NARA), Washington, DC, State Department, vol. 32, file 30-7-4, exchange of notes between The Honourable Laurence Steinharedt, Ambassador of the United States of America, Ottawa, and Lester B. Pearson, The Honourable Secretary of State for External Affairs, 12 April 1949.

148 Reg Whitaker and Gary Marcuse, *Cold War Canada: The Making of a National Insecurity State, 1945–1957* (Toronto: University of Toronto Press, 1994), 140–6.

149 State Department, vol. 32, file 30-7-4, N.R. Chappell, Canadian Executive Officer, JIMC to Ansley, 21 February 1950.

150 Ibid., exchange of diplomatic notes, H.H. Wrong, Canadian Ambassador, Washington, DC, to Dean Acheson, The Secretary of State, USA, 27 March 1951.

151 Ibid., A.R. Mackey to Millard Caldwell, 16 April 1951.

152 On 4 March US Secretary of State Dean Acheson formally rejected the Communist bloc charges. In a statement on 7 May he cited a number of prominent American scientists, notably Drs C.H. Curran (entomologist) and Rene Dubos (bacteriologist), who had labelled the germ warfare charges as a clumsy hoax.

153 The first major reference to biological weapons was made at the Warsaw Peace Congress of November 1950. LAC, Records of the Department of External Affairs (DEA), vol. 5921, file 50208-40: The Communist Germ Warfare Campaign, July 30, 1952.

154 Ibid., file 50208-40, pt 3, DEA Report, Communist Propaganda on Bacteriological Warfare, United Nations, General Assembly, Seventh Session., October 1952.

155 Joliot-Curie's statement brought an outraged response from the famous US physicist Robert Oppenheimer, who accused Joliot-Curie of violating scientific principles by allowing his name to be used "in support of these unverified charges." United States Library of Congress, Robert

Oppenheimer Papers, box 42, Joliot-Curie File, Oppenheimer to Joliot-Curie, April 1952.

156 On 13 March, the Red Cross agreed to the request of Dean Acheson, U.S. Secretary of State, to investigate the germ warfare allegations. On 20 March Secretary General Trygve Lie offered the assistance of the WHO in combating epidemics in the region. Both these initiatives were rejected on the grounds that these organizations were tools of the United States. DEA, vol. 5921, file 50208-40, Report of 18 April 1952.

157 Canadian diplomats were impressed with the analysis of George Kennan, the State Department's expert on the Soviet Union, who claimed that the 1952 "hate America" campaign was "unprecedented both in intensity and extent." Ibid., Canada's Permanent Representative to the United Nations (George Ignatieff) to Secretary of State for External Affairs (Lester B. Pearson), 21 June 1952.

158 DEA, vol. 5921, file 50208, Address by the Secretary of State for External Affairs, Lester B. Pearson, 25 May 1952; ibid., Canadian Ambassador in Washington to DEA, 20 May 1952.

159 Ibid., Canadian Representative to DEA, 18 June 1952.

160 Ibid., Canadian Permanent Representative to DEA Headquarters, 3 July 1952.

161 Pearson's 12 May speech was well received by all political parties. An irate Reverend Endicott, however, requested permission to appear before the Committee on External Affairs to refute Pearson's arguments and to justify the Chinese rejection of an investigation by the International Committee of the Red Cross and the WHO. It was rejected. Hansard, 12 May 1952; DEA papers, vol. 2412, file 102-AZW-40, pt 2, Endicott to Pearson, 13 May 1952.

162 Ibid., British Embassy, Peking to Foreign Office, 15 April 1952. *Shanghai News* 25 March 1952.

163 LAC, interim box 123, vol. 5921, file 50208-40 pt.6, Address by the Secretary of State for External Affairs, Iroquois United Church, Iroquois, Ontario, 25 May 1952.

164 Ibid. Ironically Pearson and Endicott had been close friends during their student days at Victoria College, University of Toronto, in the 1930s. Stephen Endicott, *James G. Endicott: Rebel Out of China* (Toronto: University of Toronto Press, 1980).

165 DEA, vol. 5921, file 50208-40, Pearson to Reverend Fred Reed, Sobright, Ontario, 22 March 1952; ibid., K.J. Burbridge, Legal Division, DEA, memorandum for Escott Reid, 29 April 1952.

166 Omond Solandt also advised against the prosecution of Endicott on the grounds that a trial might disrupt the DRB relationship with the US Army.

Ibid., Canadian Ambassador to the United States to Secretary State, DEA, 3 June 1952.

167 DEA Papers, vol. 2412, file 102-AZW-40, pt 3, Memorandum for Escott Reid, 26 May 1952.

168 There were also numerous comparisons between the Endicott situation and the controversy involving Hewlett Johnson, the Anglican Church's "Red" Dean of Canterbery, who also accused the US of germ warfare at meetings in the UK and in North America. Endicott, *Rebel Out of China*, 267–70.

169 In a non-circulating report, Brittain claimed that Endicott's brief demonstrated that he was a "dangerous fanatic, quite incapable of making a reasonable decision in matters of this kind. Ibid., W.H. Britain to Pearson, 28 May 1952, Enclosed, "Report: Comment on Communist Charges of Germ Warfare."

170 C.E. Atwood sent his report as a five-page letter to G. de T. Glazebrook of DEA (22 May), while A.W. Baker's submission was a brief one-page summary.

171 Ibid., A.J. Pick, Memorandum for the Minister, 27 June 1952.

172 DEA, vol. 5921, file 50208, Report of W.H. Brittain,Vice-Principal, Macdonald College, McGill University, to Lester B. Pearson, 28 May 1952. Ibid., A.W. Baker, "Report: Comment on Communist Charges of Germ Warfare." Another expert, Guilford Reed, also regarded the CPC pamphlet as seriously flawed, concluding that "the supposed method of dispersal was ridiculous." DRES, Report no. 43, 30 October 1954, "Fleas as Vectors of Plague in Bacterial Warfare," by A.S. West and G.B. Reed.

173 The three scientists also cited the comment by Rene Dubos, a world-famous scientist at the New York Rockefeller Institute, who branded the germ warfare charges as "an amateurish attempt at 'scientific fakery.'" *Ottawa Journal* 28 June 1952.

174 In the US, the *New York Herald Tribune*, and the *Christian Science Monitor* were two of the papers most interested in the DEA scientific report. Ibid., vol. 5920, file 50208-40, pt 3, Memorandum for the Minister: "A Survey of the Effects of the Work done by Professors Baker, Britain and Atwood in Countering Communist Charges of Germ Warfare."

175 A special effort was made to use the Joint Report to convince officials in Indonesia about the baseless nature of the germ warfare allegations. Memorandum for L.D. Wilgress, Under-Secretary of State for External Affairs, 10 July 1952.

176 Ibid., Colonel J. Woodall Greene, Chief, Psychological Warfare, Headquarters, United Nations and Far East Command, Tokyo, to A.R. Menzies, Chargé d'Affaires, Canadian Embassy, Tokyo, 29 July 1952.

177 Since India was a key player in the propaganda campaign about germ warfare, Canadian officials were delighted when in December 1952 Prime Minister Nehru declined to support Communist allegations. Ibid., High Commissioner (Canada) to Secretary State, DEA, 5 August 1952; ibid., George de T. Glazebrook, DEA, to Canadian Delegation, United Nations, 17 December 1952.

178 The other members of the International Scientific Commission were: Dr Andrea Andreen, laboratory director, Stockholm, Sweden; Dr Oliviero Olivo, human anatomy, University of Bologna, Italy; Dr Samuel Pessoa, parasitologist, University of Sao Paulo, Brazil; Dr N. Zhukov Verenznikov, bacteriologist, Vice-President of the Soviet Academy of Science; Dr Franco Graziosi, Institute of Microbiology, Rome, Italy. In addition, an eleven-man committee of Chinese experts was drawn from the scientific, medical, and public health fields. Ibid., DEA, file 50208-40, pt 4, "Analysis of the Report of the Scientific Commission for the Investigation of the Facts Concerning Bacterial Warfare in Korea and China," DEA, Defence Liaison, 20 October 1952.

179 Ibid., file 50208-40, pt 4, Memorandum for Mr Glazebrook, 24 September 1952.

180 Joseph Needham's role as chairman of the commission attracted considerable media comment in the UK, including criticism from the Royal Society of London, of which he was a member, that the report drew most of its conclusions "from evidence largely not scientific." Ibid., file 50208-40, Foreign Office Circular: Germ Warfare, 13 October 1952; *London Times* 1 October 1952.

181 Ibid., Canadian High Commissioner, UK, to Secretary State DEA, 16 December 1952.

182 Ibid., Canadian Delegation to SSDEA, 26 November 1952; ibid., Canadian Delegation to DEA Headquarters 25 March 1953.

183 Ibid., file 50208-40, pt 5, Chairman, Canadian Delegation, to DEA Headquarters, 22 October 1953.

184 Ibid., file 50208-40, pt 6, DEA Report, Eighth Session of the General Assembly, First Committee 24 November 1953. In December 1952 James Endicott was awarded the Stalin Peace Prize for his "outstanding contribution to the cause of the struggle for the preservation of world peace." Endicott, *Rebel Out of China*, 303.

185 Jessica Wang, *American Science in an Age of Anxiety: Scientists, Anticommunism and the Cold War* (Chapel Hill: University of North Carolina Press, 1999); Whitaker and Marcuse, *Cold War Canada.*

186 Everitt Murray to Bob Murray, 17 February 1952; ibid., Everitt to Bob Murray, 6 April 1949, cited in Avery and Eaton, *The Meaning of Life*, ciii.

187 See Martin Furmanski and Mark Wheelis, "Allegations of Biological Weapons Use," in *Deadly Cultures,* 252–61.

3 Operational Biological Weapons and Alliance Cooperation, 1955–1969

1 Daniel Gerstein, *Bioterror in the 21st Century: Emerging Threats in a New Global Environment* (Annapolis, MD: Naval Institute Press, 2009), 54–7.

2 John van Courtland Moon, "The U.S. Biological Weapons Program," in *Deadly Cultures: Biological Weapons since 1945,* ed. Mark Wheelis, Lajos Rozsa, and Malcolm Dando (Cambridge: Harvard University Press, 2006), 32.

3 Ibid., 15–17; John Hart, "The Soviet Biological Weapons Program," in Wheelis et al., eds., *Deadly Cultures,* 134.

4 In 1952, a Canada-US project carried out cold weather chemical warfare testing in the proximity of Fort Churchill, Manitoba.

5 Erika Simpson, *NATO and the Bomb: Canadian Defenders Confront Critics* (Montreal and Kingston: McGill-Queen's University Press, 2003), 25–148; Robert Bothwell, *Nucleus: The History of the Atomic Energy of Canada Limited* (Toronto: University of Toronto Press, 1988), 99–102, 139, 151.

6 In 1947 the Clement Attlee government decided to proceed with its own atomic bomb project on the grounds that the United Kingdom required its own nuclear deterrent. During the next five years, it spent almost £100 million on the A-bomb project with almost no public discussion. Margaret Gowing, *Independence and Deterrence: Britain and Atomic Energy, 1945–52,* 2 vols. (London: Macmillan, 1974).

7 One of the authors of this report was W.G. Penney, one of Britain's foremost nuclear scientists. He was also well acquainted with DRB Chairman Omond Solandt because of their mutual experiences in British operational research planning during the Second World War.

8 Department of National Defence, Directorate of History, file 94/121, "The Technical Feasibility of Establishing an Atomic Weapons Proving Ground in the Churchill Area," prepared by C.P McNamara of the DRB and W.G. Penney, of the UK Ministry of Supply (n.d.).

9 Ibid. This document remained classified until May 1994 when the UK Ministry of Defence informed the DND Directorate of History that it could be released into the public domain.

10 Ibid. Six other possible sites were considered: Port Nelson, Manitoba; Eskimo Point, Northwest Territories; Coral Harbour (Southampton Island, NWT); Belcher Islands (Hudson Bay); Frobisher Bay (Baffin Island); and Suffield, Alberta.

11 Historian Margaret Gowing has criticized the British government's decision to carry out atomic trials in Australia on the grounds that the US had

a number of well-developed test sites. Gowing, *Independence*, 50–112. In total, there were nine separate trials, beginning in March 1952 with the atomic bomb series, and then proceeding with a series of H-bomb trials between 1956 and 1958, with explosions in the megaton category. Joan Smith, *Clouds of Deceit: The Deadly Legacy of Britain's Bomb Tests* (London: Faber and Faber, 1985).

12 LAC, DRB Papers, vol. 8132, file 1287-25-2, "Official Report of the Canadian Delegation to the Thirteenth International Congress of Military Medicine and Pharmacy Held in Paris, 17–23 June 1951."

13 DRB, 1983-84/215, box 232, file 2001-91-149, Observations on Exercise Medical Rubicon, 20 October 1952.

14 Ibid., Report on Exercise Medical Broadfront III.

15 Ibid., box 233, file S-2001-41/M12, Brigadier K.A. Hunter, Director General Medical Services, memorandum, 9 November 1955. Canadian representatives were invited, along with British and NATO observers, to the March 1955 US radiological defense field exercises at the atomic tests in Nevada (Operation TEAPOT), with the proviso that they "should receive as much information concerns weapons effects on a need-to-know basis as is permitted under existing [US] law." US, Declassified Documents 37A, Report by the Joint Strategic Plans Committee to the Joint Chiefs of Staff, 21 January 1955.

16 Lawrence Wittner, *Resisting the Bomb, 1954–1970: A History of the World Nuclear Disarmament Movement*, vol. 2 (Stanford, CA: Stanford University Press, 1997), 2–3.

17 DRB, 1983-84/215, box 232, file 102-1-1, CDRA Special Round Table, 8 April 1954.

18 Ibid., Worthington to Deputy Minister, Welfare (DHW), 7 April 1954.

19 The Canadian Civil Defence organization was divided into four divisions, along with 172 basic training organizations across the country. In addition, there was advanced training at the Civil Defence College. Ibid.

20 *Debates* of the House of Commons, 19 March 1959, 2073-80.

21 DRB, 1983-84/215, box 232, file 102-1-1, Dr W.J. Riley, Chairman, Metropolitan Civil Defence Board to Andrew Currie, Provincial Civil Defence Co-ordinator, Winnipeg, 12 February 1959.

22 In December 1962 the *Canadian Medical Association Journal* 87, no. 22 (1 December 1962) devoted an entire issue to the threat of WMD.

23 Ibid., 1156–60. Suffield scientists were aware of the secret urban BW field trials being carried out in London, England during the late 1950s using anthrax simulants. Similar trials were carried out in New York City.

24 Both the US Senate and the House of Representatives considered the mobilization of scientific personnel. R. Martineau, Defence Research Member, Washington to DRB, Ottawa, 22 March 1951.

25 Ibid. Solandt to the Minister of Defence, 6 January 1951. A special meeting on the subject was held on 5 February with representatives of the DRB, the NRC, the Department of Labour's Roster Register of Technical Personnel, and the Personnel Committee of the Armed Forces before the proposal was submitted to the Cabinet Defence Committee. Otto Maass strongly endorsed Solandt's proposal, claiming that, if adopted, "our friends south of the border will be envious of the Canadian set-up." Ibid., Maass to Solandt, 2 February 1951.

26 Wilder Penfield, Director of the Montreal Neurological Institute, also proposed that all major research institutes be integrated into Canada's civil defence system. Ibid., Penfield to Solandt, 2 October 1950.

27 Brian Balmer, *Britain and Biological Warfare: Expert Advice and Science Policy, 1930–1965* (London: Palgrave, 2001), 51–3.

28 WO, 188/668, BRAB meeting, 25 April 1952. It was recommended that the British government issue a statement "that the British policy remained unaltered, i.e. that BW would in no circumstances be used except in retaliation."

29 Another group of prominent ASM members were involved with Canada's BW program, notably E.G.D Murray (McGill) and James Craigie (Toronto).

30 Archives of the American Society of Microbiologists (ASM Archives), University of Maryland, (Baltimore), file NLSAB, xiiii: 2 (1942).

31 ASM Archives, 13-II AT, folder 2, Minutes of the Committee on Biological Warfare, 15 May 1951, Room 3 E-1060, The Pentagon, Dr Ira Baldwin presiding.

32 Ibid., NLSAB 17:3, August 1951.

33 *SAB News Letter* 21, no. 3 (August 1955), and 22, no. 3 (August 1956).

34 Cited in *ASM News* 58, no. 4 (1992), ASM Archives, 13-II AT, folder 2, ASM Committee Advisory to Fort Detrick, 21 April 1967.

35 Ibid., 8 IA, folder 4, C.B. Maquand, Executive Director CBR Advisory Council Secretariat, Edgewood Arsenal, to Dr Erling Ordal, Chairman of the Committee, ASM, 13 February 1964; Erling Ordal, chairman, Annual Report, 18 February 1965.

36 DRB, 83-84/167, vol. 7359, file 170-80, C1c (vol. 2): Terms of Reference.

37 Ibid., Armand Frappier, "Role of Civilian Laboratories in BW Research and Production," ACBWR Meeting, 21–2 March 1960.

38 Ibid.

39 DND, Access Request: DRB, 89-90/203, vol. 32, file DCBRLS 202-60/45, Organization for Quadripartite CRB Conference and Associated Standing Working Groups, enclosed in letter, Chairman, DRB to Chief Superintendent DCBRL (Ottawa), 28 April 1965.

40 The TTCP soon changed its name to The Technical Cooperation Program. It also added two new members: Australia in 1965, and New Zealand in 1969. See TTCP website, www.acq.osb/ttcp.
41 DRB, 1983-84/167, vol. 1339. In 1957 when TTCP was established, it was anticipated that all CBR work would be consolidated under one TTCP subcommittee. But because Canada did not share in the exchange of atomic data, which was only on a bilateral UK-US basis, it was proposed that Sub-Group E facilitate the exchange of non-restricted data dealing with "the biological aspects of blast and heat of atomic weapons." Memorandum, 18 May 1961.
42 The medical aspects of CBW warfare was another important field of joint research, with special emphasis on how assistance could "be given to troops for dealing with B.W. contamination." DRB, 1983-84/167, vol. 740, file DRBS 171-80/C10, pt 2, Conclusions and Recommendations of the 7th Tripartite Conference on Toxicological Warfare, September 1952.
43 DRB, 83-84/167, DRBS 1800-60/141, pt 1, Minutes of a Meeting Held at Chemical Corps Biological Laboratories, Camp Detrick, 16 March 1953.
44 Ibid., Lieutenant-General G.G. Simonds, Chief of the General Staff (Canada), Memorandum: "Tripartite Meetings on Toxicological Warfare," 10 October 1952. Ibid., Conclusions and Recommendations of the 7th Tripartite Conference on Toxicological Warfare, September 1952.
45 Ibid., Statement Chiefs of Staff Committee, 23 September 1953.
46 DRB, 1983-84/167, vol. 7415, file 202-60/0, pt 1, cited in Memorandum: Tripartite Toxicological Warfare Conferences, 15 July 1955.
47 Ibid., Chairman, DRB to the Director of Defence Scientific Research, Ministry of Supply, London, 16 December 1954.
48 Ibid., G.R. Vavasour, for Chairman, DRB to Lt Colonel H.E. Staples, Canadian Army Technical Representative, U.S. Army Chemical Corps, 3 May 1956.
49 DRB, 1983-84/167, vol. 740, file DRBS 171-80/C10, pt 2, Summary Canadian Progress Report to Eleventh Tripartite Conference, October 1956.
50 Ibid.
51 Moon, "The US Biological Weapons Program," 12.
52 US Declassified Documents, 164A, J.R. Mares to J.R. Killian, National Science Adviser, 4 June 1959.
53 In 1957 American and Soviet virologists negotiated an important agreement for the exchange of medical information about the newly emergent A/H2N2 influenza virus (Asian flu). This arrangement formed the basis of expanded medical cooperation between the two countries. See chapter 6 for a discussion of this trend.

54 LAC, National Research Council Papers (RG 77), vol. 1, file 3-3-C-154-9, Lester B. Pearson, Minister of External Affairs, to President Edward Hall, University of Western Ontario, 3 February 1956. DEA officials hoped that Murray could obtain some information about the Soviet BW program when he attended the Leningrad meetings of the All Union of Microbiology. In the end, Murray's trip was cancelled. Ibid., Murray to Pearson, 9 February 1956.

55 DEA, vol. 84-85/150, file 6922-40, Canadian Ambassador, Moscow to Ottawa, 9 October 1957. Throughout the fall of 1957 Foreign Affairs officials in Moscow reported how Soviet propaganda was exploiting *Sputnik*, including accounts filed by Ottawa-based reports of *Pravda* and *Investia.*

56 Solandt was vice-president of the Canadian National Railways. He was appointed first chairman of the newly created Science Council of Canada (1966–72).

57 Pennie was educated as a chemical engineer at the Technical University of Glasgow. He joined the DRB in 1948, working first at the Valcartier Explosives Establishment, and was superintendent of the DRB Churchill (1954–6). After his Suffield stint, Pennie became Deputy Chair of Operations at DRB headquarters until 1974. Donald Avery, interview with Archie Pennie, Ottawa, 8 June 2004.

58 On 10 July 1956 the Cabinet Defence Commission cancelled Britain's offensive BW warfare program. In 1957, MRD was changed to the Microbiology Research Establishment, with David Henderson as director, rather than superintendent. In 1959 the Ministry of Supply was disbanded, and MRE was moved into a reconstituted Ministry of Defence. G.B. Carter, *Chemical and Biological Defence at Porton Down 1916-2000* (London: The Stationery Office, 2000), 84–5.

59 DRB, 89-90/203, vol. 31, file DRCLS 202-60. Draft of the proceedings of the Tripartite Toxicological Meeting, 9 September 1958.

60 Ibid, HQS 2040-1, Memorandum N.W. Morton, received DAR 15 July 1958.

61 DRB 1983-84/167, vol. 7415, file 202-60/0, pt 1, Memorandum: Toxicological Warfare R &D, Coordination of Implementation of Eleventh Tripartite Recommendations, 21 February 1957.

62 Ibid. Draft Final Agenda, BW Basic Research, Report by Joint Chairmen: Dr D.W. Henderson (UK), Dr E.V. Hill (US), Mr N.J.B. Wiggin.

63 Ibid.

64 The delegates also recommended that the United States re-enter the offensive anti-animal BW field at the earliest possible moment and that the United Kingdom reactivate its program for the possible offensive use of foot-and-mouth disease. Ibid.

65 Ibid., Chairman, DRB to Chief Supt., SES, 22 June 1958.
66 DRB, vol. 1339, file 170-3-E, J. Koop, DRB Washington Contact officer for Sub-Group E, Report of Meeting of Sub-Group E of the TTCP, 18 December 1959.
67 DND, Access Request: DRB, 83-84/167, vol. 1339, file DRBS 170-3-E, Tripartite Technical Cooperation, Report on 13th Tripartite Conference on Toxicological Warfare, enclosed in letter from Sir Frederick Brundrett, Ministry of Defence, London, 22 December 1958.
68 Another priority that required a division of scientific labour was the development of vaccines against the major BW agents, with the US working on nine items, the British three, and Canada one. Ibid.
69 Ibid. In operational terms, there was considerable interest in the use of bacterial slurries dispersed from aircraft-mounted spray devices.
70 DRB, box 31, file DRCLS 202-60/144-13, Report by A.M. Pennie, 7 August 1958.
71 Ibid., vol. 7514, C.R. Vavasour to Pennie, 28 February, 22 April 1958.
72 The criterion for biological and chemical incapacitants was that their effects had to be "instantaneous, overwhelming and of short duration." Ibid., Vavasour to Staples, 16 January 1958.
73 DRB, 83-85/167, vol. 7415, file 201-60/144, pt 2, Sixth Tripartite Conference on Army Operational Research: Memorandum and Recommendations, 28 October 1958.
74 DRB, 1982-83/167, vol. 4416, file DRBS 202-60/144-16, pt 3, G.R. Vavasour, Canadian Coordinator for 16th CBR Conference to Canadian Army Headquarters, 5 September 1962,
75 Ibid., Vavasour memorandum: CBR Tripartite Conference, 18 June 1962; ibid., J.F. Currie, Canadian Chairman, B.W. Basic Research to Dr D.W. Henderson, Director, MRE, 11 June 1962.
76 British researchers were also active in the recombinant field, with most of their work concentrating on "strains of *P. pestis* having all the known virulence determinants but not fully virulent ... [and] streptomycin resistant derivatives." In contrast, DRB scientists did not achieve the same level of innovation. Ibid., Vavasour memorandum 18 June 1962.
77 The freeze-drying process was particularly successful in the weaponization process involving Q fever, anthrax, and Venezuelan equine encephalitis. Ibid. Report of B.W. Basic Research Discussion Group Meetings (1962). Ibid., J.F. Currie, Canadian Chairman B.W. Basic Research to Chairman, DRB, 2 August 1962: Report enclosed.
78 Ibid., Report of the Basic Research Group.

79 WO, 195/15142, "Large Area Coverage by Aerosol Cloud Generated at Sea," by G.F. Collins, A.R. Laird, and R.A. Titt (6/59).
80 WO, 195/15164, "Early Warning Devices for BW Defence. The Situation in April 1961," by K.P. Norris and E.O. Powell.
81 DRB, 1982-83/167, vol. 4416, file DRBS 202-60/144-16, pt 3, J.E. Mayhood to G.R. Vavasour, 24 July 1962; ibid., Vavasour to Mayhood, 19 July 1962.
82 The confrontation over Canada's response to the Cuban missile crisis, and its refusal to accept nuclear warheads for either the Bomarc B ground-to-air missile system or for Canadian air squadrons assigned to NATO, have not been discussed here. There is, however, extensive literature on the subject. See Jocelyn Ghent-Mallet and Don Munton, "Confronting Kennedy and the Missiles in Cuba, 1962," in Donald Avery and Roger Hall, *Coming of Age: Readings in Canadian History since World War II* (Toronto: Harcourt Brace, 1996), 319–42; Erika Simpson, *NATO and the Bomb: Canadian Defenders Confront Critics* (Montreal and Kingston: McGill-Queen's University Press, 2001); Denis Smith, *Rogue Tory: The Life and Legend of John G. Diefenbaker* (Toronto: MacFarlane, Walter and Ross, 1995).
83 CMA *Journal* (1 December 1962): 1156–60.
84 The original policy was included in COS Committee Paper 1/63 of 1 May 1963. DEA, file 28-6-6, vol 5, Statement Brigadier General H. Tellier (Director General Plans Canadian Forces) to J.S. Nutt, Office of Politico-Military Affairs, DEA, 31 December 1968.
85 DRB, 1983-84/167, vol. 7358, file 110-80, C.R. Vavasour Memorandum: Matters Concerning the Next Meeting, 21 January 1963. His paper, "BW Attack against North America," was enclosed.
86 Ibid. Vavasour also speculated that the Soviet Union might use biological weapons under certain circumstances "a) to attack cities after the nuclear retaliatory forces were destroyed (b) to create a domestic catastrophe in the US in order to … discourage US intervention in Communist aggression in another part of the world (c) to create widespread illness in important industrial cities by repeated use of BW to weaken industrial output and create serious economic problems."
87 DRB, 1983-84/167, vol. 7358, file 110-80, Advisory Committee on BW Research: Terms of Reference, 14 February 1963.
88 Ibid.
89 DRB, 1983-84/167, vol. 7358, file 110-80, J.F. Currie to Chairman, DRB, 3 July 1963.
90 Ibid.

91 DRB, 1983-84/167, vol. 7358, file 110-80, Vavasour to Chairmen and Members, Advisory Committee on BW Research, 11 September 1963.
92 Ibid., Vavasour to van Rooyen, 11 September 1963.
93 Ibid., file 1800-1, H. Sheffer, Chief Supt to AK Longair, DAR, 27 February 1967.
94 North Atlantic Treaty Organization (NATO) Archives, Brussels, Civil Defence Committee (AC/023), file 1-79, First Meeting of senior civil emergency planning committee, 10 February 1956; D/499, Report of the meeting of the Restricted Working Party on Protection Against Chemical Warfare, 12 August 1965.
95 NATO file, SGM 117/62, NATO Military Committee, Standing Group CBR, 15 April 1962; ibid., Long-Term Scientific Studies for the Standing Group North Atlantic Treaty Organization.
96 NATO file, MCM 122/62, Memorandum for Members of the Military Committee, 22 October 1962; SHAPE 84/62-1450/20, "Chemical and Biological Warfare Policy," 13 April 1962.
97 NATO file, SG161/19, Report of Standing Group on CBR, "Soviet Bloc Strength and Capabilities: Part I The Soviet Bloc Threat to NATO, 1963–1967" (meeting of 27 April 1965).
98 NATO files, SG 265, Final Report, Section 11, Biological Warfare, 14 March 1966.
99 "Report by the Military Committee to the Defence Planning Committee on Overall Strategic Concept for the Defence of the North Atlantic Treaty Organization Area, 11 May 1967 (DPC/D (67) 23." It was concluded that the USSR would "continue to support their objectives from a position of impressive military strength based on nuclear, massive conventional, chemical and possibility biological capabilities." Gregory Pedlow, ed., *NATO Strategy Documents, 1949–1969* (Brussels, 2000), 353, 366.
100 Ibid., "Report by the Military Committee to the Defence Planning Committee on Overall Strategic Concept for the Defence of the North Atlantic Treaty Organization Area, 11 May 1967 (DPC/D (67) 23." Pedlow, ed., *NATO Strategy Documents*, 353, 366. DEA, CBW Disarmament, 28-6-6, vol. 3, Memorandum Assistant Deputy Minister External, 6 January 1969.
101 At this stage the Tripartite meetings included User Aspects Review, CW Agent Research Review, CW Application Review, BW Basic Research Review, BW Applications Review, Medical Aspects Review, Protection Review, and Radiological Defence Review. DRB, 82-83/167, vol. 4116, file 202-60/144-16, pt 3, Schedule of Meetings.
102 Ibid., CW Application – Mr. H.J. Fish.

103 Ibid., H.J. Fish to G. Vavasour, CBR User Aspects Discussion Group, 11 October 1962.

104 DRB, vol. 7450, file 375-4/62-6, G.R. Vavasour to Robert Heggie, Defence Research Chemical Laboratories, Ottawa, 17 December 1963.

105 Moon, "The US Biological Weapons Program," 32–3. At this stage, the US biological weapons arsenal consisted of nine anti-personnel agents: *Bacillus anthracis* (lethal), *Francisella tularensis* (lethal), *Brucella suis* (incapacitating), *Coxiella burnetii* (incapacitating), yellow fever virus (lethal), Venezuelan equine encephalitis (incapacitating), botulinum toxin (lethal), staphylococcal enterotoxin type B (incapacitating), and saxitoxin (lethal).

106 Jeanne Guillemin provides an interesting discussion of Deseret trials carried out between 1962 and 1968 in her book *Biological Weapons: From the Invention of State-Sponsored Programs to Contemporary Bioterrorism* (New York: Columbia University Press, 2005), 109–11.

107 DND, Access Request: DRB, 89-90/203, vol. 47, file DREO 1800-1-1, pt 2, "Procedural Rules for Participation by the United Kingdom, Canada, and Australia in the Deseret Test Center Program, 11 August 1964.

108 DND Access Request: DRB 83/84, vol. 7513, file 1800-60/141-1, pt 1, Brigadier J.A.W. Bennett, Commander, Canadian Army Staff, Washington DC, to Army Headquarters, Ottawa, 15 May 1962.

109 Ibid., file 1800-1, Recommendation for SES Participation in U.S. Army Field Trial WINDSOC II, 22 August 1962. Ibid., file 1800-60/141-1, G.R. Vavasour to Chief Superintendent SES, 6 September 1962. See Ed Regis, *The Biology of Doom* (New York: Henry Holt, 1999), 188–92, 200–6.

110 Ibid., Vavasour to SES, 6 September 1962. Archie Pennie was superintendent of SES between 1957 and 1964, a critical period in the Station's development.

111 DND, Access Request: DRB, 89-90/203, vol. 47, 1800-1, pt 1, Archie Pennie, Memorandum: Collaborative Trials with U.S. in BW-CW, 19 December 1962. See Guillemin, *Biological Weapons*, 110–11; Regis, *Biology of Doom*, 188–92.

112 DRB, 89-90/203, vol. 47, 1800-1, pt 1Ibid., Pennie Memorandum.

113 DRB, vol. 7513, file 1800-60/141-1, pt 1, cited in letter, A.K. Longair to Pennie, 20 December 1962.

114 Ibid. In terms of manpower, the Deseret Testing Center had its own staff and drew support from sixty-five positions at Dugway and thirty at Detrick and Edgewood. Ibid., 1800-60/141-1, pt 1, Report on a Visit to Desert Test Center, Fort Douglas, Utah by A.P.R. Lambert and H.J. Fish, 8–9 January 1963.

115 Ibid., Pennie to A.K. Longair, 18 January 1963.
116 DRB, vol. 47, 1800-1, G.K. Vavasour Memorandum: 22 January 1963.
117 Ibid., "Suggested Letter to Be Sent to Army Materiel Command."
118 Ibid., "Procedural Rules for Participation by the United Kingdom, Canada, and Australia in the Deseret Test Center Program," 18 August 1964.
119 The purpose of the Shipboard Hazard and Defence tests (SHAD) were "to identify US warships' vulnerabilities to attacks with chemical and biological weapons and to develop procedures to respond to such attacks while maintaining a war-fighting capability." Moon, "The U.S. Biological Weapons Program," 26. See also William Page, Heather Young, and Harriet Crawford, *Long Term Effects of Participation in Project SHAD* (Washington: The National Academies Press, 2007).
120 Canada had certain advantages in being involved with the joint sea vulnerability trials, since it had already carried out a series of ship-CBW hazard tests in 1960–1, and again in 1964, when the destroyed *St Croix* was involved in a series of successful trials using the stimulant BG.
121 DND Access Request: DRB 47, 1800-1, pt 2, C.E. Hubley, Canadian Joint Staff, Washington to Chairman, DRB, 12 August 1964. Guillemin, *Biological Weapons*, 110. Updates on the release of declassified information on SHAD are available at www.publichealth.va/gov/exposures/shad/basics.asp.
122 DRB, vol. 47, 1800-1-I, Hubley to Chairman DRB, 20 August 1964. Ibid., 1800-1, B.J. Perry, SES to Chairman, DRB, 21 August 1964.
123 Ibid., B.J. Perry, SES to Chairman DRB, 21 August 1964.
124 Ibid., A.P.R. Lambert to Chairman DRB, 9 November 1964.
125 Ibid., Vice Chairman DRB, Memorandum: "Deseret Test Centre, 16 August 1964." Significantly, infections disease specialists of the Department of National Health and Welfare were kept informed about Canada's involvement with the various Project 112 undertakings.
126 Ibid., A.K. Longair Memorandum, SES Trial Programme – BW, CW, 10 November 1964.
127 DRB, vol. 7513, file 1800-60/141-1, pt 2, Longair to Chief Superintendent, SES, 31 December 1964; ibid., Lambert, Report on Visit to Deseret Test Center ... 19 to 22 January 1965.
128 Ibid., R.A. McIvor to Chairman, DRB, 22 January 1965.
129 It was also argued that "because of the known innocuousness of the material, the safety of releasing BG in the San Francisco trials was never in question." DRB, vol. 47, file 1800-1-1, pt 2, Colonel L. Baker, Acting Commander DTC to Canadian Joint Staff, Washington, 15 February 1965.
130 Significantly, there was no interest in participating in Magic Sword II, scheduled for Baker Island in June 1966, which was "an evaluation of the

effectiveness of the Aedes aegypti mosquitoe as an agent vector … as a concept of entomological warfare." DRB vol. 7513, file 1800-60/141-1, pt 2, W.F. Cockburn, DAR (B &C}, Memorandum: Project Desert, Clay Pigeon and Westside II, 9 March 1965.; ibid., G.E. Hubley to Chairman, DRB, 15 March 1965.

131 In May 1965 simulated bioterrorist attacks in Washington, DC, targeted the Greyhound Station and the National Airport terminal. Moon, "The U.S. Biological Weapons program," 27.

132 Housewright joined Detrick in 1943 and continued his work as a defence scientist until the end of the war. He then obtained a PhD in microbiology from the University of Chicago while working as chief of the Microbial Physiology and Chemotherapy Branch of the Medical Bacteriology Division. In 1956 he became scientific director and then was promoted to technical director in the early 1960s.

133 DRB 83-84/167, vol. 7450, file 375-4/18/2, Longair to Lt Colonel D.N. Dalton, US Standardization Group, Washington, 8 March 1966. In many ways, this work represented a continuation of the 1965 meetings in the United Kingdom; idib., C.E. Gordon Smith, U.K. Chairman, Quadripartite Standing Working Group BW, "Report on Informal Meetings held at M.R.E. Porton," 6–10 September 1965.

134 Ibid., Longair to Lt Colonel D.N. Dalton, USA Standardization Group, Washington, 8 March 1966.

135 DND, Access Request: DRB, 83-84/167, vol. 7450, file 375-4/18/2, Canadian Position Paper, Quadripartite CBR Conference, Standing Working Group on Biological Warfare, Suffield Experimental Station, 4–6 April 1966 (hereafter Cnd Report).

136 Ibid. However, it was pointed out that no experiments with highly virulent viruses had taken place at SES "because of the lack of suitable safety facilities in the Central Laboratory."

137 Ibid. There were some problems in establishing a standard criterion for high infectivity since previous models had been based on bacterial rather than viral pathogens.

138 At this stage Detrick scientists such as William Patrick III, chief of the development program, had demonstrated that tularaemia had many advantages over anthrax as an operation BW agent, and "if disseminated by airplane, could cause casualties and sickness over thousands of miles." *Washington Post* 5 October 2010, "Obit: William C. Patrick III, Ft Detrick Biowarfare Scientist."

139 Ibid. Little was known about the Abrovirus group (dengue fever, Marburg fever, and Ebola), except that they were characterized by haemorrhagic

fever and had animal or avian reservoirs. Under the poxvirus family, vaccinia (smallpox) was already regarded as a major biological weapon.

140 Ibid. There are four types of influenza, A and B being the most common. Within the A strain, there was also human, animal, and avian subtypes.

141 DRB, 83-84/167, vol. 7450, file 375-4/18/2, Quadripartite CBR Conference, Standing Working Group on Biological Warfare, Suffield Experimental Station, 4–6 April 1966, Summary Record of Meetings.

142 The recommendations of the symposium were subsequently endorsed, without qualification, by the four armies in April–May 1966. Ibid., Colonel D.S. MacLennan, for Chief of the Defence Staff to Chairman, DRB, 16 September 1966.

143 Ibid., J.F. Currie, Memorandum: Long Range Planning, 12 October 1966.

144 Ibid., C.E. Hubley to Chairman, DRB 29 December 1967.

145 DRB, file 1800-1-1, Canadian Defence Liaison Staff Washington to Chief Superintendent SES, 3 July 1967.

146 DRB, 92-94/132, box 1, file 100/200/0, Major General N.G. Wilson-Smith to Commander, Canadian Defence Liaison Staff Washington, 16 April 1968.

147 The meetings included about twenty-five delegates from the US, UK, and Canadian intelligence agencies, and operated under strict security.

148 DEA, 28-6-6 (pt 13), HB Robinson: memorandum to the minister, 7 October 1968. The meetings included about twenty-five delegates from US, UK, and Canadian intelligence agencies.

149 During his five years as Minister of Defence, Hellyer resolutely proceeded with his unification system, which he regarded as essential for a modern and efficient military. J.F. Granatstein, *Canada 1957–1967: The Years of Uncertainty and Innovation* (Toronto: McClelland and Stewart, 1986), 241.

150 Ibid., 236.

151 DRB 1983–84/167, vol. 7329, file 100-31, Chairman DRB: "Memorandum to Defence Council, 9 April 1968." NBC was military short-hand for nuclear, biological, and chemical weapons.

152 In August 1964, the Regina chapter of the Voice of Women announced that they were going to conduct a two-hour vigil before the main gate of SES on 6 August to commemorate the nineteenth anniversary of the atomic bombing of Hiroshima and "to protest the devotion, at Suffield and other laboratories, of Canadian science and money to research for ever more efficient techniques for human slaughter." Ibid., President, Regina Voice of Women to Dr A.H. Zimmerman, Chairman, DRB, 30 July 1964.

153 On the positive side, SES's role in biological weapons had been praised by DRB Chairman H. Zimmerman in December 1964 as being "an important

and highly valued contribution to defence" that made a long-term commitment "to maintain a good program at Suffield indefinitely, well balanced between laboratory research and field trials ... [with] suitably qualified scientists." Ibid.

154 Ibid., Bobyn to Chairman, DRB, 4 January 1966.

155 Ibid., E.J. Bobyn to Chairman, DRB, 4 October 1967.

156 Ibid., pt 3, Air Vice Marshall to Chairman, DRB, 22 February 1967.

157 Ibid., pt 3, E.J. Bobyn to Chairman, DRB, 17 March 1967; ibid., B.J. Perry to Chairman, DRB, 16 April 1967.

158 Ibid., Pennie Memorandum; Note: Future Use of SES Ranges for Grazing, 2 December 1966.

159 Ibid., Robert Uffen to Minister of Defence Hellyer, 1 February 1968.

160 Ibid., Uffen to Flynn, 12 February 1968.

161 DRB, 1983-84/167, vol. 7329, file 100-31/0, pt 3, E.J. Bobyn, Director General to Chairman, DRB: Comments on PRG Submissions, 17 and 18, 19 February 1968. Special reference was made to a series of field exercises in which Canadian Forces personnel were exposed to the rigours of the CW battlefield, notably the forthcoming Exercise Vacuum, scheduled for the fall of 1968.

162 DRB, 1983-84/167, vol. 7329, file 100-31/0, pt 3, E.J. Bobyn, Director General to Chairman, DRB: Comments on PRG Submissions, 17 and 18, 19 February 1968.

163 Ibid., "Report: Reduction and Consolidation of NBC Programs, 13 December 1968."

164 Judith Miller, Stephen Engleberg, and William Broad, *Germs: Biological Weapons and America's Secret War* (New York: Simon and Schuster, 2001), 55–6, 68–9; Norman Covert, *Cutting Edge: The History of Fort Detrick* (Frederick: Public Affairs Office, Fort Detrick, 1997), 95–130.

165 ASA Archives, 8-1A folder, EX3, Housewright to Alvin Clark, 29 March 1967.

166 Ibid. Housewright claimed that Detrick scientists published "about 75–100 papers in scientific journals each year." One of his more gifted colleagues was William C. Patrick III, who joined Detrick in 1953, after he obtained a master's degree in microbiology from the University of Tennessee. *Washington Post* 2 October 2010. "William Patrick III, Expert on Germ Warfare, Dies at 84," *New York Times* 10 October 2010.

167 Donald Avery, interview with Archie Pennie, Ottawa, 8 June 2004. Pennie was previously superintendent of the DRB research establishment at Churchill, Manitoba (1954–6).

168 Ibid.

169 Ibid. Pennie had special praise for A.K. McPhail, A.P.R. Lambert, and J.R. Maltman, who were key members of the Suffield BW team. Another close contact was A.K. Longair, another Scottish-educated scientist, who joined the DRB after earning a PhD in physics at St Andrews University. He apparently died in the early 1970s in an auto accident outside of Heathrow "when he drove on the wrong side of the road."

170 Cited in Moon, "The U.S. Biological Weapons Program," 35.

171 Cited in Miller et al., *Germs*, 64.

4 Canada and BW Disarmament: National and International Developments, 1968–1975

1 NARA, Papers of Richard M Nixon, box 311, Henry Kissinger to President Nixon, "Negotiations of a Convention Banning Biological Weapons," 23 April 1971; ibid., Michael Guhin to Kissinger, 30 March 1971.

2 David Goldman, "The Generals and the Germs: The Army Leadership's Response to Nixon's Review of Chemical and Biological Warfare Policies in 1969," *Journal of Military History* 73 (April 2009): 531–69.

3 John van Courtland Moon, "The US Biological Weapons Program," in *Deadly Cultures: Biological Weapons since 1945*, ed. Mark Wheelis, Lajos Rozsa, and Malcolm Dando (Cambridge, MA: Harvard University Press 2006), 35.

4 LAC, Department of External Affairs Papers (DEA), vol. 11534, file 28-6-6, pt 3, EA, NATO to DEA, Ottawa, 8 October 1968.

5 J.L. Granatstein and Robert Bothswell, *Pirouette: Pierre Trudeau and Canadian Foreign Policy* (Toronto: University of Toronto Press, 1990), 5.

6 Ibid., 4; Erika Simpson, *NATO and the Bomb: Canadian Defenders Confront Critics* (Montreal and Kingston: McGill-Queen's University Press, 2001), 69.

7 John English, *Just Watch Me: The Life of Pierre Elliott Trudeau, 1968–2000* (Toronto: Alfred Knopf, 2009), 61. English also emphasized the influence of Trudeau's closest intellectual confidants, who were hostile to the US involvement in the Vietnam war and wary of the Canadian military.

8 Granatstein and Bothwell, *Pirouette*, 7,14.

9 English, *Just Watch Me*, 72.

10 Ibid., 62. Ivan Head was a constitutional legal expert who emerged as Trudeau's principal foreign policy adviser, in part because of the personal affinity between the two men and in part because Trudeau disliked Undersecretary of State Marcel Cadieux, whom he regarded as a rigid right-wing ideologue.

11 Granatstein and Bothwell, *Pirouette*, 28.

12 English, *Just Watch Me*, 55–72.

13 George Ignatieff, *The Making of a Peacemonger* (Toronto: University of Toronto Press, 1985), 245. M.N. (Mac) Bow, Head of the DEA Disarmament Division, was an active participant in the BW disarmament discussions at the UN, along with his colleagues G.J. Smith, D.M. Corbett, F. Pillarella, and A.D. Morgan.
14 See Boyce Richardson," Scientists Advance Genocide," *Montreal Star* 29 February 1968.
15 See the controversial "Canada and Germ Warfare," *Canadian Dimension* 4 November 1968.
16 National Library Archives (NLA), Pierre Elliott Trudeau Papers (Mg 26), vol. 24, file 155 (1968–February 1969), D.B. Privy Council Office, memorandum for Mr Morris, 23 September 1968.
17 Ibid., Morris to Mrs Elizabeth Tennant, Calgary, 4 December 1969. Other organizations submitting letters of protest included the Women's International League for Peace and Freedom, the Edmonton Chapter to End the War in Vietnam, and the Humanist Fellowship of Montreal.
18 Ibid., Doucet to Morris, 9 January.
19 Ibid., Brief for the Prime Minister, 28 January 1969. Reference was also made to the report of the special study group that was formed after the 13th (1968) meeting of the Pugwash Conference on Science and World Affairs.
20 LAC, Department of External Affairs (DEA) Papers, vol. 11534, file 28-6-6, pt 3, DEA Washington to DEA Ottawa, 24 March 1965.
21 Ibid., Legal Division: Memorandum for File, 24 March 1965.
22 Ibid., Robertson to Ignatieff, 29 March 1968.
23 Ibid., F. Pillarella to J.A. Beesley, DEA, 26 April 1968. A comparison of the arguments presented by Robertson and Pillarella were prepared as a departmental briefing paper.
24 United Kingdom National Archives (UKNA), DEFE-13/557, Memorandum for prime minister, February 1967. Ibid., Memorandum for Minister of Defence, 22 August 1967.
25 Ibid., Research Programme on Chemical and Biological Warfare, prepared by A.H. Cottrell, 8 February 1967; ibid., Memo for the Minister: Chemical Warfare and Biological Warfare – The Future of MRE and CDEE, Porton, 22 August 1967; Ibid., Chiefs of Staff Committee meeting, 13 December 1967 (124/67).
26 G.B. Carter, *Chemical and Biological Defence at Porton Down 1916–2000* (London: The Stationary Office, 2000), 127. A related accident occurred at the Porton-based Chemical Defence Experimental Establishment in May 1953 when Leading Aircraftsman Ronald Maddison of the RAF died of sarin poisoning as part of a series of control experiments with service volunteers. Ibid., 119–20.

27 *The Times* 6 June 1968.

28 UKNA, DEFE-11/672, Report Military Intelligence, 5 June 1968.

29 MRE Director General Smith recommended against releasing any data about the ocean BW trials that took place between 1949 and 1955, or the more recent Portland trials, which used a ship "spraying living organisms for both downwind survival and detection studies." DEFE, 11/672, C.E. Gordon Smith to E. Broadbent, Private Secretary for Secretary of State for Defence, 22 July 1968.

30 Carter, *Porton Down*, 128.

31 Marie Isabelle Chevrier, "The Politics of Biological Disarmament," in Wheelis, Rozsa, and Dando, eds., *Deadly Cultures*, 308.

32 There are vast numbers of books about opposition to US policies in Vietnam. But for linkages between the Pentagon and the American scientific community, see Michael Sherry, *In the Shadow of War: The United States since the 1930s* (New Haven, CT: Yale University Press, 1995); James Gibson, *The Perfect War: Technowar in Vietnam* (Boston: Atlantic Monthly Press, 1986); and Kenneth Heinenman, *Campus Wars: The Peace Movement at American State Universities in the Vietnam Era* (New York: New York University Press, 1993).

33 In February 1967, 5,000 American scientists, including seventeen Nobel laureates, signed a petition urging President Lyndon Johnson to end the use of chemical weapons in Vietnam. NARA, Records of the US Department of State (RG 59), vol. 2879, file Pol-27-10, Matthew Meselson, John Edsall, Paul Doty, and Irwin Gunsalus to President Johnson, 14 February 1967.

34 Goldman, "The Generals and the Germs," 542. Another CBW opponent was Congressman R.D. McCarthy, who described his experiences in *The Ultimate Folly: Pestilence, Asphyxiation, and Defoliation* (New York: Knopf, 1969).

35 See Seymour Hersh's *Chemical and Biological Warfare: America's Hidden Arsenal* (Indianapolis, IN: Bobbs-Merrill, 1968); Robin Clarke, *The Silent Weapons* (New York: D. McKay, 1968); and Nigel Calder, *Unless Peace Comes: A Scientific Forecast of New Weapons* (New York: Viking Press, 1968).

36 After his resignation from the U.S. Army in 1970, Housewright enjoyed considerable success in the private biotechnology sector. He died in 2003, at the age of 89. *The Scientist* 24 January 2003.

37 Archives of the American Society for Microbiology (ASM Archives), 8-IA-Folder Ex3. Over 100 Detrick scientists were members of the ASM. Ibid.

38 Ibid., Riley Housewright to Alvin Clark, 29 March 1967. Housewright claimed that Detrick scientists published "about 75–100 papers in scientific journals each year."

39 Ibid., James Moulder to S.E. Luria, 1 December 1967.

40 Ibid., Robert Austrian to Moulder, 11 December 1967.
41 "Micro-Revolt of the Microbiologists over Detrick Tie," *Science* 24 May 1968.
42 ASM Archives, 8-IA-Folder Ex3, Statement by Merrill Snyder, 11 May 1968.
43 Ibid., John W, King to D.W. Watson, vice president ASM, Department of Microbiology, University of Minnesota, 14 May 1968; Ibid., President S.E. Luria to Members of CPC, 14 May 1968.
44 Ibid., Statement of Concern by the Microbiologists Committee on Chemical and Biological Warfare, 1969 (n.d.)
45 Robert Murray and his father Everitt, head of the Department of Microbiology at McGill University, were both active in the ASM – each being presidential candidates. They were also instrumental in founding the CSM in June 1951, which originally included 173 microbiologists, with Robert being its first president. During the late 1960s Bob Murray became editor of the ASM journal *Bacteriological Reviews* (1969–79) and its president in 1972–3. Donald Avery and Mark Eaton, *The Meaning of Life: The Scientific and Social Experiences of Everitt and Robert Murray, 1930–1964* (Toronto: The Champlain Society, 2008).
46 The NPT also promises to enhance the peaceful use of nuclear energy and to prevent proliferation of nuclear weapons. Thomas Reed and Danny Stillman, *The Nuclear Express: A Political History of the Bomb and Its Proliferation* (New York: Zenith Press, 2009).
47 LAC, Department of External Affairs (DEA), vol. 11534, file 28-6-6, pt 3, Marcel Cadieu, Under-Secretary to DEA, London, 24 June 1968. Throughout the late sixties, the Soviet Union carried on a vigorous campaign for all countries to immediately ratify the Geneva Protocol, a clear attempt to embarrass the United States.
48 Ibid., Marcel Cadieu, Cadieux to Dr L.J. L'Heureux, Chairman, DRB, 28 June 1968; ibid., DEA London to DEA, Ottawa, 16 July 1968.
49 Ibid., DEA, Disarmament Division to Legal Division, 19 September 1968: "Working Paper on Some Political Aspects of Agreements on Chemical and Biological Warfare."
50 Ibid.
51 Ibid., Undersecretary of State Marcel Cadieux to Dr L.J. L'Heureux, Chairman of the Defence Research Board.
52 DEA, vol. 11534, file 28-6-6, pt 3, A.K. Longair to D.M. Corbett, Disarmament Division, DEA, 4 September 1968.
53 Ibid., Longair to Corbett, 4 September 1968; ibid., Longair to Corbett, 2 October 1968.
54 Ibid., Longair to Cornett, 2 October 1968.

55 Ibid., Marcel Cadieux to Chairman, DRB, 9 October 1968.
56 Ibid., Longair to Undersecretary of State for External Affairs (Attention M.N. Bull, Disarmament Division), 7 November 1968.
57 Ibid. Longair also cited the study of the West Nile virus by scientists in Israel as another example of dual-use research, since these were "exactly the type of laboratory studies ... if one wished to fill a munition."
58 DEA, vol. 11534, file 28-6-6, pt 3, J.S. Nutt, Defence Liaison, DEA memorandum, 2 August 1968.
59 Ibid. External Affairs officials were concerned that the popular CBC TV program *The Way It Is* gave a negative interpretation of Canada's involvement in the CBW field.
60 DEA, vol. 11534, file 28-6-6, pt 3, Canadian Embassy, Washington to DEA Ottawa, 16 August 1968; ibid., Memorandum for the Minister DEA, 6 August 1968.
61 Ibid., Leo Cadieux: memorandum to the Cabinet, 23 August 1968. On 8 October 1969 Lorne Nystrom of the New Democractic Party asked whether the Minister of Defence could assure Canadians that the results of Exercise Vacuum were "not intended to use in the war in Viet Nam." House of Commons, *Debates*, 924.
62 Ibid., Morgan memorandum to M.N. Bow, 26 November 1968.
63 Ibid., vol. 11534, file 28-6-6, pt 5, Brigadier-General Henri Tellier, Director-General Military Plans, for the Chief of the General Staff, to J.S. Nutt, Office of Politico-Military Affairs, DEA, 31 December 1968.
64 Ibid., A.D. Morgan to A.D. Bow, Disarmament Division, 8 January 1969.
65 Ibid., Disarmament Division, DEA to Nutt, Office of Politico-Military Affairs, 24 July 1969.
66 Ibid., Disarmament Division memorandum to H.B. Robinson, 25 October 1968.
67 Ibid., Orders of the Day No. 528, 30 October 1968.
68 One useful source for arms control specialists was Dr Carl Goran Heden's article, "Defences against Biological Warfare," a working paper prepared for the Pugway Study Group on Biological Warfare and republished in the *Annual Review of Microbiology* 21 (1967): 639–76.
69 DEA, vol. 11534, file 28-6-6, A.D. Morgan to Bow, 24 October 1968.
70 Ibid., Legal Division DEA to Disarmament Division DEA, 8 October 1968.
71 Ibid., A.D. Morgan DEA, Disarmament: Report of conversation with Mr Marshall of U.S. North American Defence and NATO Division, 22 October 1968.
72 Ibid., A.D. Morgan to Bow, 29 October 1968, memorandum: CBW-Non-Use or Non-Possession-Further Implications.

73 Ibid., Legal Division, DEA to Disarmament Division, DEA, 23 October 1968. It was noted that NATO's strategy of flexible response anticipated "the possibility of first use of tactical nuclear weapons against conventional [Soviet] ground attack."
74 Ibid. DEA officials were sceptical about whether the US would allow international inspectors to visit its pharmaceutical and biotechnology companies as part of any BW verification system. This would eventually prove to be a correct prediction, as was evident in 2001 when the industry vigorously opposed the proposed Verification Protocol of the BWC.
75 Under the Brussels Treaty of 1948 and its amendments, the Federal Republic of Germany agreed not to acquire either chemical or biological weapons, an obligation that was enforced by NATO. Ibid., A.D. Morgan: memorandum for file, 28 October 1968.
76 DEA, vol. 11534, file 28-6-6, pt. 12, Statement by Dr Ehrenfried Petras, included in letter by the Deputy Foreign Minister DDR to George Ignatieff, Geneva, 13 December 1968; ibid., A.D. Morgan, memorandum for file, 28 October 1968.
77 DEA, vol. 11534, file 28-6-6, pt 12, Ottawa to Ignatieff, 8 January 1969; ibid., Undersecretary of State for External Affairs to Canadian Delegation to the Conference of the Eighteen Nation Committee on Disarmament, Geneva, 17 July 1969. In July 1969 Bonn was given information about the location of American CW depots in the Federal Republic of Germany and their safety/environmental status "on a strictest need-to-know basis." NARA, State Department Records, vol. 2880, file 27-1, US Embassy, Bonn to State, 31 July 1969.
78 Ibid., DEA Ottawa to George Ignatieff (Geneva), 7 August 1968; ibid., David Miller, Legal Division DEA, to F. Pillarella, Canadian Embassy, Bonn.
79 Ibid., E.L.M. (Tommy) Burns to M.N. (Mac) Bow, Head, Disarmament Division, 21 November. MacPhail had a PhD in biochemistry and physiology from McGill and taught at a number of Canadian universities before joining the DRB in 1948. At Suffield, he was head of the Physiology section until 1962, when he was appointed Director of Biosciences Research at the Defence Chemical Biological Radiation Laboratories at Shirley's Bay.
80 DEA, vol. 11532, file 28-6-6-, pt 5, Bow to Burns, 27 November 1968.
81 DEA, vol. 11532, file 28-6-6-, pt 13, Disarmament Division to Undersecretary of State, 3 January 1969. Dr Ivan Bennett, Office of the Director, New York University Medical Center, had an impressive research and administrative record and had served in an advisory capacity with other U.S. government agencies. The lead Soviet scientist was Academician O.A Reutov, Academy of Sciences of the USSR, Division of General and Technical Chemistry.

82 Ibid., Minutes of the First Meeting of Group of Consultant Experts on Chemical and Bacteriological (Biological) Warfare, 20 January 1969.
83 Ibid. Both the Pugway Movement and the Swedish International Peace Research Institute (SIPRI) made submissions for the SG's Panel.
84 Ibid., Minutes of the Second Meeting, 21 January 1969.
85 Ibid., Minutes of the Third Meeting, 22 January 1969. The projected five chapters of the report were each placed under the direction of a team leader: Chapter One, Dr Bennett (included McPhail); Chapter Two, Dr Lars-Erik Tammelin (Sweden); Chapter Three, Dr Tibor Bakas (Hungary); Chapter Four, Mr Moulin (France); and Chapter Five (Academician Reutov, assisted by Bennett and Zuckerman). Ibid., Minutes of the Sixth Meeting, 24 January 1969.
86 Ibid., Minutes of the Fifth Meeting, 23 January 1969.
87 Ibid., DEA, vol. 11535, file 28-6-6, pt 13, G.J. Smith (Disarmament Division) DEA, to Robinson, European Division, DEA, Ottawa, 15 April 1969.
88 Ibid., DEA, New York to DEA Ottawa, 10 June 1969.
89 Ibid., United Nations, Press Release, 16 June 1969.
90 DEA, vol. 11535, file 28-6-6-, pt 16, Report D.W. Campbell (DEA, Disarmament), 12 January 1970.
91 Ibid., Ignatieff to DEA, Ottawa, 23 July 1969. This report included U Thant's statement.
92 Ibid., DEA, vol. 11535, file 28-6-6, pt 13, G.J. Smith (Disarmament Division) DEA, to Robinson, European Division, DEA, Ottawa, 15 April 1969. NARA, State, box 2880, file 27-10, US Mission Geneva to State (telegram), 16 July 1969.
93 Ibid., Canadian Ambassador, The Hague, to DEA, Ottawa, 28 February 1969. One of the authors of this SIPRI study was Julian Percy Robinson, who has subsequently enjoyed an illustrious career as a CBW arms control specialist at the University of Sussex (UK).
94 The World Health Organization, *Health Aspects of Chemical and Biological Weapons: Report of a WHO Group of Consultants* (Geneva: WHO, 1969).
95 DEA, vol. 11535, file 28-6-6, #13, United Nations Division DEA to Disarmament Division, 31 July 1969.
96 Ibid., Canadian Ambassador Moscow to DEA, Ottawa 2 June 1969. The Embassy also reported on three recent spy films involving biological warfare issues: *The President's Mistake, The Dead Season,* and *The Man from Canada.* DEA, vol. 11535, file 28-6-6, pt 13, Embassy Moscow to Undersecretary State, DEA, 19 February 1969.
97 Members of the New Democratic Party, notably David Lewis (York South), were the leading critics of Canada's BW connections with the US. House of Commons, *Debates,* 18 June 1968, 10309.

98 LAC, Trudeau Papers, vol. 54, file 155 (1965–70), Trudeau to the Honourable Leo Cadieux, Minister of National Defnce, 25 June 1969.

99 Ibid., Mitchell Sharp, memorandum for the Prime Minister, 4 November 1969.

100 Ibid., D.J. Leach, Supervisor of Cabinet Documents, Record of Committee Decisions, 9 December 1969.

101 Canada's unilateral declaration on renouncing biological weapons was tabled with the CCD in February 1970. The following year, the option of using lethal chemical weapons for retaliatory purposes, which dated back to the 1925 Geneva Protocol, was removed by Cabinet directive. The government did, however, reserve the right of Canadian Forces to use non-lethal CW devices such as tear gas. DEA, vol. 22535, file 28-6-6, pt. 17, "Chemical and Bacteriological (Biological) Weapons: Report by the Conference of the Committee on Disarmament, December 16, 1971."

102 Jonathan Tucker, *War of Nerves: Chemical Warfare from World War I to Al-Qaeda* (New York: Pantheon Books, 2006), 230–80. Critics of the US chemical warfare program did not believe Pentagon claims that all of the 1,200 field trials held since 1951 were completely safe.

103 NARA, Richard M. Nixon President Papers, National Security Council Subject Files, vol. 3, box 310, Laird memorandum for NSC, 30 April 1969.

104 Nixon Papers, box 310, Morton Halperin, memorandum for Henry Kissinger: "US Policy, Programs and Issues on CBW," 28 August 1969.

105 Ibid. Particular emphasis was placed on the fact that lethal biological and chemical weapons were not for first use, but rather for "retaliation only." In contrast, deploying CBW-incapacitating weapons was regarded as part of the US strategic plan in order "to exploit military advantage with Presidential authorization." See also news release: Office of the Assistant Secretary of Defence, 9 August 1969.

106 Ibid., Lee Du Bridge, Science Adviser, memorandum for Dr Henry Kissinger, 22 October 1969. Du Bridge was a highly respected nuclear physicist who had worked on the Manhattan Project. He strongly opposed the American BW program.

107 Ibid. The State Department added its weight to the debate by advocating that the US maintain only a defensive biological research and testing program, largely "to safeguard against technological surprise." Ibid., Ronald Spiers to The Secretary Department of State 17 November 1969;

108 At the crucial 18 November meeting of the National Security Council, General Wheeler was the only member of the Joint Chiefs of Staff who argued in favour of the US retaining a full BW retaliatory capability. Goldman, "The Generals and the Germs," 561.

109 Jonathan Tucker, "A Farewell to Germs: The U.S. Renunciation of Biological and Toxin Weapons, 1969–70," *International Security* 27 (Summer 2002): 107–48.

110 Special arrangements were made by Kissinger to consult Japanese officials because the country still had not ratified the Geneva Protocol and because of the legacy of its Second World War BW program. In February 1970, Japan became a member of the Protocol. Nixon Papers, box 310, "NSC Briefings for NSC meeting on CW/BW, 19 November 1969." The White House made special arrangements to consult Japanese officials because that country had not yet ratified the Geneva Protocol and because of the legacy of its Second World War BW program. In February 1970 Japan became a member of the Protocol.

111 Ibid., Remarks of the President on Announcing the Chemical and Biological Defence Policies and Programs, 25 November 1969.

112 The possibility of a connection between biological weapons and the 1968–9 Asian flu epidemic was raised at the press conference. Ibid., background briefing, 25 November 1969.

113 Ibid., Du Bridge to Kissinger, 22 December 1969; ibid., memorandum for Kissinger from Laird, 12 February 1970.

114 Ibid., Memorandum Secretary of Defence, 9 December 1969.

115 Ibid., box 311, Fulbright to Nixon, 19 February 1970; ibid., Kissinger memorandum, Secretary of State, 5 August 1970.

116 ASM Archives, 8-IA-folder 1, Donald Shay, Secretary ASM, memorandum, 20 June 1970.

117 Nixon Papers, box 310, US Embassy Ottawa to Department of State, 10 November 1969; State, box 2880, file 27-10.

118 Ibid., telegram, US Embassy to State Department, 25 November 1969. Newspapers across the world carried stories about Nixon's declaration. In most cases these were positive, with the notable exception of Moscow, where the media claimed that this so-called transformation was merely another cover for accelerated US offensive work. Ibid., Moscow Embassy to State, 26 November, 11 December, 22 December 1969.

119 House of Commons, *Debates*, 25 November 1969, 1199–1200.

120 There was only one other question involving Nixon's announcement during the 1969 parliamentary session. This occurred on 1 December when J.M. Forrestall (Dartmouth-Halifax East) asked whether Canada had obtained US assurance that any disposal of biological weapons would not occur in coastal regions shared by the two countries. Ibid., 1423.

121 DEA, vol. 11535, file 28-6-6, pt 12, Acting Secretary of State for External Affairs: Memorandum for the Prime Minister, 4 December 1969.

122 Trudeau Papers, vol. 24, file 155, Privy Council Office, Record of Committee Decision, Meeting of December 9th, 1969.

123 DEA, vol. 11535, file 28-6-6, pt 12, "Proposal for a Unilateral Declaration on CBW," 22 January 1970.

124 Ibid.

125 Ibid., "Check List of Steps to Achieve BW Disarmament-Unilateral or Otherwise," 17 February 1970.

126 Trudeau Papers, vol. 24, file 155, Mitchell Sharp: Memorandum for the Prime Minister, 16 February 1970. Significantly, Trudeau had previously informed George Ignatieff that he would be interested in addressing the CCD about Canada's position on biological weapons. DEA, vol. 11535, file 28-6-6, pt 10, Peter Walker to Shenstone, 24 July 1970.

127 Ibid., Ivan Head, memorandum for Trudeau: Chemical and Biological Warfare, 17 February 1970.

128 Ibid., Disarmament Division, For File: Main Points Raised at Meeting This Morning with Lord Chalfont, CBW (Ottawa), 22 May 1970.

129 Ibid., DEA, Geneva to DEA, Ottawa, 2 March 1970.

130 Ibid.

131 On 22 May 1970, Pierre Tremblay, Associate Undersecretary DEA asked Dr L.J. L'Heureux, Chairman, DRB, for permission to use the services of Dr M.K. McPhail in the preparation of a working paper on scientific methods of CBW verification, "which would constitute a distinctive Canadian contribution." DEA, vol. 11535, file 28-6-6, pt. 12.

132 Ibid., Ambassador George Ignatieff, Canadian Disarmament Delegation, Geneva to Undersecretary DEA, 26 February 1970.

133 ACDA officials also indicated that they had spent over $1,208,000 on their CBW verification research "but still considered work to be in preliminary stages." Ibid., Ignatieff, Geneva to DEA, Ottawa, 24 March 1970.

134 The list included External Affairs, the DRB, Justice, Agriculture, National Research Council, Health and Welfare, and National Defence, Ibid., M.N. Bow, Memorandum: Interdepartmental Panel, CBW, 3 June 1970.

135 See Susan Wright, ed., *Biological Warfare and Disarmament: New Problems / New Perspectives* (New York: Rowman and Littlefield, 2002); Olivert Thanert, ed., *Preventing the Proliferation of Weapons of Mass Destruction: What Role for Arms Control?* (Bonn: Friedrich-Ebert Stiftung, 1999); Nicholas Sims, *The Evolution of Biological Disarament* (Oxford: Oxford University Press, 2001).

136 Chevrier, "The Politics of Biological Disarmament," 322.

137 NARA, Papers of Richard M Nixon, box 311, Henry Kissinger to President Nixon, "Negotiations of a Convention Banning Biological Weapons, 23 April 1971; ibid., Michael Guhin to Kissinger, 30 March 1971.

138 External Affairs officials were concerned about the unwillingness of France and the People's Republic of China to support the BW Convention. For Paris, the major objections were the lack of a verification system. Beijing's refusal was based on the fact that Taiwan had already signed the agreement. DEA, vol. 11535, file 28-6-6, pt 17, DEA, Washington to DEA, Ottawa, 21 April 1972.
139 In March 1970 the House Committee on External Affairs and Defence tabled a general report on chemical and biological warfare, including a submission by the Department of National Defence outlining, in some detail, the CBW work of the Defence Research Board.
140 House of Commons, *Debates,* 5 April 1971, 4898-99.
141 DEA, vol. 11535, file 28-6-6, pt 12, D.W. Campbell to M.N. Bow, 5 June 1970.
142 Ibid., Ottawa, telegram to Missions, 8 June 1970.
143 Ibid., DEA Moscow to DEA, Ottawa, 9 June 1970.
144 Ibid., M.N. Bow, to PMO, 9 July 1970.
145 English, *Just Watch Me,* 73–123.
146 Leigh Sarty, "A Handshake across the Pole: Canada-Soviet Relations in the Era of Détente," in *Canada and the Soviet Experiment: Essays on Canadian Encounters with Russia and the Soviet Union, 1990-1991,* ed. David Davies (Waterloo, ON: Centre on Foreign Policy and Federalism, 1993), 133.
147 Cited in English, *Just Watch Me,* 166.
148 Goldman, "The Generals and the Germs," 567.
149 Tom Mangold and Jeff Goldberg, *Plague War: A True Story of Biological Warfare* (New York: St Martin's Press, 1999), 61. Significantly, Nixon did not mention his 25 November 1969 Declaration in his memoirs.
150 Judith Miller, Stephen Engleberg, and William Broad, *Germs: Biological Weapons and America's Secret War* (New York: Simon and Schuster, 2001), 62–84.
151 DEA, vol. 11536, pt 16, Report DEA Disarmament: Chemical and Bacteriological (Biological) Weapons: Report of the Conference of the Committee on Disarmament (Resolution 2827 (XXVI), of December 16, 1971. United Nations, First Committee, Twenty Seventh Session.
152 Convention on the Prohibition of the Development, Production and Stockpiling of Bacteriological (Biological) and Toxin Weapons and on Their Destruction, cited in Wheelis, Rozsa, and Dando, eds., *Deadly Cultures,* 375–80.
153 DEA, vol. 11536, file 28-6-6, pt 17, DEA, Geneva to DEA Ottawa, 24 July 1972.
154 By 2010, 188 of the 195 countries recognized by the United Nations had ratified the CWC.

5 Triple Threats: Biowarfare, Terrorism, and Pandemics, 1970–1985

1 LAC, DRB, 1989/90/205, vol. 3, file 170-3, Alex Longair to Director General Defence Research Establishment, Ottawa, 23 May 1972.
2 ASM Archives, 8-IA-folder 1 (file Fort Detrick), Robert Hungate to Ivan Bennett, 23 September 1970.
3 During the past two decades the scholarly literature on international terrorism has expanded greatly. Some of the major studies for the period 1970–2000 include: Bruce Hoffman, *Inside Terrorism* (New York: Columbia University Press, 1998); Warren Kinsella, *Unholy Alliances: Terrorists, Extremists, Front Companies and the Libyan Connection in Canada* (Toronto: Lester, 1992); Walter Laqueur, *The Age of Terrorism* (Toronto: Little, Brown, 1987); Patricia Marchak, *Reigns of Terror* (Montreal and Kingston: McGill-Queen's University Press, 2003); Kerry Noble, *Tabernackle of Hate: Why They Bombed Oklahoma City* (1998); Walter Reich et al., *Origins of Terrorism: Psychology, Ideologies, Theologies, States of Mind* (Cambridge: Cambridge University Press, 1990); Brad Roberts, ed., *Hype or Reality? The "New Terrorism" and Mass Casualty Attacks* (Alexandria VA: Free Hand Press, 2000); Jessica Stern, *Ultimate Terrorists* (Cambridge: Harvard University Press, 1999).
4 Gustav Morf, *Terrorism in Quebec: Case Studies of the FLQ* (Montreal: Clarke, Irwin, 1970); Denis Smith, *Bleeding Hearts, Bleeding Country: Canada and the Quebec Crisis* (Edmonton: Hurtig, 1971); William Tetley, *The October Crisis: An Insiders View* (Montreal andKingston: McGill-Queen's University Press, 2006); John Engish, *Just Watch Me: The Life of Pierre Elliott Trudeau, 1968–2000* (Toronto: Alfred Knopf, 2009), 73–97.
5 There are also a number of excellent studies on biological and chemical terrorism prior to 9/11. See Sidney Drell et al., *The New Terror: Facing the Threat of Biological and Chemical Weapons* (Hoover Institution Press, Stanford University, 1999); Richard Falkenrath, et al., *America's Achilles Heels: Nuclear, Biological and Chemical Terrorism and Covert Attack* (Cambridge, MA: MIT Press, 1998); Joshua Lederberg, ed., *Biological Weapons: Limiting the Threat* (Cambridge, MA: MIT Press, 1999); Jonathan Tucker, *Toxic Terror: Assessing Terrorist Use of Chemical and Biological Weapons*(Cambridge, MA: MIT Press, 2000).
6 ASM Archives, 8-IA-folder 1 (file Fort Detrick), Ivan Bennett to Dr D.E. Shea, Department of Microbiology, University of Maryland (Secretary ASM), 21 August 1970.
7 Ibid., Hungate to D.E. Shay, 16 September 1970.
8 Ibid., Hungate to Bennett, 23 September 1970.
9 Ibid., J. Roger Porter to Bennett, 28 August 1970.
10 Ibid., Housewright to Donald Shay, 4 September 1970.

11 Ibid., J. Roger Porter to Bennett, 28 August 1970.

12 Ibid., Hungate to Senator Warren Magnusson, 12 October 1970. Conversation Concerning Fort Detrick, 2 October, 14 December 1970.

13 Ibid., Housewright to J. Roger Porter, 7 January 1971. After 1 July 1970 Housewright was Vice-President and Scientific Director of Microbiological Associates, a biotechnology company based in Bethesda, Maryland.

14 NARA, National Institute of Health Papers, box 158, file Interim. Res-10-2, NCI Frederick, Cancer Research Center at Fort Detrick, 1970, Robert Lourie to President Nixon, 16 June 1971. At the local level, it was noted that while Detrick had been "much criticized by outsiders, scarcely any Frederick area residents have openly protested the facility's presence in the community." *Frederick Post* 4 November 1970.

15 NIH, box 158, file NCI Frederick, Dr Roger Egeberg, MD, Assistant Secretary for Health and Scientific Matters, HEW, to Secretary Richardson, 23 March 1970.

16 Detrick scientists were credited with over 1,500 publications and praised for their collaborative research with NIH and university scientists. In addition, they were recognized as world leaders in the design of equipment and containers for hazardous biological materials. NIH, box 158, file NCI Frederick, 1970, Redirection of the Fort Detrick Facility to Cancer Research.

17 Ibid., Elliott Richardson to the Honourable George Schultz, Office of Management and Budget, 25 July 1970.

18 Ibid., Leon Jacobs, Assitant Director for Collaborative Research NIH, Memorandum: Fort Detrick, 19 January 1971.

19 Ibid., James Cardwell, Office of the Secretary, HEW, to Dr Roger Egeberg, Assistant Secretary for Health and Scientific Affairs, 9 July 1970.

20 *Washington Post* 19 October 1971; *Frederick Post* 20 October 1971.

21 "Tour of NCI Frederick Research Center," *Frederick Post,* 3 August 1972.

22 Ibid., 15 August 1972.

23 NARA, Richard M. Nixon Presidential Papers, vol. 3, box 311, National Security Council Subject Files, White House BW/CW, The Secretary of Defense: Memorandum for the President, 6 July 1970.

24 Ibid., box 311, Report: Disposal of Biological Agents … Pine Bluff, Description of Facilities, August 1970; Ibid., Dr Jesse Steinfeld, Surgeon General (PHS) to Dr Donald MacArthur, Deputy Director, Research and Technology, Pentagon, 22 June 1970; Ibid., Richard Vanderhoof, Regional Director, Federal Water Pollution Control, to Commanding Officer, Pine Bluff Arsenal, 15 May 1970.

25 Ibid., Environmental Impact Statement for Disposal of Biological Agents and Weapons, 17 September 1970. There was a somewhat different process

for the destruction of dry agents, where after being removed from its containers, the material was transported to cubicles where formaldehyde was added to decontaminate the agent. After sixteen hours, the material was removed from the containers and pumped to a storage tank, where it was then sterilized. Finally, the containers were decontaminated and burned.

26 Ibid., Memorandum for Dr Kissinger: Completion of Plan and proposed Announcement re Destruction of Biological and Toxin Weapons, 4 December 1970.

27 There were also attempts to have the US renounce the use of hallucinogens such as BZ and LSD in warfare. Nixon Papers, box 311, memorandum for Kissinger, 7 May 1970. See also Malcolm Dando and Martin Furmanski, "Midspectrum Incapacitant Programs, " in *Deadly Cultures: Biological Weapons since 1945*, ed. Mark Wheelis, Lajos Rozsa, and Malcolm Dando (Cambridge, MA: Harvard University Press, 2006), 236–51.

28 United States Senate, Hearings of the Select Committee to Study Governmental Operations with Respect to Intelligence Operations, volume 1: Unauthorized Storage of Toxic Agents. Testimony of William Colby (Director CIA), 6-46, exhibit 2 (Washington: USGPO, 1976), 6–16.

29 DRB, DRBS, 2000-20-2, Director General B.J. Perry to Chairman, DRB (telegram), 4 July 1969. Dr Harry Sheffer, Deputy Chairman, Scientific, DRB, to Perry, 11 July 1969.

30 Ibid., Perry to Chairman, DRB, 23 March 1970; ibid., Pennie to Perry, 6 April. Significantly, it was not until 1988–9 that the remaining 300 kilograms of "nerve agent … which had been kept in protective storage, was chemically neutralized by DRES scientists." In contrast, in 1975–6, 700 tons of liquid vesicant (mustard) were destroyed using a process of hydrolysis followed by incineration. Department of National Defence, *Review of the Chemical and Biological Defence Program: May 1991–March 1992* (Ottawa: DND Public Affairs, 1992), 15.

31 DRB, 2000-20-2, Pennie to Perry, 6 March 1970.

32 DRB, 1989/90, vol. 47, file 1800-1, pt. 2, Report of the RCMP (no author) to Defence Research Board, 1 May 1970.

33 Ibid., Memorandum for the Minister, 20 May: Proposed Demonstration against CBW at Suffield, 29–31 May.

34 LAC, Trudeau Papers, vol. 54, file 155 (1970-75), Allmand to Trudeau, 25 May 1970.

35 Ibid., Cadieux to Allmand, 3 June 1970.

36 DRB, L.J. L'Heureux, Chairman, DRB to Minister of Defence, 11 June 1971.

37 DRB, file 2000-2-2, A.M. Pennie, Director of Operations, DRB, memorandum for file, 26 May 1969.

38 DRB, 1989-90/203, vol. 47, 1800-1, pt 3, Defence Research Board Advisory Committee on Defence against Biological Agents, Minutes of 70/1 meeting, 19 January 1970. The five members of the committee were Dr J.C.N. Westwood of the University of Ottawa (Chair); Dr J. Frank, Department of Agriculture; Dr R.A. MacLeod, McGill University; Dr D.M. McLean, University of British Columbia; and Dr C.A. Mitchell, University of Ottawa. For Mitchell, this was his twenty-eighth year of direct involvement with Canada's BW programs.
39 The use of military volunteers for CW testing was more extensive. After 1945, for example, live agents were used in 14 trials, involving 712 volunteers; in contrast, simulants were used in 712 trials, involving 3,906 volunteers. Clement Laforce, "The Use of Human Subjects in Chemical Warfare agent Experiments: An Ethical Perspective," unpublished paper, submitted to the Canadian Forces College, May 2006 (in author's possession).
40 DRB, 1989-90/203, vol. 47, file 1800-1, pt 3, Defence Research Board Advisory Committee on Defence against Biological Agents, Minutes of 70/1 meeting, 19 January 1970.
41 Ibid.
42 DRB, 1989-90/203, vol. 47, file 1800-1, pt 3, Joint Meeting, 6 January 1971.
43 DRB, 93-94/132, vol. 20, file DREO 4-21/0, W.A. Hoddinott, for Chairman, DRB, 2 May 1971.
44 DRB, 93-94/132, vol. 20, file DREO 965-1, Chief Superintendent and Directors, DRB, Warning Terms Other Than Security Classifications, 2 March 1964.
45 Ibid., Members of the Defence Research Council, 21 January 1971; ibid., directive 30 April 1973.
46 Ibid., file 965-2, Heggie to Chairman, DRB, 7 June 1973; ibid., file 965-2-21/0, vol. 1, J.J. Norman, DREO, to Chairman, DREO Review Panel, 23 August 1973.
47 For example, in July 1974 DRB officials questioned why scientists from East Germany, Poland, and Czechoslovakia requested full copies of their reports that had appeared in distilled form in the June edition of *Chemical Abstracts*. Ibid., file 973-1, pt 4, DREO to Directorate of Security, National Defence Headquarters, 16 July 1974.
48 NARA, Records of the Department of State (RG 59), vol. 1841, DEF 12, 3/1/70, Embassy Tokyo to State, 25 May 1970; ibid., Embassy Tokyo to State, 25 July 1970.
49 Ibid., vol. 1843, file DEF 8/1/70, State Department to Cnd Embassy, Ottawa, 29 January 1970.
50 House of Commons, *Debates* 23 January 1970 (2734).

51 Ibid., Statement by Mitchell Sharp: 2968. At an 16 April 1971 meeting Secretary of State Hillenbrand told the Canadian ambassador that the shipment would be travelling "solely through that part of the Strait which is U.S. territorial waters." State Department, vol. 1841, DEF 12-3/1/70, State to Embassy, Ottawa, 16 April 1970.

52 NARA, State Department, vol. 1841, file DEF 12, 3/1/70, H.G. Torbert Jr, Acting Assistant Secretary for Congressional Relations, to The Honourable John Dingell, Chairman, Subcommittee on Fisheries and Wildlife Conservation, House of Representatives, 14 April 1970.

53 NARA, Nixon Papers, vol. 311, J. Ronald Fox, Assistance Secretary of the Army to the Honourable William Rogers, secretary, Department of State, 23 July 1970.

54 State Department, vol. 1847, file 18-8, State to Senator J.W. Fulbright, Chairman of the Senate Foreign Relations Committee, 2 January 1970. There had been a previous test in the Aleutians in October 1965 (Long Shot) and October 1969 (Milrow).

55 *Montreal Star* 22 April 1970. At 5 megatons, this nuclear device was estimated to be 400 times stronger than the Hiroshima bomb. References were also made to the 1964 Alaska earthquake, which produced a tsunami that caused over $1.5 million in damages along the BC coastal regions.

56 State Department, vol. 1847, file 18-8, American Consul Vancouver to State, 15 March 1971. On 9 August 1971 all nine Canadian premiers announced their opposition to the CANIKIV trials. Ibid., Report of Vancouver Consul, 9 August 1971.

57 Ibid., Report of US Embassy Ottawa to State, 14 September 1971. On 23 February Marcel Cadieux, Canadian Ambassador in Washington, filed an official protest with the US State Department. Even Prime Minister Trudeau got personally involved when he told reporters that he would, if necessary, "take my cap in hand and go to the President of the United States."

58 Theodore Eliot Jr, Executive Secretary State: memorandum to Henry Kissinger, NSC, 2 September 1971; ibid., US Ambassador Moscow to State, 6 January 1971.

59 Ed Regis, *Biology of Doom: America's Secret Germ Warfare Project* (New York: Henry Holt, 1999), 75. National Security Archives, box 11, U.S. Medical Research Institute of Infectious Diseases, "Project Whitecoat: A History," 14 February 1974.

60 Select Committee to Study Governmental Operations with Respect to Intelligence Operations. Under its chairman Frank Church (Democrat / Idaho), the Committee carried out 250 executive and 21 public hearings.

61 Ibid. Intelligence Activities and the Rights of Americans (Book 11); Supplementary Detailed Staff Reports (Book 111). The involvement in these experiments of Dr Ewen Cameron, Director of the Allan Institute, remains a controversial subject. See Anne Collins, *In the Sleep Room: The Story of the Brainwashing Experiments in Canada* (Toronto: Lester and Orpen Dennys, 1988).
62 US Senate, Hearings of the Select Committee to Study Governmental Operations with Respect to Intelligence Operations, vol. 1: Unauthorized Storage of Toxic Agents, Testimony of William Colby, 6-46, exhibit 2.
63 There was no evidence that Reed was aware how the shellfish toxin was being used, although as a defence scientist he believed that this was not his responsibility. Ibid., exhibit 11.
64 Hearings before the Subcommittee on Health and Scientific Research of the Committee on Human Resources, United States Senate, Ninety-fifth Congress, First Session, 8 March, 25 May 1977. The Army reports show that 79 of the 239 tests used pathogens. DEA, vol. 11536, file 28-6-6, DEA Washington to Ottawa, 4 March 1977.
65 DEA, vol. 11536, file 28-6-6, DEA Washington to Ottawa, 4 March 1977.
66 *Ottawa Citizen*, 4 March 1977. In 1980, members of the House of Commons had the opportunity of examining the recently declassified 1953 US report, "The Behavior of Aerosol Clouds within Cities." House of Commons, *Debates*, 12 May 1980, 930.
67 *Defence Research Board of Canada, Review 1972* (Ottawa: Queen's Printer, 1972).
68 Ibid. The Defence Research Council, composed of seventeen members, assisted Chairman L.J. L'Heureux in the operation of the organization. In 1972 a new centre was established, the Defence Research Analysis Establishment, with a focus on operational research and weapons systems.
69 The practice of bringing defence and academic scientists together in an annual symposium was initiated by Omond Solandt in 1947.
70 For the most part, the Defence Research and Development Board assumed most of the former duties of the DRB. It was merely a division within the department, lacking the independence and influence of its predecessor. In 2000 the present organization, Defence Research and Development Canada (DRDC) was created.
71 The Glassco Commission cited the Avro Arrow (CF-105) as a classic example of the dichotomy between technological capability and market realities. Bruce Doern, *The Structure of Policy-Making in Canada* (Toronto: Macmillan, 1971), 91–6.
72 Omond Solandt was the first president of the Science Council. In 1992 the Mulroney government curtailed its operation. Science Council of Canada, *Report No. 4: Towards a National Science Policy for Canada* (Ottawa, October

1969); John de La Mothe, "A Dollar Short and a Day Late: A Note on the Demise of the Science Council of Canada," *Queen's Quarterly* 99, no. 4 (Winter 1992): 873–86.

73 Douglas Bland, *The Administration of Defence Policy in Canada 1947 to 1985* (Kingston: Ronald Frye, 1987), 142–3. As a result of these changes, the resources of the Chief of Research and Development (CRAD) were available only to the Minister of Defence after going through two levels of civil advisers.

74 Cited in Gordon Watson, "Defence Research Board: Policies, Concepts and Organization," in *Perspectives in Science and Technology: The Legacy of Omond Solandt, Proceedings of a Symposium Held at the Donald Gordon Center, Queen's University of Kingston, Ontario, 8–10 May 1994,* ed. C.E. Law, G.R. Lindsey, and D.M. Grenville (Kingston: Queen's Quarterly, 1994), 68–73.

75 The Military Training Area occupied the northern three quarters of the Suffield Block, with the CBW area renamed Experimental Proving Grounds. Under this dual system, DRES and CFB Suffield shared the site, with British military officials dealing almost exclusively with the military base commander.

76 House of Commons, *Debates,* 28 February 1974, 25.

77 Richardson's 22 February 1974 statement specified that the Winnipeg-based DND research establishment would require a capital expenditure of $11.5 million, along with an annual maintenance cost of $6.5 million. He suggested, however, that these expenditures would be offset by the closing of Suffield. House of Commons, *Debates,* 7 March 1974, 280.

78 Ibid., Statement by Bert Hargrave, the member for Medicine Hat, 7 March 1974, 280.

79 Interview with Robert Heggie; *Debates,* 1 April 1974 (1031); ibid., 5 June 1975 (6495).

80 *Debates,* 22 October 1976 (365); ibid., 3 May 1977 (5229).

81 DRB, 1989-90/205, box 2, file 170-3 E1, pt 3, Longair to Chiefs of Establishment and members of Former DRC, 21 November 1974.

82 Ibid., Hayward to Perry, 17 March 1971.

83 MOU Report: private source in author's possession. This MOU was extended in 1984, 1989, 1999, and 2004. In June 2000 the focus was altered somewhat with the addition of radiological defence, making it a CBR arrangement.

84 On the medical side of the equation were programs for the exchange of personnel, CBRD loans, and special projects such as vaccine development. The ITF focused on short-term problems, while the Working Groups had broader terms of reference. Ibid.

85 Between 1991 and 2001, of the 41 completed ITFs, 15 dealt with the threat of biological weapons. There was increased reference to "CB Counterterrorism." Ibid.

86 William Tetley, *The October Crisis, 1970: An Insider's View* (Montreal and Kingston: McGill-Queen's University Press, 2006).

87 DRES experts also provided briefing for the Canadian Forces Nuclear, Biological, and Chemical School and the Canadian Forces Battle Group and Medical Support Unit in Europe. *DRES Annual Report, 1971*, unpublished report.

88 DRB, 93-94/132, vol. 20, file 965-1, W.A. Hoddinott for Chairman, DRB, Members of the Defence Research Council: "Bomb Threats in Relation to DRB Headquarters and Establishments, 2 November 1970." The report was submitted to the Security Panel on 2 December 1970.

89 Ibid.

90 Since over 80 per cent of the bombs were placed outside the buildings, it was argued that it was actually safer for employees to remain in their offices rather than accidentally move into a bomb location. Ibid.

91 Ibid.

92 Ibid.

93 DRB 24, 89/90, vol. 23, file 9700-1-2, F.A. Tate (for Director General DREO) to Chairman, DRB, 16 September 1971; ibid., vol. 7, file 962-2, pt 2, P.J. Armstrong, Security and Safety Officer, DREO: Memo, "Visitors to the Shirley Bay Site, 12 July 1971."

94 Ibid., vol. 17, file 2000-1, M.C. Hamblin, for Chief of the Canadian Defence Research Staff to Chairman, DRB, 13 March 1972. Report enclosed.

95 NARA, Nixon Papers, EX-FG 355, box 1, file Cabinet Committee on Terrorism, Samuel Hoskinson and Fernando Rondon to Henry Kissinger, 17 March 1972.

96 Ibid., Rogers to Nixon, 21 September 1972.

97 Ibid., Colonel Richard Kennedy to Kissinger, 23 September 1972; ibid., Kissinger and Ehrlichman to President Nixon, 24 September 1972; ibid., "Suggested Statement on Terrorism, 25 September 1972."

98 The major obstacle in obtaining a Terrorist Convention was the insistence by Arab nations that before they would support any UN action "the underlying causes of terrorism be studied." Ibid., Theodore Eliot, Jr, Executive Secretary State, memorandum for Kissinger, 21 October 1972.

99 Ibid., F.E. Rondon and R.T. Kennedy, memorandum for Henry Kissinger, 25 October 1972.

100 Of the eighty letter bombs sent internationally, six were intercepted in the United States, with no casualties. It was also agreed that the FBI would be the lead agency for investigating terrorism acts perpetrated in the United States. Ibid., Rogers to Nixon, 7 November 1972.

101 Ibid., Richard Kennedy, memorandum for Henry Kissinger, 25 November 1972.
102 Another welcomed development was the Department of Transportation's stated goal of having "100 percent screening of passengers and hand luggage by 5 January [1973]."
103 Ibid., Brent Scowcroft to David Parker, NSC, 11 March 1974.
104 In terms of the 250 letter bomb attacks, only 30 occurred after December 1972. This was partly offset by a surge of kidnappings, notably the sensationalist abduction of American diplomats in Khartoum (Sudan) and Guadalajara (Mexico). Ibid., William Rogers, Memorandum for the President, 27 June 1972.
105 Ibid., Rogers to Nixon, 8 January 1973.
106 DND, Access Request, LAC, DRB 89/90, vol. 23, file 9700-1-2, J.H. Meek, Terrorist Use of Chemical/Biological Agents, 29 January 1976.
107 Ibid., "Plausibility of Use of Nuclear, Biological or Chemical Agents for Terrorist Coercion." After 1971, Canada no longer had nuclear weapons on its soil because of the policies of the Trudeau government.
108 Ibid.
109 Ibid.
110 Ibid., "Terrorists Use of Weapons of Mass Destruction."
111 Ibid. The fact that DRES was destroying its stocks of residual chemical weapons was perhaps another reason why the DND report concluded that the theft of CW munitions seemed unlikely.
112 DRB 89/90, vol. 23, file 9700-1-2, Major General McClachlan to Commander, Mobile Command, St Hubert, Quebec, 16 March 1976.
113 Ibid., Establishment of a CBR Incident Response Capability for Operation Gamescan 76, 16 March 1976. The CBR platoon was to consist of eight to twelve trained officers and NCOs "drawn from the staff of the CF NBC School and from other CF bases where full time NBC specialists are establishment." It was expected to be able to respond "on a 24/7 basis."
114 Ibid., N.G. Trower, DGPO to CRAD, 22 March 1976.
115 Ibid., G.T. Pulman, Director General Plans and Programs, DND, to Chief, DRES, 24 March 1976.
116 Ibid., Lieutenant Colonel T.D. Nelson, Directorate Operational Training Coordination, circular letter, 24 March 1976.
117 Ibid., G.T. Pullan, memorandum, 9 April 1976.
118 Colonel A.E. Fox, for Chief of the Defence Staff, circular letter, 13 May 1976.
119 In 1984 a bioterrorist attack by the Rajneesh sect occurred in Oregon. Over a number of months, the sect was responsible for a salmonellosis

outbreak in Waco country, with the major targets being restaurant salad bars. There were many illnesses, but no deaths. Mark Wheelis and Masaaki Sugishima, "Terrorist Use of Biological Weapons," in Wheelis et al., *Deadly Cultures*, 286–92.

120 LAC, Defence Research Board Records, vol. 7358, file 170-80/B1, pt 2, C.E. van Rooyen, M.D. Professor and Head of Department, Dalhousie University, to Dr H. Zimmerman, Chairman, DRB, 18 June 1964.

121 Ibid., G.R. Vavasour to Chairman, DRB, 30 June 1964.

122 Ibid., W.F. Cockburn, Secretary ACBWR, to Committee Members, 5 November 1964.

123 "Timeline of Human Flu Pandemics, 1918–2007," National Institutes of Health and Department of Human Services. Canada, Dominion Bureau of Statistics, *Influenza in Canada, Some Statistics on Its Characteristics and Trends.* Ottawa: The Bureau, 1958.

124 Christoph Scholtissek, "History of Research on Avian Influenza," in *Avian Influenza*, ed. H.D. Klenk et al. (Basel: Karger, 2008), 101–17.

125 Jeffrey Taubenberger, Ann Reid, et al., "Characterization of the 1918 Influenza Virus Polymerase Genes," *Nature* (6 October 2005): 889–93.

126 D.A. Henderson et al., "Public Health and Medical Responses to the 1957–58 Influenza Pandemic," *Biosecurity and Bioterrorism* 7, no. 3 (2009): 270–1.

127 David Morens, Jeffrey Taubenberger, and Anthony Fauci, "Predominant Role of Bacterial Pneumonia as a Cause of Death in Pandemic Influenza: Implications for Pandemic Influenza Preparedness," *Journal of Infectious Diseases* 198 (1 October 2008): 962; Albert Osterhaus et al., "Epidemiology of Avian Influenza," *Journal of Infections Diseases* 198 (1 October 2008): 1–10; Paul Torrance ed., *Combating the Threat of Pandemic Influenza: Drug Discovery Approach* (New York: John Wiley and Sons, 2007).

128 LAC, Records of the Department of National Health and Welfare (DNHW), vol. 1193, file 311-J2-1, Dr G.D.W. Cameron, Deputy Minister of National Health to Dr A. Somerville, Deputy Minister of Public Health, Edmonton, 7 June 1957.

129 Ibid., E.H. Losing, Chief Epidemiology Division to Dr D.S. Puffer, Assistant Chief Medical Officers of Ontario, 21 June 1957.

130 Henderson et al., "Public Health and Medical Responses," 270.

131 DNHW, vol. 1193, file 311-J2-1. Ibid., First Meeting of Advisory Committee on Influenza, 11 July 1957.

132 Ibid., Report on Influenza Vaccine Production, 11 August 1957.

133 Ibid. G.D.W. Cameron, Deputy Minister of National Health, Memorandum to The Acting Minister, 13 August 1957.

134 NIH, box 150, file 1-3-C, Influenza 1958–72, Joseph Smadel, Associate Director NIH, to Dr Geist, 25 September 1957.

135 Significantly, Canada's level of morbidity and mortality from the H2N2 virus was lower than that in the United States. Among North American urban centres, New York City was particularly hard hit. Daniel Widelock and Sarah Klein, "Laboratory Analysis of 1957–58 Influenza Outbreak in New York City," *American Journal of Public Health* 50, no. 5 (May 1960): 649–60; Cecile Viboud et al., "Influenza Epidemic: England and Wales, Canada and the United States," *Emerging Infectious Diseases* 12, no. 4 (April 2006): 661–8.

136 This U.S.-Canadian influenza exchange system benefited from the joint poliomyelitis surveillance reports that had already been established. DNHW, vol. 1194, file 331-J2-15, Alexander Langmuir, Chief Epidemiology Branch, to E.H. Lossing, 25 September 1957. NIH, box 150, file Influenza Research Committee, Associate Director NIH, memorandum for files, 3 September 1957.

137 The list of laboratories was prepared by the federal Civil Defence Health Services. DNHW, vol. 1194, file 331-J2-15, G.D.W. Cameron to Dr Leonard Willer, Deputy Minister of Health, The Province of Newfoundland, 7 August 1957.

138 Ibid., Cameron: Report to the Dominion Council of Health, 7 November 1957.

139 NIH Papers, box 150, file 1030C-1 (Influenza 1957-58), Memorandum: Study of "Asian Flu" Epidemic, The Office of the Assistant Secretary of Defense, Washington, 9 October 1957.

140 In 1968 Fort Detrick scientists assisted the NIH and the Public Health Service in analysing the character of the H3N2 influenza virus. Archives of the American Society for Microbiology (University of Maryland–Baltimore), file 8-1A, folder Ex3, Housewright to Alvin Clark, 29 March 1967.

141 NIH, box 150, file Hong Kong Flu 1968, Director, NIAID to Surgeon General, PHS, 27 September 1968.

142 Claude Hannoun with Susan Craddock, "Honk Kong Flu (1968) Revisted 40 Years Later," in *Influenza and Public Health: Learning from Past Pandemics*, ed. Tarmara Giles-Vernick and Susan Craddock (London: Earthscan, 2010), 180–9.

143 United States General Accounting Office (GAO/NSIAD 92-33), December 1991, Biological Warfare, Role of Salk Institute in Army's Research Program.

144 Jeff Widmer, *The Spirit of Swiftwater: 100 Years at the Pocono Labs* (Scranton, PA: University of Scranton Press, 1998), 54–65. In 1977 the Toronto-based Connaught Medical Laboratories acquired control of the influenza

production plant as part of its renewed interest in vaccine production. Paul Bator, *Within the Reach of Everyone: A History of the University of Toronto's School of Hygiene and the Connaught Laboratories*, vol. 2 (Ottawa: Canadian Public Health Association, 1990).

145 Memorandum of Dr David Sencer, included in letter from Assistant Secretary for Health (Theodore Cooper) to Secretary of Health (David Matthews), 18 March 1976, cited in Richard Neustadt and Harvey Fineberg, *The Epidemic That Never Was: Policy-Making and the Swine Flu Affair* (New York: Vintage Books, 1983), 198, 30.

146 Cited *Globe and Mail* 31 March, 10 December 1976.

147 The Guillain-Barré syndrome was described as a form of muscular seizure that could be fatal if it affected the respiratory system. In the vast majority of cases, however, symptoms were minor and temporary, and most of the 300 to 700 cases that occurred annually did not result in permanent disability. In the United States, it was stated that "swine flu has become a minor industry." *Globe and Mail* 10 December 1976.

148 By this stage, Connaught was controlled by the Canadian Development Corporation and would soon become part of the French medical multinational Rohome-Poulne Group. See Bator, *Within the Reach of Everyone*, 75–97.

149 *Toronto Star* 3 October 1977.

150 The official story, *Smallpox and Its Eradication*, was prepared by Henderson and his WHO associates in 1982. This 1,500-page tome, known as the "Big Red Book," remains the most authoritative source. See also Henderson's 2009 book *Smallpox: The Death of a Disease* (Amherst: Prometheus Books) and his article, "Smallpox: Dispelling the Myths," *Bulletin of the World Health Organization* 86, no. 12 (December 2008): 917–19.

151 Bator, *Within the Reach of Everyone*, 116–20. Connaught Archives (Toronto), Robert Wilson Papers, box 8350. Smallpox – Henderson file, D.A. Henderson, "Smallpox Eradication Program, 18 October 1965.

152 In 1966, Connaught was designated a World Smallpox Vaccine Reference Centre by the WHO. The following year it entered into a thirteen-year contract with the Pan American Health Organization for improving vaccine quality.

153 Wilson Papers, D.A. Henderson to Wilson, 23 March 1967. In contrast, the Canadian government was initially reluctant to provide the 10 million doses requested by the WHO, despite strong pressure from Connaught officials. Ibid., Ken Ferguson (President) to the Honourable Paul Martin, Minister Health and Welfare, 8 January 1968.

154 Henderson offered three major reasons why smallpox could be eradicated: its symptoms were easily recognized, it only required one shot to

establish immunity for ten years, and it had no animal reservoir. Wilson Papers, Henderson to Wilson, 26 November 1976.

155 On 8 May 1980 the World Health Assembly officially announced that the variola virus had been eradicated.

156 Connaught Collection, box 8352 … 47, file WHO Report on Laboratories Retaining Variola Virus, WHO/SE/79.137, "Meeting of Officials from Laboratories Retaining Variola Virus and National Control Authorities Concerned (23–4 April 1979).

157 Ibid. The number of laboratories retaining variola declined from seventy-six in 1976 to eight in 1978, with each of the remaining institutions being forced to justify their reasons for retention.

158 David Fidler and Lawrence Gostin, *Biosecurity in the Global Age: Biological Weapons, Public Health and the Rule of Law* (Stanford, CA: Stanford University Press, 2008), 2.

159 ASM Archives, 8-1A-folder 1 (file Fort Detrick), Housewright to Donald Shay, 4 September 1970.

6 Preventing Germ Warfare in the Age of the Biotechnology Revolution

1 Cited in Mark Wheelis et al., eds., *Deadly Cultures: Biological Weapons since 1945* (Cambridge, MA: Harvard University Press, 2006), 376.

2 John Hart, "The Soviet Biological Weapons Program," in Wheelis et al., eds., *Deadly Cultures*, 132–57; Roger Roffey et al., *Support to Threat Reduction of the Russian Biological Weapons Legacy: Conversion, Biodefence and the Role of Biopreparat* (Stockholm: Swedish Defence Research Agency, NBC Defence, April 2003); Dany Shoham and Ze'ev Wolfson, "The Russian Biological Weapons Program: Vanished or Disappeared?" *Critical Reviews in Microbiology* 30 (2004): 241–61.

3 Estimates about the actual numbers of fatalities vary considerably, with some studies suggesting that as many as one thousand people died from anthrax exposure.

4 Many of these AIDS conspiracy theories focus on the 1969 testimony of Dr MacArthur, Director of the US Army's Advanced Research Project Agency, who informed Congress that there was considerable interest in biological agents that could attack the human immune system. United States Senate Library, Hearings before a Sub-Committee of the Committee on Appropriations, House of Representatives, 1 July 1969.

5 Libya's dictator Moammar Ghadhafi was also suspected of having BW intentions, given his ambitious nerve gas program. There were also fears that Israel might launch a pre-emptive strike to eliminate Libya's CW

threat. DEA, file 21-4, NATO representatives to DEA, 22 December 1988; Washington to DEA, 4 January 1989.

6 The first meeting of the Australia group, an informal forum of fifteen countries seeking to ensure that exports did not contribute to the development of biological and chemical weapons, took place in Brussels in June 1985. Today the organization has forty members. See www.australiagroup.net.

7 While the South African BW program has attracted little attention, compared with its nascent nuclear ambitions, it was quite an extensive operation under the Ministry of Defence. Code-named Pacific Coast, it was directed by Dr Wouter Basson, who was able to enlist the services of a number of academic and industrial scientists with the goal of producing BW weapons for military use, crowd control, and political assassination. Chandre Gould and Alastair Hay, "The South African Biological Weapons Program," in Wheelis et al., eds., *Dangerous Cultures*, 191–212; Jeffrey Richelson, *Spying on the Bomb: American Nuclear Intelligence from Nazi Germany to Iran and North Korea* (New York: W.W. Norton, 2006), 370–3.

8 In 30 November 1979 Canada had its own toxic accident when a Canadian Pacific freight train, transporting tons of potentially deadly chlorine gas, derailed in the middle of a densely occupied residential district of Mississauga, Ontario, forcing the evacuation of more than 240,000 people. Fortunately, there were no deaths. *Toronto Star* 10 November 2009.

9 Gary Zweiger, *Transducing the Genome: Information, Anarchy, and Revolution in the Biomedical Sciences* (New York: McGraw-Hill, 2001), 204. By 2001 there was a massive increase in the scale of genomic sequencing (over 200 a month), with estimates that over a million human genomes would be processed in the near future. "DNA Sequencing Is Caught in a Deluge of Data," *New York Times* 18 December 2011.

10 National Security Archives (Washington, DC), box 11, "An Evaluation of Biological Treaties," August 1976.

11 Ibid., U.S. Army Dugway Proving Ground, Bruce Grim, "Biological Agent Delivery by ICMB," April 1981.

12 Ibid., U.S. Dugway Proving Ground, Dr Frank Armstrong, Dr A. Paul Adams, and Mr William Rose, "Recombinant DNA and the Biological Warfare Threat," May 1981.

13 Donald Frederickson, *The Recombinant DNA Controversy: A Memoir* (Washington, DC: ASM Press, 2001), 14.

14 Ibid., 25.

15 Ibid., 40–7. Raymond Zilinskas and Barbara Zimmerman, eds., *The Gene-Splicing Wars: Reflections on the Recombinant DNA Controversy* (New York: Macmillan, 1986).

16 See Susan Wright, *Molecular Politics* (Chicago: University of Chicago Press, 1994), and J. Goodfield, *Playing God: Genetic Engineering and the Manipulation of Life* (New York: Random House, 1977).

17 National Institutes of Health (NIH), box 158, file NCI Frederick, Deputy Director for Science, NIH, to Director NIH, 24 March 1976.

18 NIH, box 159, file IRM Res 10-2-B, P4, Facilities for DNA Res. at Fort Detrick, 1975–76, Preliminary Report of the Special P4 Containment Capability Ad Hoc Committee, enclosed in letter Richard Krause, Director NIAID to Deputy Director for Science, NIH, 12 August 1976.

19 Ibid., Deputy Director, NIAID to Director NIH, 18 January 1977.

20 Ibid., Director NIAID to Director NIH, 3 April 1978: "Revised NIH Guidelines on Recombinant DNA Research, 25 July 1979."

21 B.D. Cohen, "DNA LAB: Fort Detrick Room Set for Genetic Engineering," *Washington Post* 18 March 1978.

22 DEA, 26-6-6, vol. 20, telegram, Washington to External/IDA, 18 June 1986.

23 *New York Times* 4 July 1982; DEA, 26-6-6 (BW), telegram, J.W. Currie to Washington, 30 July 1982. The Biotechnology Committee of the Canadian Society of Microbiologists formally endorsed the NIH protest.

24 D.A. Henderson, *Smallpox: The Death of a Disease* (New York: Prometheus Books, 2009), 273.

25 Ken Alibek, *Biohazard: The Chilling True Story of the Largest Covert Biological Weapons Program in the World – Told from Inside by the Man Who Ran It* (New York: Random House, 1999), 234.

26 DEA, 26-6-6- (BW), vol. 3, Defence Intelligence Agency Report, "Foreign Technology Weapons and Systems," 3 March 1980.

27 Ibid., US State Department Report, 18 March 1980.

28 Ibid., telegram, Washington to DFD/IDA, 4 September 1981.

29 Ibid., telegram, DEA Moscow to External/DFD, 23 September 1981.

30 Ibid., R.P. Cameron, Director-General, Bureau of International Security Policy and Arms Control Affairs, to John Anderson, Assistant Deputy Director, Policy, DND, 2 October 1981.

31 Department of External Affairs Archives, "Study of the Possible Use of Chemical Warfare Agents in Southeast Asia," by Dr H.B. Schiefer, Toxiocology Group, University of Saskatchewan (Ottawa, 1982). The report also endorsed the previous report on yellow rain by the United States Mission to the United Nations (Special Report #98, 1982).

32 Ibid., "An Epidemiological Investigation of Alleged CW/BW Incidents in South Eastern Asia" by Lt Colonel G.R. Humphreys and Major J. Dow (Ottawa, 1982).

33 For a recent scholarly debate about the yellow rain incident see Mathew Meselson and Julian Perry Robinson, "The Yellow Rain Affair: Lessons from

a Discredited Allegation," and Rebecca Katz, "'Yellow Rain' Biological Warfare Agent Use: Evidence and Remaining Questions," in Anne Clunan et al., eds., *Terrorism, War, or Disease: Unravelling the Use of Biological Weapons* (Stanford, CA: Stanford University Press, 2008), 72–96, 97–120.

34 National Security Archives (Washington), box 11, chemical and biological weapons: "Soviet Influence Activities: A Report on Active Measures and Propaganda, 1986–87" (United States Department of State, August, 1987), 33.

35 Ibid., 35, 39. This cloning theory about the origins of AIDS was developed by Jacob Segal, a 76-year-old East German biophysicist, formerly of the Humboldt University of East Germany. It became an integral part of Soviet propaganda in January 1987.

36 Ibid., 42. Several prominent members of the US Congress praised the State Department for its creative attempts to counter Soviet AIDS propaganda. NSA, box 11, record number 58485, Dan Burton, House of Representatives to Secretary George Schulz, Department of State, 6 October 1987.

37 Ibid., 42. Zhdanov was a prominent member of the USSR Academy of Sciences and had been its leading spokesman at the 1962 conference of the International Association of Microbiology in Montreal. In December 1985 he was the first Soviet scientist to acknowledge that AIDS had spread to the Soviet Union. See Donald Avery and Mark Eaton, *The Meaning of Life: The Scientific and Social Experiences of Everitt and Robert Murray, 1930–1964,* (Toronto: Champlain Society, 2008), 70–85.

38 Despite this rapid movement towards détente, hard-liners still carried on the AIDS propaganda campaign. Indeed, in 1987, there were at least thirty-two citations in Soviet print or broadcast media.

39 According to the Soviet defence scientist Igor Domaradskij, Pasechnik had excellent connections with the major power brokers of Biopreparat. This included General Kalinin, head of the organization, who ironically approved Pasechnik's participation at an international life sciences conference in Switzerland, thereby facilitating his defection. I.V. Domardadskii and W. Orrent, *Biowarrior: Inside the Soviet/Russian Biological War Machine* (Amherest: Prometheus Books, 2003), 271–81. See also Judith Miller et al., *Germs: Biological Weapons and America's Secret War* (New York: Simon and Schuster, 2001).

40 One of the sites was the massive facility at Obolensk, near Moscow, where about 4,000 scientists, technicians, and supported staff worked on anthrax, tularaemia, brucellosis, glanders, and genetically modified plague. In contrast, the Vektor Centre, near Novosibirsk, Siberia specialized in viral BW agents, including Marburg, Ebola, Lassa fever, and apparently 140 strains

of smallpox. Tom Mangold and Jeff Goldberg, *Plague Wars: A True Story of Biological Warfare* (New York: St Martin's Press, 1999), 126–37.

41 In 1993–4 there was a second round of trilateral visits of former Soviet BW facilities. This was followed by a tour of US facilities, including Fort Detrick, in 1994–5. John Hart, "The Soviet Biological Weapons Program," in Wheelis et al., eds., *Deadly Cultures*, 148.

42 DEA, file 26-6-6 (BW), pt 8, "Report from Netherlands on Soviet Union Exchange of Information under BTWC," 21 February 1990.

43 LAC, Department of Defence Records (RG24), vol. 44, file 3754-1, pt 2, NDHQ Policy Directive P3/85.

44 William H. Barton, "Research, Development and Training in Chemical and Biological Defence within the Department of National Defence and the Canadian Forces: A Review" (Ottawa, Canada, 31 December 1988), 25. ABCA was an acronym for America, Britain, Canada, Australia.

45 Barton reserved special praise for the research and testing work carried out at DRES, while questioning DND's small budget allocation for this important work. Ibid., 5–6.

46 Ibid. The use of chemical and biological incapacitants was, however, regarded as justified, since they were non-lethal ways "to neutralize hostile elements."

47 LAC, DEA, file 28-6-6 (BW), vol. 6, "USA Paper on Technological Developments of Relevance to the BTW Convention," 8 September 1986.

48 Canadian involvement with the MOU/CANUKUS primarily involved scientists from DRES, DREO (Ottawa), and DCIEM (Toronto). In order to coordinate MOU policies among the three countries, a semi-annual meeting of Program and Requirements officers was held at the U.S. Aberdeen Proving Ground, Maryland.

49 DND Access Request: LAC, DND (RG24), 2001-00003-0, vol. 46, file 338, pt 10, MOU Steering Group Meeting, 5 June 1986: Summary Record.

50 In determining the relationship between the MOU and the TTCP, it was agreed that "CB collaboration should be arranged by the MOU." Ibid. STAG was the Canadian anti-terrorist body organized by the Health Protection Branch, Department of Health and Welfare.

51 Ibid., National Defence Headquarters, Acting Director, International Research and Development, memorandum, 22 January 1986.

52 Ibid., DREO memorandum, R andD / Intelligence Interface 13 May 1986.

53 Ibid., Chief, DREO to National Defence Headquarters, 7 August 1986.

54 Ibid., Director, Nuclear, Biological, Chemical Coordination for Chief of the Defence Staff, memorandum, ITF 4 After Action Report, 8 August 1986.

55 Ibid., DREO Memorandum to DRES and NDHQ (telegram), 30 August 1986.

56 In July 1993, Graham Pearson, Chief Executive of the Chemical and Biological Defence Establishment, Porton Down, informed members of the British House of Commons that his establishment had received a number of BW toxins and agents from the United States during the previous ten years. *Debates,* 16 July 1993 (vol. 228), 708–17.

57 DND (RG24), 2001-00003-0, vol. 46, file 338-pt 10, CA Program Officer, US/UK/CA MOU on CB Defence to NDHQ, 19 February 1986; DREO memorandum, 18 February 1986.

58 Ibid., Chief, DREO to NDHQ 2 January 1986; ibid., DREO, Report to CA Steering Group Member CRAD 24 March 1986.

59 Ibid., DREO to CDE Porton, 2 November 1986.

60 Ibid., Responsibility for NBC Threat Agents, n.d. (probably November 1986).

61 DEA, 26-6-6 (BW), vol. 8, DEA Headquarters to Tehran, 14 August 1989.

62 Ibid., Gordon Vachon, IDA to Schiefer, 20 October 1989.

63 Three of the more controversial studies were Leonard Cole, *Clouds of Secrecy: The Army's Germ Warfare Tests over Populated Areas* (Savage, MD: Rowman and Littlefield, 1990); Marc Lappe, *Broken Code: The Exploitation of DNA* (San Francisco: Sierra Club Books, 1984); Susan Wright, ed., *Preventing a Biological Arms Race* (Cambridge: MIT Press, 1990).

64 Cited in Wright, ed., *Preventing a Biological Arms Race,* 52–3.

65 Science for Peace was established at the University of Toronto in 1981 by a diverse group of academics and peace activists. The first president of the organization was Dr Eric Fawcett (1981–4), a well-respected physicist. He was succeeded by George Ignatieff, formerly one of Canada's most influential diplomats and an expert on CBW arms control. See www.scienceforpeace.ca.

66 The controversial CBC television series *The Valour and the Horror* claimed that Canadian soldiers and airmen paid a high price for advancing British military goals during the Second World War. See David Bercuson and S.F. Wise, eds., *The Valour and the Horror Revisited* (Montreal and Kingston: McGill-Queen's University Press, 1994).

67 John Bryden, *Deadly Allies: Canada's Secret War, 1937–1947* (Toronto: McClelland and Stewart 1989), 260.

68 See Donald Avery, "Canada's Biological Weapons Program and Field Trials at DRDC Suffield during the Second World War: The Anthrax Controversy," DRDC Suffield (unpublished report, November 2005.

69 Barton, "Research, Development and Training in Chemical and Biological Defence."

70 Department of Defence, *First Annual Report of the Biological and Chemical Defence Review Committee* (Ottawa, 1990).

71 Jean Krasno and James Sufferlin, *The United Nations and Iraq: Defanging the Viper* (Westport, CT: Praeger, 2003).
72 Graham Pearson, "The Iraqi Biological Weapons Program," in Wheelis et al., eds., *Deadly Cultures*, 169–90.
73 Background papers by the Defence Intelligence Agency (1987), prepared for Secretary of Defence Richard Cheney, cited in Gary Matsumoto, *Vaccine A: The Covert Government Experiment That's Killing Our Soldiers and Why GI's Are Only the First Victims* (New York: Basic Books, 2004), 42.
74 While Salman Pak and Al Hakam were the main BW production centres, there were at least nine other facilities. Pearson, "The Iraqi Biological Weapons Program," 175–89.
75 The 1995 defection of Saddam Hussein's son-in-law, General Hussein Kamel, confirmed UNSCOM suspicions about the scope of the Iraqi BW program by July 1996. The final inventory showed 19,000 litres of botulinum toxin, 8,500 litres of B. anthracis, 2,200 litres of aflatoxin, and 340 litres of C. perfringens. Gregory Koblentz, *Living Weapons: Biological Warfare and International Security* (Ithaca, NY: Cornell University Press, 2009), 86, 94.
76 In 1990–1 the US deployed approximately 697,000 troops in the Persian Gulf, with 293 deaths (0.042). In the case of the United Kingdom, there were about 50,000 troops, with 24 fatalities (0.048). Both countries administered the anthrax vaccine. Matsumoto, *Vaccine A*, 59–85.
77 The specially appointed US Tri-Service Vaccine Task Force (Project Badger) had primary responsibility for securing supplies of anthrax vaccine. Most of its members were located at the United States Army Medical Research and Development Command at Fort Detrick, including Dr Bruce Ivins, who would subsequently be involved in the 2001 anthrax letter bomb controversy. Matsumoto, *Vaccine*, 84–9.
78 LAC, DND (RG 24), 2001-0003-0, vol. 44, file 3754-1, pt 2, S. Myles, Director, Defence Sciences Division, DRES and B.H. Harrison, Director, Protective Sciences division, DREO, Joint Memorandum: Strategic Planning for CBD Program, 8 August 1990.
79 David Kelly was one of the first UNSCOM scientists to realize that the Al Hakam facility was a key component of Iraq's BW program. He had previously been a member of the UK-US scientific delegation that visited the USSR / Russia during the 2001–5 Trilateral exchanges. His tragic suicide in 2007, after charges that he had revealed top-secret information to the British media, remains a matter of considerable controversy.
80 About half of the Canadian BW experts came from DRDC Suffield, with most of the others, including Ken Johnson, coming from National Defence

Headquarters. Summary of Defence Research and Development Branch (DRDB) Personnel Involvement in UNSCOM Mission, June 1991–8.

81 Johnson spent 293 days during his nine stints in Iraq and five tours at UNSCOM headquarters in New York City. Interview with Ken Johnson (Ottawa, DRDC, 28 February 2002). Koblenz, *Living Weapons*, 98.

82 In 2006 Australia joined this organization, which meant the designation was changed to AusCanUK US (Chemical Biological Radiological Memorandum of Understanding (CBR MOU). In addition, New Zealand has affiliated status.

83 DND, 2001-00003-0, vol. 46, file 338, pt 10, International Task Force (ITF) 23: "Development of Trinational Biodefence Concepts." Other important reports include ITF 27, "Biological Warfare Defence Strategic Issues," and ITF 24: "Emerging Technologies for the Detection and Identification of Biological Agents.

84 Ibid., ITF 32: "Impact of BW on Operations."

85 Ibid., Report: Medical Countermeasures Coordinating Team-CB Working Group (March 1998).

86 Under the World Health Assembly Resolution of May 1999 (WHA52.10) authorization was given for the temporary retention of variola virus (smallpox) strains "for the purpose of further international research into antiviral agents… and to permit high-priority investigations into genetic structure and pathogenesis of variola." Orthopox Working Group, Terms of Reference (2000).

87 During Operation Desert Storm, 41 per cent of US combat troops and 57 per cent of UK troops were vaccinated against anthrax.

88 Throughout the 1990s there was an intense controversy over so-called the Gulf War syndrome, whose symptoms included fatigue, headaches, memory problems, muscle / joint pains, and neurological problems.

89 Minutes of Proceedings, Standing Court Martial for the trial of Ex-Sergeant Michael Richard Kipling, Canadian Forces, Regular Force, held at 17 Wing, Winnipeg, Manitoba: 15 February–5 May 2000. Military Judge Colonel G.L. Brais, Office of the Chief Military Judge. See www.mvrd.org/AVN/sonnie/law%5Ccanadatranscript.htm.

90 During his twenty-one years at Fort Detrick, Friedlander became an expert on anthrax, publishing articles on this subject in the leading scientific journals on infectious diseases. Ibid.

91 Minutes of Proceedings, Standing Court Martial for the trial of Ex-Sergeant Michael Richard Kipling; interview with Colonel Ken Scott, Director of Medical Policy, Department of National Defence, Ottawa, 24 February 2002.

92 Defence Research Establishment Suffield publications: "Canadian Integrated Bio / Chemical Agent Detection System (CIBADS)"; "Automated Microchip Platform for Biochemical Analysis"; Jim Ho, "The Future of Biological Aerosol Detection" (DRES, 2001).

93 Research on CBW detectors has remained a priority with Canadian defence scientists, as is evident in the recent development of the Vital Point Biological Sentry System (VPBSS), and the chemical warfare Compact Atmospheric-Sounding Interfeometer Development Model, and Research Model (CATSIDM). Both these systems proved effective in detecting CBW agents approximately two kilometres away, thereby giving time for defensive countermeasures. DRDC unpublished report: "The Dragons Din" (Spring 2010).

94 DRES publications: "CB Decontamination Research"; "Chemical and Biological Incident Management"; "CB Demilitarization and Site Remedial Projects."

95 Bill Kournakakis, Joan Armour, Cam Boulet, M. Spense, and B. Barsons, "Risk Assessment of Anthrax Threat Letters," DRES, Technical Report TR-2001-048 (September 2001).

96 David Brown, "Agency with Most Need Didn't Get Anthrax Data: CDC Unaware of Canadian Study Before Attacks," *Washington Post* 11 February 2002. While attending the International Conference on Emerging Infectious Diseases in Atlanta, Georgia, in March 2002 I interviewed several scientists from the CDC, who were quite puzzled about why the DRES anthrax report was not utilized.

97 Between 1981 and 1995 South Africa's BW program was directed by Dr Wouter Basson, one of the country's leading defence scientists. During the late 1990s he was prosecuted, but later acquitted, for these activities. Chandre Gould and Alastair Hay, "The South African Biological Weapons Program, in Wheelis et al., eds., *Deadly Cultures*, 191–212.

98 Jez Littlewood, *The Biological Weapons Convention: A Failed Revolution* (Aldershot, Hants: Ashgate, 2005), 70–3; Nicholas Sims, *The Diplomacy of Biological Disarmament: Vicissitudes of a Treaty in Force, 1975–1985* (New York: St Martin's, 1988).

99 University of Sussex Archives (UK), National Biological Defence Research and Development Program. Declaration, United States (CBM Measure, 1996), 155. Reference was also made to the fact that by 1968 the US was in a position to weaponize *Brucella suis, Pasteurella tularensis, Coxiella burnetii, Bacillus anthracis,* Staphylococcal enterotoxin, and Venezuelan equine encephalitis. Ibid., 160.

100 Ibid., 404–5. See also US Department of Defence, *Chemical and Biological Defense Program: Annual Report to Congress* (March, 2000). One notable omission from the US declaration was an account of the projects carried out during the 1960s under Project 112.

101 University of Sussex Archives, Federation of Russian Republics, Confidence Building Measure F: Declaration of Past Activities in Offence and / or Defensive Biological Research and Development Programs (1992).

102 President Yeltsin's 11 April 1992 Edict 390, "On Ensuring the Implementation of International Obligations Regarding Biological Weapons," prohibited any BW activities in Russia that would violate the BWC. These measures were strongly opposed by defenders of the former BW system, who were determined to concede only what the US and UK already knew, either from defectors or through Tripartite visits. Koblenz, *Living Weapons*, 123–4.

103 *Washington Times* 27 July 1992. Koblenz, *Living Weapons*, 124.

104 University of Sussex Archives, Government of Canada, Confidence Building Measure F: Declaration of Past Activities in Offence and / or Defensive Biological Research and Development Programs (1992).

105 Ibid. Canada's 1992 Declaration was also contradictory since it claimed that "no work was aimed at the suitability of specific agents as weapons," while in the previous section of the paragraph there is reference to "studies of weapon-produced aerosols of potential BW agents … and development of M. mallei and M.1 pseudomallei as new potential BW agents." Ibid.

106 Ibid. All twenty-nine DRES laboratory personnel were civilians, twelve of whom were scientists with expertise in bacteriology, immunology, microbiology, virology, biochemistry, biotechnology, veterinary medicine, medicine, and pharmacology.

107 In the case of the Connaught Medical Research Laboratories, its major vaccines / toxoids related to diphtheria, pertussis, typhoid, staphyloccus, botulinum toxin, polio (Salk and Sabin vaccines), yellow fever, and influenza, the latter two produced at its Swiftwater, Pennsylvania plant. Ibid.

108 Malcolm Dando, *Preventing Biological Warfare: The Failure of American Leadership* (New York: Palgrave, 2002), 116, 124, 132, 173–7.

109 Littlewood, *Biological Weapons Convention*, 86–120.

110 Ronald Atlas and Michael Goldberg, "Biological Warfare: Examining Verification Strategies," *ASM News* 59, no. 8 (1993): 393–5.

111 Federation of American Scientists, "Proposals for the Third Review Conference of the Biological Weapons Convention," *Contemporary Security Policy* 12, no. 2 (September 1991): 240–50; Federation of American

Scientists Working Group on the Biological Weapons Convention, "Preliminary Paper: The Cost and Structure of a BWC Organization," June 1998.

112 Eric Myjer, ed., *Issues of Arms Control Law and the Chemical Weapons Convention* (The Hague: Martinus Nijhoff, 2002), 62.

113 Dando, *Preventing Biological Warfare*, 33.

114 Malcolm Dando and Martin Furmanski, "Midspectrum Incapacitant Programs," in Wheelis et al., eds., *Deadly Cultures*, 236–51; Editorial, "'Non-Lethal' Weapons, the CWC and the BWC," *Chemical and Biological Weapons Convention Bulletin* 61 (2003): 1–2.

115 In 1997 the US pharmaceutical was producing about 45 per cent of the world's drugs, with high research and development costs. Dando, *Preventing Biological Warfare*, 70–1.

116 Miller et al., *Germs*, 304–20.

117 Jonathan Tucker, *Scourge: The Once and Future Threat of Smallpox* (New York: Atlantic Monthly Press, 2001), 192.

118 Historian of science Susan Wright has provided an important critique of what she calls bioterrorism as "received knowledge" in her article, "Terrorists and Biological Weapons, Forging the Linkage in the Clinton Administration," *Politics and the Life Sciences* 25, no. 1–2 (15 February 2007): 57–115. In particular, she points out how prominent scientific advisers, such as Nobel laureate Joshua Lederberg and former bioweaponeer Bill Patrick, helped convince the White House and prominent congressmen of the need for extensive spending for US biodefence projects.

119 Cited in Wright, "Terrorists and Biological Weapons," 83. See also Richard Preston's sensationalist account of US biodefence activities after 9/11: *The Demon in the Freezer: A True Story* (New York: Random House, 2002).

120 Mangold and Goldberg, *Plague Wars*, 384. New York City had its CBW counterterrorism exercises in September 1997, authorized by Mayor Rudolph Giuliani, that involved more than 4,000 members of the city's police, fire, and emergency units, along with 1,500 doctors and nurses. It focused on the vulnerability of air exchange systems in large office buildings and in the NYC subway. Ibid., 352–6.

121 Most of these exercises were divided into three categories: Tabletop, Command Post, and Field, the latter being the most extensive and realistic, with activities taking place in a number of different environments over a series of days.

122 TOPOFF One began on 17 May with the simulated release of an aerosol of Y. pestis at the Denver Performing Arts Centre. It ended four days later, with projections that 3,700 people would have contacted pneumatic

plague, resulting in 950 deaths. Thomas Inglesby, Rita Grossman, and Tara O'Toole, "A Plague on Your City: Observations from TOPOFF," *Clinical Infectious Diseases* 32 (2001): 436–45.

123 Ibid. Interview with Dr Huffmann, Chief Medical Officer of Denver (1988–2001), International Conference on Emerging Infectious Diseases, Atlanta, 10 March 2002.

124 Thomas Inglesby, "Lessons from TOPOFF, The Second National Symposium on Medical and Public Health Response" (11/30/2002), www.hopkins-biodefense-org/sympcast/transcripts/trans_ingl.html. A year later, on 22–3 June 2001, another important bioterrorism exercise, called Dark Winter, took place, involving D.A. Henderson, director of the newly established biodefence center at Johns Hopkins University. Predictably, the pathogen involved was smallpox.

125 Elinor Sloan, *Security and Defence in the Terrorist Era* (Montreal and Kingston: McGill-Queen's University Press, 2005), 32, 340–6.

126 George Tenet, *At the Center of the Storm: My Years at the CIA* (New York: Harper Collins, 2007), 123. During the Atlanta Olympics a bombing killed one person and injured scores. Richard Clarke, *Against All Enemies. Inside America's War on Terror* (New York: Free Press, 2004), 97, 108–10.

127 Tenet, *Center of the Storm,* 260.

128 Clarke, *Against All Enemies,* 212.

129 Tenet, *At the Center of the* Storm, 126.

130 There had been three attacks on Turkish diplomats in Ottawa during previous years. But the March 1995 assault on the embassy was the most serious incident, since it resulted in the death of a security guard. Three men were subsequently convicted of first-degree murder.

131 Of those accused in the Air India bombing, only bomb maker Inderjit Singh Reyat was ever convicted. Kim Bolan, *Loss of Faith: How the Air-India Bombers Got Away with Murder* (Toronto: McClelland and Stewart, 2005).

132 *Commission of Inquiry into the Investigation of the Bombing of Air India Flight 182,* vol. 1: *The Overview* (Ottawa: Canadian Government Publishing, January 2010) (electronic resource).

133 This plan replaced the 1967 arrangement and greatly expanded the scope for cooperation. "Agreement between the Government of Canada and the Government of the United States of America on Cooperation in Comprehensive Civil Emergency: Planning and Management," declassified document in possession of author.

134 Office of Solicitor General of Canada, "Developing Options to Strengthen National Consequence Management Response Capabilities for Terrorist

Incidents," cited in Discussion Paper, 18 April 2001 (declassified document in possession of the author).

135 Ron Purver, "The Threat of Chemical / Biological Terrorism," *Commentary* 2, no. 60 (August 1995): 2, 10. STAG was originally set up during the counterterrorist response to the 1976 Montreal Summer Olympics. Because of its focus on medical prevention and treatment, the chairman of the group was usually an official of Health Canada.

136 Canadian Security Intelligence Service, Report #2000 / 02, "Chemical, Biological, Radiological and Nuclear (CBRN) Terrorism," *Perspectives* (18 December 1999). See also Report #20005/05, "Biological Weapons Proliferation," *Perspectives* (9 June 2000).

137 In 1987, and again in 1989, special studies were carried out by the Senate Special Committee on Terrorism and Public Safety, but these were essential ad hoc measures. "The Report of the Special Senate Committee on Security and Intelligence (Chairman: The Honourable William M. Kelly)," January 1999, chapter 1.

138 There were also serious questions why Parliament was not kept informed about major counterterrorist projects involving foreign intelligence. In this debate, reference was made to the situation in the United States, where Congress was extensively involved with intelligence matters, making it a "more effective critic of the executive branch." Ibid., chapter 4.

139 Ibid., chapter 3.

140 It was also noted that Canada lagged behind the United States in providing protection for critical infrastructure, although the work of the government's Interdepartmental Information Operations Working Group appeared quite promising. Ibid., chapter 2.

141 Significantly, six of the committee's thirty-three recommendations involved weapons of mass destruction. It was also recommended that CSIS be given the authority and resources to obtain "relevant foreign intelligence" (26) as well as access to the RCMP's Violent Crime Linkage Analysis System (29). Ibid., chapter 5.

142 Solicitor General Canada, "Developing Options," Discussion Paper, 18 April 2001.

143 Ibid. Equally disruptive were the large number of hoaxes such as the January 2001 incident in Ottawa in which the appearance of an envelope, apparently containing anthax spores, closed down a federal building for two days while HAZMAT teams carried out a thorough examination.

144 Ibid. Reference was also made to the possibility that terrorists might use nerve gas agents as well as toxic industrial chemicals against civilian targets. The impact of this type of incident was revealed on 3 December 1984

when the accidental leak of a deadly pesticide from the local Union Carbide plant in Bhopal, India killed over 10,000 people.

145 The Canadian Forces provided most of the CBW expertise through its special base at Camp Borden, Ontario, reinforced by its direct linkages to DRDC Suffield. In 2001 the team consisted of sixteen instructors and nine specialists. DRDC, "2001 Annual Report: Biological and Chemical Review Committee," unpublished document.

146 Health Canada maintained the National Emergency Services Stockpile (NESS) of medical materials, drugs, and equipment, which, it was assumed, could "be immediately shipped and used as required upon request of a provincial authority."

147 Bill Kournkakis, a DRES expert on anthrax, profited from his contact with the celebrated David Kelly when they were members of the same UNSCOM team in Iraq. Interview with Bill Kournkakis, DRDC Suffield, 31 March 2005.

148 After 1993 there was a 37 per cent reduction of scientific personnel at the Department of National Defence; Suffield lost many senior personnel through transfers or early retirement. During this period many DRES scientists were involved with UNSCOM. Interview with Dr Kent Harding, DRDC Suffield, 23 August 2003.

149 Interview with Mike Thielmann, American University, Washington, DC, 7 December 2006.

150 Ibid. Additional information about the CSIS bioterrorist operation after 2001 was obtained from other sources: Interview with Katie Tolan, CSIS liaison officer, Canadian Embassy, Washington, DC, 20 December 2006.

151 After obtaining his MD at Columbia and a MPh in public health from the Harvard Medical School, St John worked for the Surgeon General's Office and CDC, specializing in AIDS/HIV work. Interview with Dr Ronald St John, Manotick, Ontario, 23 January 2009.

152 Interview with Ronald St John. By the end of October 1994, when the plague epidemic was contained, fifty-two deaths were reported in Surat. *New York Times* 29 September 1994. See See Ron Barrett, "The 1994 Plague in Western India: Human Ecology and the Risks of Misattribution," in Anne Clunan et al., eds., *Terrorism, War, or Disease? Unraveling the Use of Biological Weapons* (Stanford, CA: Stanford University Press, 2008), 49–71.

153 Interview with Ronald St John.

154 At the beginning CEPR had only a handful of professionals, including Dr Frank Welsh, formerly of the Population and Public Health Branch, who became responsible for the biodefence training programs and

liaison with provincial and local health officials. Interview with Frank Welsh, Office of Emergency Preparedness, Planning and Training, Ottawa, 25 October 2002.

155 Interview with Dr Marc-André Beaulieu, Ottawa, 24 October 2002.

156 On the morning of 11 September 2001, the system had its first real test, when news of the terrorist assaults on American targets reached Ottawa. "I was in a meeting when the attack occurred," Beaulieu recalled, "but within two hours our Plan was off and running." Ibid.

157 Interview with Ronald St John.

158 Donald Avery, tour of CSCHAH/NML facility, 13 October 2010.

159 "Winnipeg's Fortress of Deadly Disease," CBC News, 13 May 2009. Roger Loftstedt, "Good and Bad Examples of Siting and Building Biosafety Level 4 Laboratories: A Study of Winnipeg, Galveston and Etobicoke," *Journal of Hazardous Materials* 93, no. 1 (1 July 2002): 46–66.

160 The level-4 laboratory had originally been scheduled to be built in the Etobicoke region of Toronto during the late 1970s, under the jurisdiction of the Ontario Ministry of Health, which spent $5.8 million on a feasibility study. However, because of local protests, the Ontario government cancelled the project on the grounds that this type of facility was a federal responsibility. The Centre is now just one of fifteen level-4 laboratories in the world. "Winnipeg's Fortress of Deadly Disease," CBC News, 13 May 2009.

161 Donald Avery, tour of CSCHAH/NML facility, 13 October 2010. The NML consists of six sections: Zoonotic Diseases and Special Pathofens; Bacteriology and Enterics; Viral Diseases; Prion Diseases; National HIV and Retrovirology; and Science, Technology and Core Services. Only about 4 per cent of the laboratory space is devoted to Containment Four (CL-4) work, with 35 per cent given to CL-3 and 61 per cent to CL-2. Of the 400 NML employees, about 100 are scientists; of these, about 25 are qualified to enter the CL-4 laboratory. Interview with Hank Krueger, Director Business Operations, and Dr Jim Strong, tour of NML, 13 October 2010.

162 Dr Francis (Frank) Plummer was born in Winnipeg on 2 December 1952. He studied medicine at the University of Manitoba, becoming an MD in 1976. During the early 1980s he began to focus his work on infectious diseases, and carried out research on HIV/AIDS in Nairobi, Kenya for seventeen years. Interview with Dr Frank Plummer, 13 October 2010.

163 Given its advanced research facilities, the NML was able to attract a group of outstanding researchers such as Harvey Artsob (Director, Zoontoic

Diseases and Special Pathogens Program), and Heinz Feldman, one of the world's authorities on Ebola, whose work in understanding the molecular biology and pathogenesis of the virus has attracted much scientific attention. Interview with Frank Plummer, 13 October 2010.

164 The National Microbiology Laboratory had a rather rocky beginning. On 23 June 1999, over 2,000 litres of waste water from the laboratory was inadvertently discharged into the Winnipeg sewer system, creating the possibility that dangerous pathogens had been released into the community. Fortunately, this did not occur, and the laboratory subsequently allayed local fears by creating a special Community Liaison Committee. Loftstedt, "Good and Bad Examples of Siting and Building Biosafety," 46–66. Mia Rabson and Dan Lett, "A Super-Lab to Fight Super-Bugs," *Winnipeg Free Press* 6 February 2010.

165 Miller et al., *Germs*, 295.

166 Ibid., 295–8.

167 Dando, *Preventing Biological Warfare*, 33.

168 The West Nile virus, closely related to the Japanese encephalitis virus, is a vector-carried zoonotic disease. In 1999 a major epidemic in New York City resulted in 20,000 reported cases and 770 deaths. Since that time it has spread throughout North America, peaking at 1,481 cases in Canada in 2003.

169 The AIDS/HIV pandemic has been the most serious. Approximately 24 million people have perished from this disease since 1981, with over half a million of these deaths occurring in the United States and approximately 14,800 in Canada. Sigall Bell, Courtney McMickens, and Kevin Selby, *AIDS* (Santa Barbara: Greenwood, 2011); Elisabeth Kubler-Ross, *AIDS: The Ultimate Challenge* (New York: Macmillan, 1987).

170 Problems of coordinating US biodefence and emergency response strategies were dramatically displayed in August 2005 when Hurricane Katrina hit New Orelans, killing over 1,500 people and disrupting the lives of thousands more. Ronald Daniels, Donald Ketti, and Howard Kunreuter, *On RisK and Disaster: Lessons from Hurricane Katrina* (Philadelphia: University of Pennsylvania Press, 2006); Lynn Goldman and Christina Coussens, eds.; *Environmental Public Health Impacts of Disasters: Hurricane Katrina, Workshop Survey* (Washington: National Academies Press, 2007).

171 Canada's most serious natural disaster of recent times was the ice storm of January 1998, which knocked out power to a million homes. H. James Miskel, *Disaster Response and Homeland Security. What Works, What Doesn't* (Westport, CT: Praeger Security International, 2006).

7 Biodefence after 9/11: Old Problems and New Directions

1 *Globalization, Biosecurity, and the Future of the Life Sciences* (Washington: National Academies Press, 2006), 2.
2 S.J. Armour, B.V. Kournikakis, E. Yee, and K. Brooks, "Hogtown Disaster: A BW Terrorist Attack on a Major Canadian City," DRDC Suffield, Technical Report TR 2004-265, December 2004.
3 Ibid., The original report was first presented at a meeting of the Scientific and Technical Intelligence Group in Washington, DC, 22–4 February 1999.
4 A.F. Kaufmann et al., "The Economic Impact of a Bioterrorist Attack: Are Prevention and Post Attack Intervention Programs Justifiable?" *Emerging Infectious Diseases* (1997): 3, 83–94.
5 Interview with Dr Ron St John, 23 January 2009, Manotick, Ontario. Ron St John and Brian Finlay, "Bioterrorism in Canada: An Economic Assessment of Prevention and Post Attack Response," *Canadian Journal of Infectious Diseases* 12, no. 5 (2001): 275–84.
6 On 1 October, Bob Stevens, a Florida-based photo-journalist, fell ill after opening one of the first anthrax letters. He died on 5 October of pulmonary anthrax. Leonard Cole, *The Anthrax Letters: A Medical Detective Story* (New York: Joseph Henry Press, 2003); W. Seth Carus, *Bioterrorism and Biocrime: The Illicit Use of Biological Agents since 1900* (Amsterdam: Fredonia Books, 2002).
7 Gregory Koblentz, *Living Weapons: Biological Warfare and International Security* (Ithaca, NY: Cornell University Press, 2010), 203.
8 It is estimated that in October and November 2001 more than 750 anthrax hoax letters were circulated in the United States. "Tracing Anthrax Hoaxes and Attacks," CNS James Martin Center for Non-Proliferation Studies, 20 May 2002, http://cns.miis.edu/pubs/index.htm.
9 J.A. Jenigan et al., "Bioterrorism Related Inhalational Anthrax: The First Ten Cases Reported in the United States," *Emerging Infectious Diseases* (2001), 7, 933–944; T.V. Inglesby, D.A. Henderson et al., "Anthrax as Biological Weapon 2002: Updated Recommendations for Management," JAMA (2002), 287, 2236–2252; C.J. Peter and D.M. Hartley, "Anthrax Infection and Lethal Human Infection," *The Lancet* (23 February, 2002), 710–11; F.R. Sidell et al., (eds.), *Textbook of Military Medicine: Medical Aspects of Chemical and Biological Warfare* (Washington DC: Office of the Surgeon General, Department of the Army, 1997).
10 The standard US military doctrine emphasized that between 8,000 and 10,000 spores were typically required to cause pulmonary anthrax. In the

2001 incident, however, some of the victims died with very low exposure, prompting some scientists to suggest that even one spore could cause inhalation anthrax." United States Congress House of Representatives Committee on Homeland Security, Subcommittee on Prevention of Nuclear and Biological Attack, *Engineering Bio-Terror Agents: Lessons from the Offensive U.S. and Russian Biological Weapons Programs* (Washington: U.S. Government Printing Office, 2006), 31–4.

11 For example, U.S. scientific groups were involved extensively in the various hearings involving the Bioterrorism Preparedness Act of 2000 since it provided extensive guidelines for the possession, use, storage, and transportation of category A pathogens. Interview with Ronald Atlas (president of ASM in 2001), Washington, DC, October 2005.

12 The author carried out an extensive search of the House of Commons debates between September 2001 and March 2002 on the subject of bioterrorism.

13 There was particular interest in the claim by US Senator Patrick Leahy that the anthrax letter he received was potent enough "to kill 100,000 people." *Toronto Star* 26 November 2001. See also Peter Calamai, "Terrorism in a Test Tube: Canada's Vaccine Supplies Woefully Inadequate to Counter a Biological Attack," *Toronto Star* 21 September 2001.

14 CRTI was launched in May 2002 to deal with potential CBRN terrorist attacks by assisting in the development of laboratory clusters and new defensive technology. In March 2006 it became a larger organization as a result of a cooperative arrangement between the Department of National Defence and Public Safety Canada. Plans were also made to establish close connections with the US Department of Homeland Security through the CA-US Public Security Technical Program. Centre for Security Science, *News* 1, no. 2 (May 2007).

15 Dr James Young, Chief Coroner of Ontario, was appointed Assistant Deputy Minister of Public Safety in 2000. He was subsequently Commissioner of Emergency Management. Interview Dr James Young, Toronto, 11 August 2005. See also Ontario, Backgrounder, "Ontario Response to September 11, 2001" (Toronto: Ministry of Public Safety and Security, 5 September 2002).

16 *Securing an Open Society: Canada's National Security Policy* (Ottawa, March 2004), 10.

17 The 15 October evacuation of the Hart Building after the discovery of the contaminated letter sent to Senator Tom Daschle was followed by the closure of many other government buildings in Washington and scores of mail sorting facilities across the country. R. William Johnston, *Bioterror: Anthrax, Influenza and the Future of Public Health Security* (Westport, CT: Praeger International, 2008), 5–30.

18 For many years, the FBI concentrated on Steven Hatfield, a former Detrick microbiologist, who appeared to have the motive, opportunity, and skills necessary for the task. In turn, Hatfield successfully sued the FBI for harassment. See Koblenz, *Living Weapons,* 200–12; "FBI Anthrax Investigation under Scientific Review," *ScienceInsider* (Science AAS), 6 May 2009; "Was Ivins the Anthrax Killer? *New Scientist* 1 August 2008.
19 Koblenz, *Living Weapons,* 211.
20 There were two cases in Florida (both inhalation, with one death), eight in New York (one fatal inhalation case, seven cutaneous), seven in New Jersey (five inhalation and two cutaneous), five in Washington, DC (inhalation, both fatal), and one in Connecticut (inhalation, fatal).
21 By 2005 the FBI Amerithrax team of thirty-one experts had carried out over 9,000 interviews. Jeanne Guillemin, *American Anthrax: Fear, Crime and the Investigation of the Nation's Deadliest Bioterror Attack* (New York: Henry Hold and Company, 2010), 58, 172, 214.
22 Ibid., 59, 73; David Willman, *Mirage Man: Bruce Ivins, the Anthrax Attack, and America's Rush to War (*New York: Bantam, 2008), 149.
23 Cited in Peter Andreas and Thomas Biersteker, eds., *The Rebordering of North America: Integration and Exclusion in a New Security Context* (New York: Routledge, 2003), 14. Richard Holbrook, former US ambassador to the United States, and Senator Hillary Rodham Clinton were other prominent critics of security at the northern border.
24 The allocations consisted of $2.1 billion for airport security, $1.2 billion for border security, $1 billion for screening entrants to Canada, $396 million for emergency preparedness, and $513 million to chemical, biological, and nuclear threats. By 2007, the amount of money allocated to counterterrorism in Canada reached $10 billion. Kent Roach, *September 11, Consequences for Canada* (Montreal and Kingston: McGill-Queen's University Press, 2003), 190.
25 Canada, Report of the Standing Senate Committee on National Security and Defence, 1st Session, 38th Parliament, December 2004; Roach, *September 11,* 160–202.
26 In February 1995 President Clinton and Prime Minister Chrétien had signed the Shared Border Accord, which was further enhanced by the 1999 Canada-US Security Partnership Forum. Victor Konrad and Heather Nichol, *Passports for All: Canadian-American Public Policy* (Orono: The University of Maine, May 2008), Occasional Paper #74.
27 Among the agreement's thirty-three points were a number of references to the problems of CBRN terrorism, which was an essential priority for the special joint border investigative (IBET) teams. Peter Andreas and Thomas

Biersteker, *The Rebordering of North America: Integration and Exclusion in a New Security Context (*New York: Routledge, 2003).

28 In February 2006 Ottawa created a special branch of the military, Canada Command, similar in function to NORTHCOM. See www.forces.gc.ca/site.

29 CEPR expanded from 34 to 190 positions between 2000 and 2003, with an accompanying surge in budget allocations. Interview with Ron St John, 23 January 2009.

30 In October 2002, Health Minister Alan Rock attempted to obtain cheaper supplies of the antibiotic cirpflaxin by challenging the patent rights of Bayer A.G., the international pharmaceutical giant. "Patent Wars Looming over Drug for Anthrax," *Globe and Mail* 19 October 2001; House of Commons, *Debates,* 23 October 2001, 15–16.

31 This project also received $500,000 from the Gates Foundation. *Global Security Network,* 17 November 2001; "GHSI Background," Global Health Security Initiative, www.ghsi/ca/enlgish/background/asp.

32 "Canadian Institutes of Health Research (CIHR) and Health Canada Conference on Biological Terrorism: Canadian Research Agenda, January 18–20, 2002, Toronto, Ontario" (unpublished report). The session on viral agents attracted considerable attention since it featured Peter Jahrling of USAMIIRID, D. Butler of St Louis University, and Grant McFadden of the University of Western Ontario, three of the world's leading experts on smallpox.

33 Ibid. Cooperation between Canadian and American pharmaceutical companies was also encouraged, particularly since the Toronto-based Aventis Pasteur still had a stockpile of freeze-dried smallpox vaccine.

34 C-36 passed the House of Commons on 29 November 2001 by a margin of 189 to 47 votes. Roach, *September 11,* 21.

35 Cited in Roach, *September 11,* 22.

36 Ibid., 33. The Patriot Act, or, "Uniting and Strengthening America by Providing Appropriate Tools Required to Intercept and Obstruct Terrorist Acts,' passed the House of Representatives 356 to 66, and the Senate 98 to 1. Nancy Chang, *Silencing Political Dissent: How Post-September 11 Antiterrorism Measures Threaten Our Civil Liberties* (New York: Seven Stories Press, 2002); Brian Hook et al., "The USA Patriot Act and Information Sharing between the Intelligence and Law Enforcement Communities," in *Homeland Security and Terrorism: Readings and Interpretations,* ed. Russell Howard et al. (New York: McGraw-Hill, 2006), 384–99.

37 Government of Canada, Department of Foreign Affairs and International Trade, press release, "The BTWCIA: What It Will and Won't Do," May 2004.

38 Donald Avery, "The North American Plan for Avian and Pandemic Influenza: A Case Study of Regional Health Security in the 21st Century," *Global Health Governance* 3, no. 2 (Spring 2010): 1–26.

39 One of the reasons why the United States was so insistent on including Mexico in the GHSI was the creation of the United States-Mexico Border Health Commission in 2000 to coordinate binational actions that would improve the quality of life on the border. Website for the United States-Mexican Border Health Commission.

40 Congressional Research Service (Library of Congress), *High Containment Laboratories: National Strategy Is Needed* (Report GAO-09-574, September 2009), 19–22.

41 Ministerial Statement, "Health Minister Launches Initiatives to Improve Health Security Globally," Mexico City, December 2002, www.ghsi.ca/english/statementmexicocityDec2002.asp (accessed 16 November 2008); interview with Dr Frank Plummer, 13 October 2010, NML.

42 Both the WHO and the European Commission also participated in the event. *Exercise Global Mercury: Post-Exercise Report* (Ottawa, 2002); Fourth Ministerial Meeting on Health Security and Bioterrorism, Berlin, November 2003.

43 Sixth Ministerial Meeting on the Global Security Initiative, Rome, November 2005.

44 Jonathan Tucker, "Updating the International Health Regulations," *Biosecurity and Bioterrorism*, 3, no. 2 (2005): 338–47.

45 Department of Justice official press release, 8 October 2002, "Justice Department, State Department to Conduct Exercises Combating Weapons of Mass Destruction."

46 US Department of State, "Fact Sheet," Office of Counterterrorism: TOPOFF 24 July 2002.

47 Department of Homeland Security, *Top Officials (TOPOFF) Exercise Series: TOPOFF2. After Action Summary Report for Public Release* (19 December 2003). www.dhs.gov/xlibraryassets/T2_Report_Final_Public.doc.

48 Use of a Virtual News Network helped make the disaster scenario appear more realistic, while keeping the 8,500 participants fully informed about the progression of the crisis. *Exercise Series: TOPOFF2.*

49 According to Mark Mes of DFAIT, who participated in the Canadian sessions, once the foreign trawler entered Canada's 200-mile limit, Foreign Affairs Minister Bill Graham argued strongly that the ship should be sunk, claiming "that if we don't do it, the United States Air Force will." Interview with Mark Mes, 11 June 2003.

50 The NLE series is somewhat different from its TOPOFF predecessor because

of its emphasis on prevention rather than response and recovery. It also adopts a much broader approach in preparing for catastrophic crises, including both CBW attacks and natural disasters.

51 TOPOFF 3 included a bioweapons attack on two regions of New Jersey and a CW attack on New London, Connecticut. In addition, the UK carried out its own exercise (Atlantic Blue).

52 Interview with Ron St John, 23 January 2009.

53 Many of these issues were discussed at the Canadian Conference on Counter-Terrorism and Public Health (29 October–1 November 2003), jointly sponsored by CEPR and the Canadian Public Health Association. The author presented a paper in the session entitled "Understanding Terrorism – What Is Needed to Prevent It."

54 *Eurosurveillance Weekly*, 1997–2005: European Commission, www.Eurosurveillance.org/Default/.

55 At this stage, biodefence research expenditure in the EU was 1.9 per cent of GDP, while in the US it was 2.6 per cent. *The Lancet* 733 (7 September 2002).

56 *Toronto Star* 14 February 2003; *Global Security Newswire* 8 November 2002.

57 Interview with John Mattisussi, British Nuclear Section, UK Delegation, NATO headquarters (Brussels), 26 April 2002.

58 Among NATO countries France was the least interested in having the alliance become involved with biodefence matters, on the grounds that these were domestic concerns. Interview with Stephen Orosz, Director CEPR, NATO headquarters, 27 April 2002.

59 *Public Health Response to Biological and Chemical Weapons: WHO Guidance, Second Edition* (Geneva: WHO, November 2001), Executive Summary. In 2000, the WHO reissued its 1970 report on the *Public Health Responses to Biological and Chemical Weapons*, which provided new insights into how molecular biology and genetic engineering could make biological weapons even more formidable.

60 Ibid. This resolution was based on the 17 January 2002 report of the WHO Executive Board.

61 Jonathan Tucker, "Updating the International Health Regulations," *Biosecurity and Bioterrorim* 3, no. 4 (2005): 346. On 25 May 2005, all 192 members of the WHO approved this sweeping expansion of the original 1951 IHR.

62 Kenneth D. Ward, "The BWC Protocol: Mandate for Failure," *The Non-Proliferation Review* 11, no. 2 (Summer 2004): 260–72; Malcolm Dando, *Preventing Biological Warfare: The Failure of American Leadership* (New York: Palgrave, 2002), 163.

63 It was reported that the Japanese representative was so outraged by Bolton's actions that he denounced him as "a man without honour." Interview with Trevor Smith, DFAIT arms control officer, Pearson Building, 10 February 2002.
64 Scholars associated with the Bradford Peace Studies program such as Malcolm Dando and Graham Pearson prepared a number of reports on attempts to improve the Biological Weapons Convention. See www.brad.ac-uk/ssis/peace-studies/research.
65 One of the more important of these meetings was the Tokyo Seminar in July 2002, which included a group of non-government BW specialists from the United Kingdom, Australia, France, Germany, Russia, South Africa, and the United States. The author represented Canada. Foreign Office of Japan, Special Meeting to Prepare for the Review Conference BWC.
66 The Bush administration was not prepared to support the findings of the Blix Commission, established in 2003 as an independent non-governmental organization, which submitted its report, *Weapons of Terror*, in 2006. See also Jonathan Tucker, "Preventing the Misuse of Pathogens: The Need for Global Biosecurity Standards," *Arms Control Today* (June 2003).
67 On 18 May 2002 the Fifty-Fifth World Assembly (WHA55.16) passed the resolution "Global public health response to natural occurrence, accidental release or deliberate use of biological and chemical agents or radio-nuclear material that affect health."
68 Jez Littlewood, *Managing the Biological Weapons Problem: From the Individual to the International*, Report no. 14 (Stockholm: Weapons of Mass Destruction Commission, 2005).
69 Cited in the webpage of the Global Partnership Program, Department of Foreign Affairs, www.dfait-maeci.gc.ca/foreign. Considerable progress was achieved in decommissioning Soviet-era nuclear submarines, assisting in the destruction of Russia's vast stockpiles of chemical weapons, and improving nuclear and radiological safety.
70 In 2004, Canada agreed to provide $18 million to the Moscow-based International Science and Technology Centre (ISTC). By 2005, thirty-eight projects were being funded, about half in the biosciences.
71 Donald Avery, "Reconciling the Goals of the Biological and Toxin Weapons Convention and the Global Partnership Program: Confidence Building Measures, Biosecurity and Scientific Codes of Conduct," Rome, 26–7 April 2005, Advancing International Cooperation on Bio-Initiatives in Russia and the CIS, organized by the Russian American Nuclear Security Advisory Council.
72 The major recipients for this funding were HSS ($31.7 billion), DHS ($9.01 billion), and DoD ($5.5 billion). Crystal Franco, "Billions for

Biodefence: Federal Agency Biodefence Funding, FY2008–FY2009," *Biosecurity and Bioterrorism: Biodefense Strategy, Practice and Science* 6, no. 2 (2008): 131–45.

73 Michael Mair, Beth Maldin, and Brad Smith, "Passage of S.3678: The Pandemic and All-Hazards Preparedness Act," *Biosecurity and Bioterrorism* 5, no. 2 (2007): 72–4.

74 Barry Kellman, *Bioviolence: Preventing Biological Terror and Crime* (New York: Cambridge University Press, 2007), 50, 137.

75 Ronald Atlas and Malcolm Dando, "The Dual-Use Dilemma for the Life Sciences: Perspectives, Conundrums, and Global Solutions," *Biosecurity and Bioterrorism* 4, no. 3 (2006): 276–86.

76 David Fidler and Lawrence Gostin, *Biosecurity in the Global Age: Biological Weapons, Public Health and the Rule of Law* (Stanford: Stanford University Press, 2008), 76–7. "From Red Peril to Red Tape," editorial, *Nature* June 2003.

77 National Academy of Sciences, Committee on Research Standards and Practices to Prevent the Destructive Application of Biotechnology, Gerald R. Fink Chair, *Biotechnology Research in an Age of Terrorism: Confronting the Dual Use Dilemma* (Washington: National Academies Press, 2003).

78 Constraints on certain types of research and the possession of type A pathogens had already been established by the October 2001 Patriot Act (Section 817) and the June 2002 Public Health Security and Bioterrorism Preparedness Act (Title II-Enhanced Controls for Dangerous Biological agents and Toxins).

79 Minutes of the First Meeting of the National Science Board for Biosecurity, NIH offices, Bethesda, Maryland, 30 June–1 July 2005.

80 Ibid., 89.

81 Ibid. On his USAMIRID experiences, Franz emphasized that only about 5 per cent of the scientific staff were allowed into the BL-4 laboratories and that it was a supervised, gradual process. On the other hand, he admitted that unlike nuclear or chemical plants it was difficult to prevent the removal of dangerous material (191, 215–17).

82 Ibid., Dr Michael Osterholm, Director, Center for Infectious Disease Research and Policy, University of Minnesota (207).

83 National Academy of Sciences, *Globalization, Biosecurity and the Future of the Life Sciences* (Washington, DC: NAC Press, 2006).

84 Ibid. The report criticized Washington's obsession with security, claiming that it impeded scientific advancement, particularly since the researchers in the life sciences did not have the same culture of secrecy as nuclear physicists (35).

85 Some of the eight major scientific developments cited were: RNA interferences-reverse genetic engineering; DNA synethesis-and bio-prospecting; fusion of enabling technologies; molecular biology and bioinformatics; gene therapy and genomic medicine; and synthetic biology-CDNA. Ibid.

86 National Science Advisory Board for Biosecurity, *Addressing Biosecurity Concerns Related to the Synthesis of Select Agents* (October 2006), 13.

87 Ibid., 2.

88 Ibid. Viral genomes are routinely reconstructed using recombinant DNA methods and reverse engineering techniques, having access to a natural source of the virus of interest.

89 There was also a strong appeal for a reassessment of biosafety procedures on the grounds that many of the biotechnology companies who were involved in synthetic genomics had not adopted the necessary biosafety precautions. Ibid., 4, 12, 13.

90 Ibid., 9–11; Jan van Aken, "When Risk Outweighs Benefit: Dual-Use Research Needs a Scientifically Sound Risk-Benefit Analysis and Legally Binding Biosecurity Measures," *EMBRO Reports* (European Molecular Biology Organization), 7 July 2006.

91 In October 2007 the NSABB's Working Group on International Collaboration hosted a meeting in Washington of molecular biologists and virologists from a variety of countries. No Canadian life scientists were present.

92 Grant McFadden, "Dual Use Biotechnology: A Canadian Survey," unpublished study (March 2005).

93 Ibid., 24.

94 Ibid., 23, 25.

95 *Report on the National Forum on Dual Use Biotechnology*, Ottawa, 1–2 March 2006 (unpublished report).

96 Donald Avery, "Notes of the Working Group Session," 28 March 2006.

97 Among US laws, the most relevant was the Public Health Security and Bioterrorism Preparedness and Response Act of 2002. In the UK the most pertinent legislative measure was the revised Anti-Terrorism, Crime and Security Act of 2001.

98 House of Commons *Debates*, 40th Parliament, 2nd Session, Second Reading of Bill C-11 (8 February 2009). Statement by Mr Colin Carrie, Parliamentary Secretary to the Minister of Health, 1110.

99 According to the CDC, biosafety level-4 laboratories contained pathogens that posed a high individual risk of life-threatening disease and for which no treatment was available. The Canadian category four group included nineteen viruses, while category five, labelled "Prohibited Human

Pathogens and Toxins," was left deliberately vague. Bill C-11: Human Pathogens and Toxins Act.

100 Laboratory safety was institutionalized in the United States in 1984 with the first edition of the *Biosafety in Microbiological and Biomedical Laboratories*, published by the CDC. According to the CDC, biosafety level-4 laboratories contained pathogens that posed a high individual risk of life-threat ening disease and for which no treatment was available.

101 During the debate over C-11, Pat Martin, the NDP member from Winnipeg Centre, reminded the House that the Winnipeg National Laboratory had a pattern of unfortunate incidents, ranging from the 1999 accident involving polluted water to the 2009 crash of a FedEx truck carrying laboratory samples of "anthrax, the Newcastle disease virus and a number of other serious toxins." House of Commons, *Debates*, 28 February 2009, 1120.

102 House of Commons Committees-HESA (40-2), Minutes of Proceedings, Meeting No. 8, 5 March 2009.

103 Ibid., Meeting No. 9, 10 March 2009.

104 Ibid., 10 March 10, 2009, Testimony of Dr Vivek Goel, President and Chief Executive Officer, Ontario Agency for Health Protection and Promotion.

105 "The Human Pathogens and Toxins Act-Bill C-11," *Canadian Society for Medical Laboratory Science* 16, no. 3 (Fall 2009): 19–20.

106 *Toronto Star* 9 July 2005. The terrorist attack on London's transit system left fifty-six dead and thousands injured.

107 The eight accused, all young men of Pakistani background, devised an innovative attack plan of smuggling liquid explosives in plastic sport-drink bottles, which would then be armed with detonators once the plane was in the air. *Toronto Star* 3 April 2008.

108 *Toronto Star* 27 December 2009.

109 Members of the so-called Toronto 18 were arrested in July 2006 under the federal Anti-Terrorism Act of 2001. *Globe and Mail* 6 June 2006; Canadian Press 29 December 2009.

110 *Securing an Open Society: Canada's National Security Policy* (Ottawa: Queen's Printer, April 2004), vii.

111 Ibid., 41–2. *Report of the Event Relating to Maher Arar. Commission of Inquiry into the Actions of Canadian Officials in Relation to Maher Arar*, 3 vols. (Justice Dennis O'Connor) (Ottawa, 2006).

112 Between 1992 and 2007 twenty people were charged under Security Certificate legislation, most for national security reasons. *Macleans* 7 March 2007; *Toronto Star* 23 October 2007.

113 Canadian Press 2 February 2009; *Toronto Star* 7 March 2009.

114 Kerry Pither, *Dark Days: The Story of Four Canadians Tortured in the Name of Fighting Terror* (Toronto: Viking Canada, 2008), 62; Robert Diab, *Guantanamo North: Terrorism and the Administration of Justice in Canada* (Halifax: Fernwood, 2008), 9–12, 92–107.

115 Arar Commission, *Analysis and Recommendations*, 13–15.

116 A number of highly placed government officials were badly tarnished by the Arar Inquiry, including RCMP Commissioner Guiliano Zaccardelli (forced to resign in December 2006) and CSIS Director General Jack Hooper (retired April 2007). Pither, *Dark Days*, 440–6.

117 The December 1999 apprehension of Ressam provoked some US commentators to call Canada "the Club Med for terrorists ... and the weak link in North American security." Pither, *Dark Days*, 34; *Toronto Star* 28 July 2005.

118 *The 9/11 Commission Report: Final Report of the National Commission on Terrorist Attacks upon the United States* (New York: W.W. Norton, 2006).

119 Pither, *Dark Days*, 41.

120 In March 2007 the British Home Office released statistics showing that 1,126 arrests had been made under the Terrorism Act 2000; 117 were subsequently charged with terrorist-related activities. Steve Hewitt, *British War on Terror: Terrorism and Counter-Terrorism on the Home Front since 9/11* (London: Continuum, 2008), xv–xxiv; Jon Coaffee, *Terrorism, Risk and the Global City: Towards Urban Resilience* (London: Ashgate, 2009), 248, 303. In 1995, the 9/11 mastermind Sheikh Mohammed, had devised a plan to bring down eleven airlines over the Pacific. *London Times* 10 August 2006.

121 *Toronto Star* 2 July 2010; *Toronto Star* 12 January 2010.

122 *Toronto Star* 2 June 2006, 26 March 2008, 27 September 2008.

123 *Toronto Star*, 2 April 2008.

124 *Toronto Star* 24 April 2008; *Globe and Mail* 6 June 2006; *Toronto Star* 2 June 2009. There were also reports that Osama bin Laden had named Canada an enemy of Islam.

125 Tanya Primiani and Christopher Sands, "Terrorists in Toronto: Is Canada Secure? Are We?" Center for Strategic and International Studies (Washington, DC), 20 June 2006.

126 Canadian Press 29 December 2009.

127 *Toronto Star* 23 June 2009.

128 "The Toronto 18," *Toronto Star* 2 July 2010.

129 *Toronto Star* 24 June 2010.

130 *Canadian Security Guidebooks, 2005 Edition: An Update of Security Problems in Search of Solutions*, a Report of the Standing Senate Committee on National

Security and Defence, 1st session, 38th Parliament, December 2004 (Chairman Colin Kenny).

131 Ibid., 64.

132 2009 November Report of the Auditor General of Canada, Chapter 7, "Emergency Management-Public Safety Canada," 6–7, www.oag-bvg.gc.ca/.

133 Ibid., 13, 9, 14.

134 *Emergency Preparedness in Canada*, Report of the Standing Senate Committee on National Security and Defence, vol. 1, Second Session, Thirty-ninth Parliament, 2008, 6.

135 The central depot of NESS is located in Ottawa, but 1,300 pre-positioned caches are located across Canada, under the joint management of the provinces and the federal government. Auditor General's 2009 Report, 195–207.

136 Ibid., 113, 213–19.

137 Fidler and Gostin, *Biosecurity in the Global Age*, 2.

138 Almost all SARS cases occurred in the Greater Toronto Area (GTA). *Learning from SARS: Renewal of Public Health in Canada*, a report of the National Advisory Committee on SARS and Public Health (Nayor Committee), October 2003, 37.

139 SARS originated in southern China, spread to Hong Kong, and eventually reached twenty-seven countries: 8,500 were diagnosed with the virus, resulting in over 900 deaths. *Learning from SARS*, 1.

140 The first SARS outbreak in Toronto lasted from 13 March to 17 May, when Ontario premier Ernie Eves lifted the emergency status; the second phase began on 23 May, with a series of cases at the North York General Hospital, and lasted until 30 June. *Learning from Sars*, 23–42. Donald Low, "SARS: Lessons from Toronto, in *Learning from Sars: Preparing for the Next Disease Outbreak: Workshop Summary*, Institute of Medicine of the National Academies (Washington: National Academies Press, 2004), 67.

141 James Young, "My Experience with SARS," in *SARS in Context: Memory, History, Policy*, ed. Jacalyn Duffin and Arthur Sweetman (Montreal and Kingston: McGill-Queen's University Press, 2006), 29.

142 Ibid., 37.

143 Although SARS patients were treated at twenty hospitals in the GTA, most of the cases were located at five major centres: the Scarborough Grace / West Park, the North York General, Sunnybrook and Women's College Health Sciences Centre, Mount Sinai, and the Toronto Western site of the University Health Network. Chad Marley et al., "SARS and Its Impact on Current and Future Emergency Department Operations," *Journal of Emergency Medicine*, 26, no. 4 (2004): 415–20. *Learning from SARS*, 28.

144 Matthew Muller et al., "Clinical Trials and Novel Pathogens: Lessons Learned from SARS," *Emerging Infectious Diseases* 10, no. 3 (March 2004): 389, 392; Laura Hawryluck et al., "SARS Control and Psychological Effects of Quarantine, Toronto, Canada," *Emerging Infectious Diseases* 10, no. 7 (July 2004): 1206–12; Zoutman, "Remembering SARS," 30–4.

145 The Naylor Report (*Learning from SARS*); *Spring of Fear: The SARS Commission Final Report, the Honourable Mr. Justice Archie Campbell* (Toronto: SARS Commission, 2006); and *For the Public's Health: A Plan for Action: Final Report of the Ontario Expert Panel on SARS and Infectious Disease Control* (Toronto: Ontario Ministry of Health and Long Term Care, 2004).

146 *The Interim Report of the SARS Commission, Issued by the Honourable Mr. Justice Archie Campbell* (Toronto: SARS Commission, 15 April 2004). Many Toronto health care workers experienced severe psychological stress, with about 30 per cent reporting some emotional trauma. See Robert Maunder et al., "The Experience of the 2003 SARS Outbreak as a Traumatic Stress among Front-Line Health-Care Workers in Toronto: Lessons Learned," in *SARS: A Case Study in Emerging Infections*, ed. Angela McLean et al. (Oxford: Oxford University Press, 2005), 96–106.

147 Interview with Dr James Young, Toronto, 11 August 2005.

148 *For the Public's Health* (Walker Report), Executive Summary, 15.

149 *Learning from SARS*, 220; *Toronto Star* 1 April 2004.

150 Young, "My Experience with SARS," 19. Interview with Frank Welsh, 12 July 2003, 8 June 2004. Interview with Howard Njoo, Director General, Centre for Communicable Disease and Infection Control, PHAC, Ottawa, 22 January 2009.

151 Shelley Hearne et al., "SARS and Its Implications for U.S. Public Health Policy: 'We've Been Lucky,'" *Biosecurity and Bioterrorism* 2, no. 2 (2004): 131.

152 "Seeking Security: Pathogens, Open Access and Genome Databases," Committee on Genomics Database for Bioterrorism Threat Agency, National Research Council (Washington: National Academy of Sciences, 2004).

153 By 2007 the H5N1 Eurasian strain of the influenza A virus had infected birds in over fifty-nine countries, resulting in the deaths, through illness and culling, of 240 million birds. Mike Davis, *The Monster at Our Door: The Global Threat of Avian Flu* (New York: Henry Holt, 2006).

154 The countries most affected by H5N1 were Indonesia (141), Vietnam (109), Egypt (58), China (38), Thailand (25), and Turkey (12).

155 H.D. Klenk et al., *Avian Influenza* (Freiburg: Krager, 2008).

156 Health Canada, "Preparing for an Influenza Pandemic" (September 2006; revised 2009), www.hc-sc-gc-ca/hc-ps/ed-ud/prepar/flu-pandem/index-eng.php.

157 David Rosner and Gerald Markowitz, *Are We Ready? Public Health since 9/11* (Los Angeles: University of California Press, 2006).

158 Joint Statement by President George W. Bush, Prime Minister Stephen Harper of Canada and President Vicente Fox of Mexico – the Security and Prosperity Partnership of North America, www.spp-psp-gc-ca.

159 Ibid.

160 The six authors of this report included Dr Julio Frenck Mora (Secretario de Salud) and Dr Pablo Kuri Morales (Director General de la Direccion General de Epidemiologia).

161 President Bush's plan called for Congress to authorize $7.1 billion, which would be used for increased preparedness in vaccine and drug production, disease surveillance at home and abroad, and assistance to state and emergency response operations. *New York Times* 1 November 2005.

162 Donald Avery, "The North American Plan for Avian and Pandemic Influenza: A Case Study of Regional Health Security in the 21st Century," *Global Health Governance* 3, no. 2 (Spring 2010).

163 *North American Plan for Avian and Pandemic Influenza* (Ottawa / Washington, August 2007), chapter 4: Pandemic Influenza), 21.

164 "Joint Statement by Ministers Responsible for the Security and Prosperity, 28 February 2008," www.spp.gov/news/news_02282008.asp.

165 "Concern over Flu Pandemic Justified," Dr Margaret Chan, Director-General of the World Health Organization, Address to Sixty-second World Health Assembly, Geneva, Switzerland, 18 May 2009.

166 Associated Press, "WHO Chief Does not Raise Swine Flu Alert Level," 18 May 2009, www.google.com/hostednews/ap/article/AL.

167 *Nature News* 30 April 2009. Dr Yi Guan, a Hong Kong-based authority on influenza, was among those considering the remote chances of a "super-bug."

168 "Flu Pandemic Underway," *Nature News* 11 June 2009,

169 "Vaccine Surveillance Report, 28 January 2010," Public Health Agency of Canada, www.phac-aspc/gc-ca/publicat/ccdr-rmtc/10vol36/acs-6/index-eng.php.

170 Susan Sherman et al., "Emergency Use Authority and 2009 H1N1 Influenza," *Biosecurity and Bioterrorism* 7, no. 3 (2009): 245–50; Sandra Quinn et al., "Public Willingness to Take a Vaccine or Drug under Emergency Use Authorization during the 2009 H1N1 Pandemic," idem, 275–91.

171 Sando Cinti, "Bacteria Pneumonias During an Influenza Pandemic: How Will We Allocate Antibiotics?" ibid., 311–16.

172 *The Lancet* 374 (10 October 2009), 1215.

173 *Nature News*, 19 June 2009. Interview with Dr Yan Li, NML, 13 October 2010. The NML also sent a team of seven virologists and epidemiologists

to Mexico City for two months to assist Mexican health officials in dealing with the pandemic. "Exclusive: Interview with Head of Mexico's Top Swine Flu Lab," *Science* 1 May 2009.

174 Joint Statement by North American leaders, 10 August 2009, Guadaljara, Mexico, Canada News Centre. Prime Minister Harper also issued his own assessment of the NAPAPI, praising the "shared and effective response thus far to H1N1 – a cross border threat to all of us." Statement by the Prime Minister of Canada, 10 August 2009, www.pm.gc.ca.

175 "Special Section: Virus of the Year," *Science* 18 December 2009, 1607.

176 *Prevention of WMD Proliferation: Report Card*, US Commission on the Prevention of Mass Destruction Proliferation and Terrorism (Washington, January 2010), www.pharmathene.com/WMD_report_card.pdf.

177 Joint Statement by Senators Lieberman and Collins, in support of Bill S.1649, *Congressional Record*, 8 September 2009, S9135–39.

178 Washington, DC: National Intelligence Council, November 2008. The NCI produced a similar study, *Strategic Implications of Global Health* (Washington, December 2008).

179 Another important report was the *Report of the Defense Science Board Task Force on Department of Defense Biological Safety and Security Programs* (Ottawa, May 2009).

180 The US continued to argue that access to the CDC collection of variola was "essential" for developing advanced therapeutics and vaccines. "Highlights and Happenings," *Biosecurity and Bioterrorism* 7, no. 3 (2009): 235–6.

181 *Globe and Mail* 14 January 2010.

182 Interview with Frank Plummer and Hank Krueger, 13 October 2010.

183 Ibid.

184 Donald Avery to Clement Laforce, Deputy Director-General, DRDC Suffield, 12 March 2012.

185 Marie-Helene Brisson, Public Affairs Officer, Defence Research and Development Canada to Donald Avery, 3 May 2012.

186 National Academy of Sciences, *Biotechnology Research in an Age of Terrorism.* See also National Academy of Sciences, *Globalization, Biosecurity and the Future of the Life Sciences.*

187 Malcolm Dando, *Bioterrorsm: What Is the Real Threat?* Carnegie Endowment for Science and Technology, Report no. 3 (New York, March 2005).

188 Kellman, *Bioviolence*, 16. At a special session on 15 July 2005 representatives of Congress and the biodefence community endorsed the view that, because of advances in biotechnology and the widespread dissemination of dangerous information, international terrorist organizations could

develop operational bioweapons. Subcommittee on Prevention of Nuclear and Biological Attack, *Engineering Bio-Terror Agents,* 1-53.

189 Brian Michael Jenkins, *Unconquerable Nation: Knowing the Enemy, Strengthening Ourselves* (Santa Monica: Rand Corporation, 2006), 1.

190 Milton Leitenberg, *Assessing the Biological Weapon and Bioterrorism Threat,* (Strategic Studies Institute, U.S. Army War College, December 2005), 43.

191 Koblentz, *Living Weapons,* 214–15.

192 Interview with Frank Plummer, Scientific Director NML, 13 October 2010.

193 Interview with Dr Ron St John, former director general CEPR, 23 January 2009.

Conclusion

1 Queen's University Archives, G.H. Ettinger Collection 5116, box 1, file 14. Dr Wilder Penfield, Montreal Neurological Institute, to Dr Harold Ettinger, Faculty of Medicine, Queen's University, 5 July 1974.

2 Major General J.V. Young, Master-General of the Ordnance, DND, to Minister of Defence R.L. Ralston, June [n.d.] 1943, Cited in Donald Avery and Mark Eaton, *The Meaning of Life: The Scientific and Social Experiences of Everitt and Robert Murray, 1930–1964* (Toronto: Champlain Society, 2008), 121.

3 "Some Historical Comments and Background on TTCP," prepared for the 25th Anniversary Meeting of the NAMRAD Principals, Washington, DC, 12–13 October 1983.

4 One of the strengths of the TTCP system was that the NAMRAD Principals, the executive body of the organization, was composed of high-profile defence scientists and had very little turnover. In the case of Canada, for instance, between 1958 and 1983 there were five representatives, three being DG of the DRB (Zimmerman, Uffen, L'Heureux), while E.J. Bobyn was a former Superintendent of DRES. Ibid.

5 David Fidler and Lawrence Gostin, *Biosecurity in the Global Age: Biological Weapons, Public Health and the Rule of Law* (Stanford: Stanford University Press, 2008), 2.

6 Canada was not alone in concealing its BW programs, both out of fear of negative public opinion and concerns that terrorists groups might use knowledge about bioweapons experiments and testing for nefarious purposes. Olivier Lepick, "The French Biological Weapons Program," in *Deadly Cultures: Biological Weapons since 1945,* ed. Mark Wheelis, Lajos Rozsa, and Malcolm Dando, *Deadly Cultures* (Cambridge, MA: Harvard University Press, 2006), 108.

7 Stacey Gibson et al., "Terrorism Threats and Preparedness in Canada: The Perspective of the Canadian Public," *Biosecurity and Bioterrorism* 5, no. 2 (2007): 134–44.

8 According to one report, the $8.01 billion allocated for civilian biodefence programs in the 2009 budget brought the total to $49.66 billion since 2001, with HHS and DHS receiving the bulk of the funds. Crystal Franco, "Billions for Biodefence: Federal Agency Biodefense Funding, FY2009–FY2009," *Biosecurity and Bioterrorism* 6, no. 2 (2008): 131. In Canada, the original $280 million allocated for counterterrorism in 2001 was later expanded to about $7.7 billion by 2008, about one-third of which was used for biodefence.

9 In early 2001 BW scientists at DRDC Suffield carried out a sustained test within the aerosol test chamber on so-called anthrax letters "to measure the actual aerosol release resulting from the passive dissemination of the BG spores." Paper in possession of the author. See also the paper by University of Victoria biologists David Levin and Giovana V. de Amorim, "Potential for Aerosol Dissemination of Biological Weapons: Lessons from Biological Control of Insects," *Biosecurity and Bioterrorism* 1, no. 1 (2003): 37–42.

10 The author was fortunate to have been involved in a number of leading biosecurity meetings, notably the January 2002 conference in Toronto organized by the Canadian Institutes of Health Research, "Biological Terrorism: Canadian Research Agenda," and the October 2003 Canadian Conference on Counter-Terrorism and Public Health, sponsored by Health Canada and the Canadian Public Health Association.

11 The NML Ebola / Marburg program was started in 2001 by Heinz Feldmann, a German virologist who built up an impressive research team that quickly gained international recognition. This work has continued under the leadership of James Strong, both in the laboratory and in the mobile field laboratory that assumed a lead role during the WHO emergency response to the 2009 Marburg outbreak in Angola. In April 2009, a German laboratory worker who was accidentally infected with the Ebola virus received the NML experimental vaccine and survived the experience. *ScienceDaily* March 2006; BIOEd Online 21 January 2008; CTV.ca 6 April 2009; interview with Dr James Strong, NML, 13 October 2010.

12 Testimony of Anthony Fauci, Director, National Institute of Allergy and Infectious Diseases, NIH, HSS, "The Role of NIH Biomedical Research in Preparing for Emerging Public Health Threats," Committee on Homeland Security, Subcommittee on Emerging Threats, Cybersecurity, and Science and Technology, US House of Representatives, 18 April 2007.

13 While the US Department of Defense vaccinated over 630,000 military personnel, only about 40,000 civilian public health personnel participated in the program, even though the Bush administration targeted over 10 million for the vaccine. "Lessons Learned from the Smallpox Vaccination Program, *Institute of Medicine of the National Academies* (Washington: National Academy of Sciences, 2005), 81, 101. See also Dale Rose, "How Did the Smallpox Vaccination Program Come About?" in *Biosecurity Interventions: Global Health and Security in Question*, ed. Andrew Lakoff and Stephen Collier (New York: Columbia University Press, 2008), 89–240.

14 Dominic Murphy et al., "Why Do UK Military Personnel Refuse the Anthrax Vaccination?" *Biosecurity and Bioterrorism* 6, no. 3 (2008): 237–42; "Highlights and Happenings," *Biosecurity and Bioterrorism*, 7, no. 4 (December 2009).

15 Kathlee Carr et al., "Implementation of Biosurety Systems in a Department of Defense Medical Research Laboratory," *Biosecurity and Biodefense* 2, no. 1 (2004): 1–10; Gigi Kwik Gronwall et al., "High Containment Biodefense Research Laboratories: Meeting Report and Center Recommendations," *Biosecurity and Biodefense* 5, no. 1 (2007): 75–85.

16 The new DHS laboratory facilities at Camp Detrick include the National Forensics Center and the National Biodefence Analysis and Countermeasures Center. Jonathan Tucker and Gregory Koblentz, "The Four Faces of Microbial Forensics," *Biosecurity and Bioterrorism*" 7, no. 4 (2009): 389–97.

17 Lynn Klotz and Edward Sylvester, *Breeding Bio Inseucity: How U.S. Biodefence Is Exporting Fear, Globalizing Risk, and Making Us All Less Secure* (Chicago: University of Chicago Press, 2009), 84. There has been vigorous criticism of this expansion, with one study denouncing the American "bloated, largely secret biodefence program [which] increases the risk of accidents and theft by terrorists." Ibid., 4.

18 The Winnipeg facility is shared by the National Microbiology Laboratory for human pathogens and the Canadian Food Agency. The development of the $775 million US National Bio and Agro-Defense Facility has been more controversial because of its mandate to examine dangerous foreign animal diseases and due to delays in its opening, now scheduled for 2018. *Government Security News* 13 December 2010, www.gsnmagazine.com.

19 Maureen Best, "Bioterrorism: Impact and Implications for Biosafety," *American Biosecurity* 8, no. 4 (2003): 166–8; Peter Jahrling et al., "Triage and Management of Accidental Laboratory Exposures to Biosafety Level-3 and -4 Agents," *Biosecurity and Bioterrorism* 7, no. 2 (2009): 135–43.

20 David Franz and James Le Duc, "Balancing Our Approach to the Insider Threat," *Biosecurity and Bioterrorism* 9, no. 3 (2009): 205.

21 "Lone Suspect in the Amerithrax Investigation Identified, but Skepticism Remains," *Biosecurity and Bioterrorism* 6, no. 4 (2008), 285–7. Jeanne Guillemin, *American Anthrax: Fear, Crime, and the Investigation of the Nation's Deadliest Bioterror Attack* (New York: Henry Holt, 2010), 58, 172, 214, 245–61.
22 *Boston Globe* 24 February 2005.
23 *Winnipeg Free Press* 3 March 2005.
24 *Winnipeg Free Press* 14, 15 May 2009; *Toronto Star* 14 May 2009; *Nature News* 26 May 2009. Public Health Agency Canada the Audit Services Division, "Audit of Security of Laboratories," June 2009.
25 While biosafety is generally linked with laboratory practices, the meaning of biosecurity is much more diverse. In Canada, the US, and the UK it is usually associated with attempts to ensure that the public's safety is not threatened either by the actions of sinister groups or by unsafe research practices. See Brian Rappert and Chandre Gould, eds., *Biosecurity: Origins, Transformations and Practices* (London: Palgrave-Macmillan, 2009), 1–19, 25–6.
26 Another controversy involved the work of scientists at the State University of New York (Stony Brook), who used the genetic map of the polio virus to contruct a synthetic version capable of infecting humans. Fidler and Gostin, *Biosecurity in the Global Age,* 42.
27 Arturo Casadevall et al., " Biodefense Research: A Win-Win Challenge," *Biosecurity and Bioterrorism* 6, no. 4 (2008): 291.
28 In June 2011, after a strong recommendation from the NSABB, category I of the select agent list was reduced from 82 to 11, using such criteria as the agent's ability to produce a mass casualty event, its communicability, and its history of use as a biological weapon. "Panel Recommends Changes to the Select agent Program," *Biosecurity and Bioterrorism* 9, no. 1 (2011): 202–3.
29 Interview with Dr Grant McFadden, 12 February 2003, University of Western Ontario.
30 "Bioterrorist Fears Could Block Crucial Flu Research," *New Scientist* 21 November 2011.
31 "An Engineered Doomsday," *New York Times* 15 January 2012. This article also rejected proposals that about 100 laboratories be authorized globally to study the H5N2 virus since "the consequences should the virus escape, are too devastating to risk."
32 "Scientists Must Decide on Bird Flu Research Risks, "*Toronto Star* 15 February 2012.
33 "Researcher at Heart of Bird Flu Studies Controversy Reveals Details of His Findings," *Winnipeg Free Press* 25 January 2012. The views of the US-based virologist Yoshi Kawaoka and his Dutch counterpart, Ron Fournier of the Erasmus Medical Centre in Rotterdam received an extensive critique at he

Royal Society (London), 3–4 April 2012. This special session, "H5N1 Research: Biosafety, Biosecurity and Bioethics: An International Symposium," included many of the world's leading influenza and biodefence experts. "Latest on Killer Virus," *New York Times* 29 April 2012.

34 "Canada Confines Mutant Flu to Maximum-Security Facilities," *NatureNews Blog,* 23 February 2012, www.nature.com/nature/about/indix/html; "The Truth about the Doomesday Virus?" *Toronto Star* 11 March 2012.

35 For instance, the 2002 US Public Health and Bio-Preparedness Act provided $4.6 billion for detection systems, long considered a serious weakness in North American biodefence capabilities. Gregory Koblentz, *Living Weapons: Biological Warfare and International Security* (Ithaca: Cornell University Press, 2009), 28–30.

36 See Meeting Report, "Prevention of Threats: A Look Ahead," *Biosecurity and Bioterrorism* 7, no. 4 (2009): 433–42.

37 Committee on Determining a Standard Unit of Measure for Biological Aerosols, National Research Council, *A Framework for Assessing the Health Hazard Posed by Bioaerosols* (Washington: NASP, 2008), Executive Summary, 2.

38 Milton Leitenberg, Ambassador James Leonard, and Dr Richard Spertzel, "Biodefence Crossing the Line," *Politics and the Life Sciences* 22, no. 2 (2004): 40–6. See also Jonathan Tucker, "Biological Threat Assessment: Is the Cure Worse than the Disease?" *Arms Control Today* (October 2004); Laura Kahn, "Biodefense Research: Can Secrecy and Safety Coexist?" *Biosecurity and Bioterrorism* 2, no. 2 (2004): 1–8.

39 Mark Wheelis, "Will the New Biology Lead to New Weapons?" *Arms Control Today* (July / August 2004).

40 This subject was extensively discussed at the 2008 meeting organized by the Centre for Emergency Preparedness and Response, "National Biological Counter-Terrorism Seminar: Integrating Intelligence, Policing and Health," Ottawa, May 2008.

41 Interview with Frank Plummer (NML), 13 October 2010. Extensive use was made of Vital Points Bio Sentry, developed by General Dynamics Canada as part of a $30 million contract awarded in April 2007 for six full and twenty-three partial systems. It combined the advantages of the focused stand-off approach and the more complex point detection strategy. *The Devil's Din* (Spring and Summer 2010).

42 In 2006 the CBRNE Research and Technology Initiative (CRTI) was merged into the Centre for Security Science, jointly operated by the Department of National Defence and the Department of Public Safety. Security Science, *News,* May 2007.

43 This clash of professional cultures was clearly described at the February 2003 conference at Wilton Park on Bioterrorism when police authorities from the city of London described the difficulties of coordinating their CBRN response teams with hospital and scientific experts.

44 *Globalization, Biosecurity and the Future of the Life Sciences* (Washington: NAS, 2006), Executive Summary, 10.

45 Commission of Inquiry into the Actions of Canadian Officials in Relation to Maher Arar, 3 vols. (Ottawa: Ministry of Public Works, 2006); *Air India Flight 182: A Canadian Tragedy, Remarks by Commissioner John C. Major, June 17, 2010,* www.majorcomm.ca/em/reports/finalreport.

46 Interview with Frank Plummer, 13 October 2010.

47 Dr Boulet obtained his PhD in organic chemistry from the University of Victoria in 1986 and is a 2005 graduate of the Canadian Forces College National Security Studies Course. Since his 2009 appointment as Director General, he has assumed many important duties, including Special Advisor to the Canadian Forces for CBRN personal protection and Canadian Program Officer for the AS/CA/UK/US Chemical, Biological and Radiological Defence Materiel (CBR MOU).

48 Suffield has also upgraded its containment level 3 (BL-3) laboratory and its open-air sites to develop new technology and techniques for aerosol detection and analysis. See www.drdc-rddc.gc/drdc/en/centre/drdc-suffield-rddc-suffield/.

49 Marc Osfield, "Intersectoral and International Cooperation on Combating Bioterrorism," US Department of State, Office of International Health Affairs, 14 September 2005.

50 Kellman, *Bioviolence,* 240. In 2003 the Bush administration launched the Proliferation and Security Initiative, with like-minded allies including Canada, to seize CBRN components on the high seas. In addition, there has been considerable US involvement with the special United Nations Committee created by Security Council Resolution 1540 prevent WMD proliferation to non-state parties.

51 *North American Plan for Avian and Pandemic Influenza* (Ottawa / Washington, August 2007), chapter 4 (Pandemic Influenza), 21.

52 Donald Avery, "The North American Plan for Avian and Pandemic Influenza: A Case Study of Regional Health Security in the 21st Century," *Global Health Governance* 3, no. 2 (Spring 2010): 13–19.

53 Louise Lemyre et al., " A Psychosocial Risk: Assessment and Management Framework to Enhance Response Response to CBRN Terrorism Threats and Attacks," *Biosecurity and Bioterrorism* 3, no. 4 (2005): 316–30; Charles

DiMaggio et al., "The Willingness of U.S. Emergency Medical Technicians to Respond to Terrorist Incidents," *Biosecurity and Bioterrorism* 3, no. 4 (2005): 331–7.

54 Daniel Kollek, "Canadian Emergency Department Preparedness for a Nuclear, Biological or Chemical Event," *Canadian Journal of Emergency Medicine*, www.caep.ca/004/1/9/2004.

55 In March 2005, 750 research scientists signed a petition against the funding policies of the National Institutes of Health, which included a fifteen-fold increase in grants for biodefence projects between 1998 and 2005, while other infectious diseases studies lost 27 per cent of their financial support. *New Scientist* 1 March 2005.

56 Nicholas King, "The Influence of Anxiety: September 11, Bioterrorism, and American Public Health," *Journal of the History of Medicine and Allied Sciences* 58, no. 4 (2003): 433–41; Louise Lemyre et al., "A Psychosocial Risk Assessment and Management Framework to Enhance Response to CBRN Terrorism Threats and Attacks," *Biosecurity and Bioterrorism* 3, no. 4 (2005): 316–30.

57 NATO–Russia Advanced Research Workshop on Social-Psychological Consequences of Chemical, Biological, Radiological Terrorism, 25–7 March 2002, NATO Headquarters, prepared by Dr Simon Wessely, GKT School of Medicine and Institute of Psychiatry, King's College, London.

58 For the medical aspects of the Tokyo attacks see Azik Hoffman, "A Decade after the Tokyo Sarin Attack: A Review of Neurological Follow-Up of the Victims," *Military Medicine* 172 (June 2007): 607–10; Ezra Susser et al., "Combating the Terror of Terrorism," *Scientific American* August 2002, 74.

59 Ian Lustick, *Trapped in the War on Terror* (Philadelphia: University of Pennsylvania Press, 2006), 121–4. See also Kenneth King, *Germs Gone Wild: How the Unchecked Development of Domestic Biodefence Threatens America* (New York: Pegasus Books, 2010).

60 Remarkably, over 50 per cent of those given antibiotics because of possible exposure to anthrax spores in 2001 soon discontinued their treatment, either because of concerns about side-effects, or because of a belief that they could not develop inhalation anthrax. Gillian Steel Fisher et al., "Public Response to an Anthrax Attack: Reactions to Mass Prophylaxis in a Scenario Involving Inhalation Anthrax from an Unidentified Source," *Biosecurity and Bioterrorism* 9, no. 2 (2011): 239–50.

61 There have been a number of studies of the experiences of postal workers at the Washington, DC, Brentwood facility (two of whom died), who were part of the 10,000 exposed to anthrax during the 2001 attacks. See Guillemin, *American Anthrax*, 80–160.

62 *Toronto Star* 25–6 February 2009. In February 2010, the courts rejected the claims of all fifty-three of these victimized hospital workers.

63 John Mueller, "Fear Not: Notes from a Naysayer," *Bulletin of the Atomic Scientists* (March / April 2007): 32,

64 Despite projections that there was only a 1 in 75,000 chance of being killed by terrorists, during these years the US public appeared "to have chosen … to wallow in a false sense of insecurity." John Mueller, "Inflating Terrorism," in *American Foreign Policy and the Politics of Fear: Threat Inflation since 9/11*, ed. A. Trevor Thrall and Jane K. Cramer (New York: Routledge, 2009), chapter 11. See also Lynn Klotz and Edward Sylvester, *Breeding Bio Insecurity: How U.S. Biodefence Is Exporting Fear, Globalizing Risk, and Making Us All Less Secure* (Chicago: University of Chicago Press, 2009).

65 Dan Gardner, *Risk: The Science and Politics of Fear* (Toronto: McClelland and Stewart), 2008), 28–36; Lustick, *Trapped in the War on Terror*, 5–8.

66 *Toronto Star* 10 September 2003.

67 "Remarks at the 7th Biological and Toxin Weapons Convention Review Conference, 7 December 2011," Hillary Rodham Clinton, Secretary of State, Palais des Nations, Geneva, Switzerland, www.state.gov/secretary/rm/2011/12/178409.htm.

Glossary of Terms

Amerithrax	FBI operation during 2003 anthrax letter bomb attacks
B. anthracis	*Bacillus anthracis* (anthrax): Also called N
BCBW	British Cabinet Committee on Biological Weapons (1939–45)
BDP	Biology Department, Porton: BW research unit
Bursting chamber	Used at SES for small-scale bomb trials – usually with simulants
"Bot Tox"	Botulinum toxin was one of the major BW agents: Also called X
BTWC	Biological Toxin Weapons Convention (1972)
BW	Biological warfare
C-1 Committee	Canadian BW organization during Second World War
CAB	Cabinet Records (UK)
Camp Detrick	Centre of the US biological warfare program – near Washington, DC. In 1956 it was renamed Fort Detrick.
CANUKUS	Canadian, United Kingdom, United States CBW organization
CBW	Chemical and Biological Warfare
CDC	Centers for Disease Control and Prevention (US)
CIA	Central Intelligence Agency (US)
CSCHAH	Canadian Science Centre for Human and Animal Health (Winnipeg)
CSIS	Canadian Security Intelligence Agency
CWC	Chemical Weapons Convention (1993)
CWISB	Chemical Warfare Inter-Service Board
CWS	Chemical Warfare Service (US): became Corps in 1947
DCW&S	Directorate of Chemical Warfare and Smoke (Cdn Army)

DEA	Department of External Affairs
DFAIT	Department of Foreign Affairs and International Trade
DH&W	Department of Health and Welfare
DND	Department of National Defence (Canada)
DOD	Department of Defence (US)
DRB	Defence Research Board (1947–75)
DRDC S	Defence Research and Development Canada Suffield (2000–
DRES	Defence Research Station Suffield (1967–2000)
DTC	Deseret Test Centre (Utah)
ESS	Experimental Station Suffield (1942–7)
FBI	Federal Bureau of Investigation (US)
GIN	Grosse Ile anthrax project
Granite Peaks	Part of the Dugway Utah CBW testing facility
Gruinard Island	Site of the 1942–3 UK anthrax trials – northern Scotland
Horn Island	American BW testing facility 1943–5.
ISSBW	Inter-Services Sub-Committee on Biological Warfare (UK)
LAC	Large-Area Coverage (BW offensive strategy)
M-1000 Committee	The first organization to coordinate Canada's BW program
NARA	National Archives and Records Administration (US)
NASC	National Academy of Sciences Collection (Washington, DC)
NATO	North Atlantic Treaty Organization
NAUK	National Archives United Kingdom (Kew, England)
NIH	National Institutes of Health (US)
NLA	National Library Archives (Ottawa)
NML	National Microbiology Laboratory (Winnipeg)
NRC	National Research Council
NSABB	National Science Advisory Board for Biosecurity
NSAC	National Academy of Sciences Archives
OSG	Office of the Solicitor General (Cdn)
OSS	US Office of Strategic Services
PHAC	Public Health Agency of Canada
Porton Down	UK research centre for chemical and biological warfare
PRO	Public Record Office (UK): Now National Archives UK
RCAMC	Royal Canadian Army Medical Corps
RCMP	Royal Canadian Mounted Police
SES	Suffield Experimental Station (1948–67)
UNSCOM	United Nations Special Commission (Iraq)
USAMRIID	United States Army Medical Research Institute of Infectious Diseases (Detrick)

VIGO Plant	BW production facility in Indiana during Second World War
WDCS	War Disease Control Station (Grosse Ile)
WMD	Weapons of mass destruction
WHO	World Health Organization
WRC	Special American BW Committee in 1942
WRS	War Research Service under George Merck (1943–4)

Note on Sources

Obtaining information about biological warfare programs, past and present, is not an easy undertaking. In part, this is because of the sensitive nature of the subject, which despite its inclusion in the military policies of Canada, the United States, and the United Kingdom retains a negative image with the general public. In turn, this means that both elected government officials and bureaucrats are often unwilling to provide relevant BW documents. This was certainly the experience of this author, although use of the federal access-to-information process did produce some previously restricted material.

Despite these obstacles, this book draws on a wide variety of documentary sources for all aspects of the Canadian bioweapons saga. For the Second World War, for instance, the author consulted the copious records of the Army Directorate of Chemical Warfare and Smoke that administered the wartime program, along with other useful material from the National Research Council. The personal papers of E.G.D. Murray, Colonel Ralston, and C.J. Mackenzie were also of value in understanding the broader contours of the subject. Equally important were a range of British sources, made available during the 1990s, which provided extensive information about how Paul Fildes and his Porton colleagues attempted to utilize Canadian scientific resources and the Suffield testing grounds in the development of their anthrax retaliatory weapon. Collections at the United States National Archives and American Academy of Sciences were also crucial in understanding the relationship between the American biowarfare program and that of its two allies.

In assessing the activities of Canadian BW scientists during the Cold War the author was able to utilize the voluminous records of the Canadian Defence Research Board, including the personal records of Director

General Omond Solandt. Also valuable were the files of the arms control division of the Department of External Affairs for the light they shed on Canada's role in the establishment of the Biological Weapons Convention. The book's discussion of the BW disarmament movement of the 1960s benefited from a variety of US documents, notably the records of the American Society of Microbiologists and the National Institutes of Health. The British perspective on these developments was obtained from declassified material from the Foreign Office and the Ministry of Defence.

The book also deals with a number of specialized topics during the Cold War years. One of these is bioterrorism, which is explored in various sections of this book and is based on information obtained from various documentary sources, notably DRB records and the proceedings of various government committees. The records of President Richard Nixon's Cabinet Committee on Counter-Terrorism helped place Canada's response to terrorism within a broader context and demonstrate a pattern of ongoing cooperation between security agencies of the two countries. Another major theme was the link between biological weapons and natural disease pandemics. Here, valuable information was obtained from the records of the Canadian Department of Health and Welfare, the Connaught Medical Research Laboratories, and the US National Institutes of Health. Information on Canada's involvement in the CANUKUSA defence science alliance system is based on sporadic reports obtained through access-to-information requests. The great array of documentary sources has been greatly enhanced through a series of interviews that were conducted over the last fifteen years, with an impressive group of defence scientists and health security officials.

For the historian, there are many frustrations in trying to analyse major trends in Canadian biodefence policies after 9/11, since most relevant documents are not yet available. As a result, this section of *Pathogens for War* is based largely on government reports, congressional / parliamentary testimony, and scholarly articles, most of which was generated in the United States. Where possible, the author has attempted to relate the major arguments of these studies to the Canadian situation, both through his own research and through interviews with many key participants in Canada's biodefence system. Without these insights, this section of the book would not have been possible.

Many of the themes developed in this book have been explored in a range of conferences and workshops, at the national and international level, during the past fifteen years. Most of the author's publications associated with these scholarly events have been cited in the endnotes of the book, although several deserve special mention. These include "Canadian Biological and Toxin Warfare Research, Development and Planning, 1925-45," in *Biological and Toxin Weapons: Research, Development and Use from the Middle Ages to 1945,* edited by Erhard Geissler and John Ellis van Courtland Moon (Oxford: Oxford University Press, 1999), and "Canada, Alliance Warfare and the Biological Arms Race," in *"Deadly Cultures": Bioweapons from 1945 to the Present,* edited by Mark Wheelis et al. (Cambridge, MA: Harvard University Press, 2005). In both these projects the leading experts in this scholarly field exchanged ideas and cooperated in the publication of outstanding surveys of the global impact of biological warfare during the twentieth century.

Primary Sources

Canada

Connaught Medical Research Laboratories (Toronto)
  Records on research during the Second World War
  Files on influenza vaccine research, 1944–76
  Files on eradication of smallpox, 1965–85
Library and Archives Canada (Ottawa)

Private Papers

Brock Chisholm
Hon. Brooke Claxton
Hon. Clarence Howe
Rt. Hon. William Lyon Mackenzie King
C.J. Mackenzie (wartime diaries)
General A.G.L. McNaughton
E.G.D. Murray
Archie M. Pennie
Hon. J.L. Ralston

Government Records

Cabinet Conclusions (on-microfilm)
Cabinet War Committee, Minutes and Records (PCO)
Defence Research Board Records, 1947–74
Department of National Defence Records
  DND Chemical and Biological Warfare Files on Microfilm (C-5001–5019)

General Records: Solandt Correspondence 1946–50
Files on CW trials with US in Churchill, Manitoba 1952
Files on BW Working Group (Advisory Committee) 1946–65
Reports from Research Establishments: Suffield and Kingston
Isolated reports of Tripartite Meetings: 1947–69 (Access to Information)
Occasional Files on cooperation between Suffield and Deseret Testing Centre, 1962–68 (Access to Information)
Occasional Reports on Canadian–United Kingdom–United States CBW research programs, 1986–92 (Access to Information)

Department of External Affairs:
Files on the Korean War Germ Warfare Controversy
CBW Arms Control Records, 1965–90

Department of National Health and Welfare: Civil Defence Records, 1946–65
National Research Council: CBW files, 1939–45
Defence Research Establishment Canada Suffield
Field Reports of BW Trials, 1943–60 (Declassified)
Cooperative Projects with the Dugway Proving Grounds (Declassified)

National Defence Headquarters: Directorate of History
Chemical Warfare Records Inter-Service Board
Files on proposed UK nuclear trials at Churchill, 1950

National Research Council
Frederick Banting Papers

McGill University Archives
E.G.D. Murray Papers (Department of Microbiology)
Otto Maass Papers

Queen's University Archives
Guilford B. Reed Papers

University of Toronto Archives
Omond Solandt papers

University of Toronto Rare Book Room
Wartime Diary of Sir Frederick Banting

United Kingdom

National Archives of the United Kingdom, Kew (formerly Public Records Office)
War Office Records on BW warfare (1940–70): Paul Fildes Correspondence
Cabinet Records, 1940–55
Foreign Office Records, 1965–75

Centre for Applied Microbiology and Research (CAMR) Porton, Down
Reports of the Biology Department, 1945–6

United States

Archives of the American Microbiology Society (University of Maryland, Baltimore)
ASM Committee Advisory to Fort Detrick, 1951–65
Correspondence of Executive Committee, 1965–72
Archives of the National Academy of Sciences (Washington, DC)
Files on biological warfare, 1942–46
Library of Congress
Robert Oppenheimer Papers
National Archives and Records Administration, College Park, MD
Records of the US Joint Chiefs of Staff
Records of the Joint Committee on New Weapons
Records of the Office of Scientific Research and Development
Records of the War Department: General and Special Staffs
New Developments Division (George Merck correspondence re: biological warfare)
Records of the National Institutes of Research, 1955–82
Records of the Department of State, 1946–75
Records of the State / War / Navy Coordinating Committee, Lot 57
Richard Nixon Correspondence: 1968–83
National Security Archives (Washington)
Files on biological and chemical warfare, 1945–90

European Sources: Archives of the North Atlantic Treaty Organization (Brussels)

Civil Defence Committee
Working Committee on Protection against Chemical and Biological Warfare
Reports of the Military Committee

Personal interviews (conducted by author)

Kelly Anderson, Canadian Embassy, Washington DC, 2 February 2007
Joan Armour, 30 September 2000, 14 August 2003 (Defence)
Marc-André Beaulieu, 24 October 2002 (Health)
Bill Bide, DRDC Suffield, 15 August 2003 (Defence)
Peter Boehm, Ottawa, 13 November 2007 (Foreign Affairs)
Cam Boulet, Ottawa, 26 July 2006 (Defence)
Jeff Bozworth, American University, Washington, DC, 11 January 2007 (US-security)
Kathryn Clout, Ottawa, 20 December 2004 (Solicitor General)

Elaine Feldman, Ottawa 23 January 2009 (Foreign Affairs)
David Franz, Frederick, Maryland, 4 January 2007 (US-Biodefence)
Daniel Goodspeed, Brussels, 25 April 2002 (UK-diplomatic)
Kent Harding, DRDC Suffield, 16 August 2003 (Defence)
Robert Heggie, Ottawa, 9 June 2004 (Defence)
George Ignatieff, Toronto, 13 November 1987 (Foreign Affairs)
Ken Johnson, Ottawa, 28 February 2002 (Defence)
Bill Kournkakis, DRDC Suffield, 31 March 2005 (Defence)
Clement Laforce, DRDC Suffield, 30 September 2000, 28 March 2005 (Defence)
George Lawrence, Deep River, 20 November 1981 (Nuclear)
Stuart MacPherson, Ottawa, 12 September 2006 (Border Agency)
J. Carson Mark, Los Alamos, 15 November 1993 (US Nuclear)
Grant McFadden, London, Ontario: seven session, 2006–2010 (Scientists)
Marc Mes, Ottawa, 1 June 2003 (Foreign Affairs)
Gail Miller, Ottawa, 23 January 2009 (Health)
Randy Murch, Washington, DC, 27 December 2006 (US-Biodefence)
Robert Murray, London, Ontario, 25 January 1994, 19 October 2004 (Scientist)
Stephen Orosz, Brussels, 3 April 2002 (NATO)
Archie Pennie, Ottawa, 8 June 2004 (Defence)
Frank Plummer, National Microbiology Laboratory (Winnipeg), 13 October 2010
Ken Scott, Ottawa, 24 February 2002 (Health-DND)
Trevor Smith, Ottawa, 12 July 2005 (Foreign Affairs)
Omond Solandt, Bolton, Ontario, 12 October 1989 (Defence)
R.B. Stewart, Queen's University, 10 July 1998 (Scientist)
Ronald St John, Manotick (Ottawa), 23 January 2009 (Health)
James Strong, National Microbiology Laboratory (Wpg.), 13 October 2010
Michael Thielmann, American University, Washington, DC, 7 December 2006 (Solicitor General-Biodefence)
Harry Thode, Hamilton, 6 May 1992 (Nuclear Science)
Katie Tolan, Canadian Embassy, Washington, DC, 20 December 2006 (CSIS)
James Young, Toronto, 11 August 2005 (Public Health-Biodefence)
Frank Welsh, 7 February 2003,12 July 2003, 8 June 2004 (Health)
Ted Whiteside, Brussels, 28 April 2002 (NATO-WMD)
Yan Li, National Microbiology Laboratory (Wpg.), 13 October 2010

Scholarly Conferences Dealing with Major Themes of This Book

2002

Biological Terrorism: Canadian Research Agenda, King Edward Hotel, Toronto, 18–20 January. This international conference, jointly organized by Health Canada and the Canadian Institutes of Health Research, involved 130 of the leading national and international experts on bioweapons and biosecurity.

International Conference on Emerging Infectious Diseases, Atlanta, Georgia, 24–7 March.

International Symposium on the Biological and Toxin Weapons Convention, 15–17 July.

September 11, 2001: The Impact and Aftermath for Canada and Canadians, Canadian Studies Conference, Ottawa, 12–14 September.

2003

International Symposium, Meeting the Threat of Biological Terrorism, held at the Wilton Park Conference Centre, England, February.

Canadian Conference on Counter-Terrorism and Public Health. Sponsored by the Centre for Emergency Preparedness and Response, Health Canada, and the Canadian Public Health Association, Toronto, 29 October–1 November.

2004

Ethics and Weapons of Mass Destruction: A Comparison of the Response of Biological and Nuclear Scientists. Invited lecture at the 16th HSP London Seminar (UK, Foreign and Commonwealth Office), March.

Special seminar: Biological Weapons: Coping with Current Threats, hosted by Simons Centre for Disarmament and Nonproliferation Research, Liu Centre (Vancouver), 10–12 November.

Conference on The Future of the Life Sciences: Reaping the Rewards and Managing the Risks, Washington, DC, 8–10 December. International Institute for Strategic Studies and the Chemical and Biological Arms Control Institute.

2005

How Canada Can Assist in the Transformation of Former BW Laboratories and Their Scientists in Russia and the Commonwealth of Independent States. Department of Foreign Affairs meetings, Government Consultations with Civil Society on Issues Related to International Security, Nuclear Weapons and Other Weapons of Mass Destruction and Their Delivery Systems, Ottawa, 8–9 March.

Conference: Advancing International Cooperation on Bio-Initiatives in Russia and CIS, RANSAC, Rome, Italy, 26–7 April.

The National Biological Counter-Terrorism Seminar: Integrating Intelligence, Policing and Health. Ottawa, 13–14 June.

2006

Symposium: Ethical Issues Relating to Dual Use Technologies in the Life Sciences. Sponsored by Health Canada and Defence Research and Development Canada, Ottawa, 7–8 February.

2011

"The 1957 Influenza Pandemic and Biological Warfare Planning: An Unexplored Relationship," paper presented at conference: After 1918: History and Politics of Influenza in the 20th and 21st Centuries, Rennes, France, 24–6 August.

Index